VISIT
CR

WORKSHOPS

IN THE ENGLISH COUNTRYSIDE

Editor: John Cole-Morgan

Compiler: Carol Turner

Foreword by His Royal Highness The Prince of Wales

MPC

in association with
Rural Development Commission

sponsored by
Barclays Bank plc

Published by
Moorland Publishing Co Ltd
Moor Farm Road, Ashbourne
Derbyshire DE6 1HD

in association with
Rural Development Commission
141 Castle St, Salisbury SP1 3TP

ISBN 086190 195 9
© Rural Development Commission
1992

Printed by: The Cromwell Press Ltd
Broughton Gifford, Wiltshire

Front cover: Glass blowing at Derwent Crystal Craft Centre, Ashbourne, Derbyshire

Rear Cover: Jonathan Chiswell Jones in his workshop at Drusillas, Alfriston, East Sussex (photo: Leigh Simpson)

A considerable effort has been made to contact all those eligible for inclusion in this book but inevitably there will be some craftspeople who feel they should have been included but whom we were unable to reach. Any craft business wishing to be considered for inclusion in any subsequent edition of this book should write to the editor immediately in order that the necessary information can be added to the database and used for reference purposes.

Vineyards
It was hoped that it would be possible to include in this publication vineyards open to the public. However, so many entries were recieved from craft workshops the vineyards had to be omitted. A list of vineyards which provided information for inclusion in this directory is available from the address below.

> The Editor
> *Visiting Craft Workshops in the English Countryside*
> Rural Development Commission
> Communications Branch
> 141 Castle Street, Salisbury, Wilts SP1 3TP
> ☎ 0722 336255

CONTENTS

RURAL DEVELOPMENT COMMISSION AND CRAFTS

Crafts of many kinds have traditionally played an important part in the economy of rural areas and the Rural Development Commission encourages and supports rural craft workshops as part of its general objective of maintaining and fostering a living, working countryside.

For many years the Commission has been directly engaged in teaching craft skills especially in areas where there is a danger of them otherwise dying out. Many new entrants into, for example saddlery, forgework and furniture restoration have done part of their training in the Commission's workshops in Salisbury. In this way it has helped to maintain and improve standards in many traditional crafts and skills.

The introduction of National Vocational Qualifications based on standards of competence set by industry and commerce itself will for the first time give craftspeople and others access to nationally recognised qualifications. This has been warmly welcomed by the Commission which has provided considerable support to the body organising the development of these standards in the Craft Section, the Crafts Occupational Standards Board. This body is responsible for arranging for practicing craftspeople to define the standards.

NVQs are already changing the ways people think about training in the crafts. The assessment process is very flexible and removes many of the hurdles of written examinations. It makes competence in the job the deciding factor and should encourage more people to gain occupational qualifications.

At the higher levels, NVQ qualifications equate to GCE 'A' levels and look not only at the craft skills but at the level of business competence which is equally necessary to make any craft workshop a success. Through NVQs the Commission believes craftspeople will be able to raise not only their standards but also their status as certification of competence will be recognised throughout the Single European Market.

In a world in which miraculously efficient and sensitive robots are taking over many of the jobs which required highly skilled people there is something immensely satisfying about watching the craftsman at work.

Whilst we all enjoy the enormous benefits to be derived from economies of scale, it is hard to match the lasting pleasure of owning useful and beautiful objects which have taken shape before our eyes. Increasingly, people today look for and are even prepared to pay a little more for the individuality which craftsmen and women can bring to their lives and surroundings. The work of the thatcher and the dry stone waller enhances our countryside; a hand thrown casserole dish may become a family treasure.

With the decline in agriculture the importance of nurturing alternative and appropriate work in our rural areas has never been greater and what can be more appropriate than craft workshops? Our countryside was moulded by work; without enduring economic activity it would become a much poorer place.

This guide will not necessarily help people to find what they are looking for, but it will hopefully lead its users into the rural and working heart of England. If they derive as much pleasure as I do from visiting the craft workshops on the Duchy of Cornwall estates their journeys will certainly not be wasted.

I, therefore, hope this guide will help craftsmen and women find new customers and lead the public to an even greater appreciation of the tremendous wealth of art and skill which exists in every corner of our rural communities.

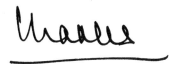

THE IMPORTANCE OF RURAL BUSINESSES

This book guides its readers into some of the hidden corners of rural England — to places where they can watch beautiful and useful things take shape in the hands of craftsmen and craftswomen. However, it is important to remember that this is only one side of their work.

For many people, as well as providing a living, self-employment gives them the opportunity to express their independence and though many of them did not go into business primarily for the money, they could not survive unless their small businesses were soundly based financially.

All the small businesses listed in this book have gone through the trauma of starting up, with the associated problems of planning the business, finding and funding premises and seeking the right advice. Only when such important issues are properly addressed can these businesses succeed and add richness to our lives and surroundings.

At Barclays we believe small businesses are a fundamental part of the economic and social scene. That is why we attach so much importance to providing relevant advice and why we have published a series of Business Guides aimed at those thinking of setting up or running a business. One set of Barclays' Business Development Guides, which we have produced in association with the English Tourist Board, provides advice for tourism businesses, many of which are based in rural areas.

We have also worked closely with the Rural Development Commission to produce a special loan package to assist businesses in rural areas.

When you visit craft workshops, as well as admiring the artistry and skill of these creative people, remember also the considerable skill which is required to make any small business work.

Brian Carr
Head of Barclays Community Enterprise

INTRODUCTION

Visiting Craft Workshops in the English Countryside is not only a guide to the many craft workshops which are open to the public but also a way of finding people who can provide very specialised products and services. For this reason we have also included those who practise their craft in situ such as thatchers, dry-stone wallers and hedge-layers.

Those who wish to use the book to enhance a visit to an area should turn first to the county section and then to the nearest large town under which the locally based craft workshops are listed. Those looking for a particular craft or skill will find a cross-reference under separate craft headings in the index at the back; again the county is indicated.

Some workshops are in private homes or have limited space; in others the machinery, tools and methods may require special safety measures. These workshops can be visited by appointment only and visitors are asked to be especially considerate and give as much notice as possible. Many small craft businesses rely upon craft fairs as an outlet for their work and may be away from their workshop from time to time, so it may be advisable to telephone before visiting.

Respect the Craftsmen

Some of the crafts people who have submitted entries to this guide have done so with hesitation. Most craftspeople are self-employed and whilst they may welcome visitors watching them at work, it has to be remembered that they are not in business to give demonstrations. We would therefore ask all those visiting the craftsmen and women listed in this directory to respect the fact that they are visiting a place of work, and not to be too demanding nor outstay their welcome.

Most of the craft workshops listed in this book depend on direct sales to the public. They naturally hope that visitors will become customers and tell others about their work.

Acknowledgements

This guidebook has been compiled with the financial help of Barclays Bank. Their generous support meant that it has not been necessary for the compilers to impose an entry charge on craft workshops and it can therefore be seen as a direct subsidy to the craftspeople whose businesses appear and to crafts in the English countryside in general. The Rural Development Commission is grateful to Barclays Bank PLC and in particular to Mr Brian Carr, Head of Barclays Community Enterprise, for the interest and enthusiasm he has shown in this project.

The editor and compiler are grateful to Joyce Dingley for collating and indexing all the information received and to the many craftspeople who offered illustrations for inclusion in this book.

AVON CRAFT WORKSHOP CENTRES

Church Farm Business Park
Corston, Nr Bath BA2 9EX
Farm buildings converted for business use. Several craft workshops.
Directions: just off A4 in Corston village.

Clevedon Craft Centre
Newhouse Farm, Moor Lane,
Clevedon BS21 6TD
☎ *0275 879519*
Twelve studio workshops.
Open all year round, some workshops closed on Monday.
Free entry and parking
Directions: from M5 Exit 20, follow signs to Clevedon Court (National Trust) on B3130 Bristol Road. Turn right opposite the Court, proceed over the motorway and turn right into Moor Lane.

Ivy House Farm Craft Workshops
Wolvershill
Banwell BS24 6LB
Workshops include: French polishing, textile conservation, upholstery, handmade button guitars, patchwork & quilting, graphics, Western Steam Model Engineers.
Directions: from M5 Junc 21 follow signs to West Wick and Banwell, proceed over motorway, Ivy House Farm on right.

AVON CRAFT WORKSHOPS

Bath

JEWELLERY/FINE METALWORK
Ursula Cook
14E Church Farm Business Park
Corston, BA2 9EX
☎ *0225 423667*
Copper and silverwork with bird imagery as a theme. Awards & exhibitions. C/W/E
Mon-Fri 2-6pm
Directions: see under Craft Centres.

STONE CARVING
Nick Moore
Church Farm Business Park,
Corston BA2 9EX
☎ *0225 873842*
Experienced restoration mason, now producing all types of stone carved garden sculpture, as well as undertaking corporate and public commissions. Tuition for all abilities, seminars for trade and public. C/E
Mon-Fri 9am-5pm, Sat by appointment.
Directions: see under Craft Centres.

CERAMIC SCULPTURE
Peter Hayes Contemporary Ceramics
2 Cleveland Bridge, Bath BA1 5DH
☎ *0225 466215*
Ceramic artist producing indoor and landscape sculpture. Work commissioned for Broadgate Development and Stanhope

---

Abbreviations used in this book
C   commissions undertaken
R&R repairs and restoration
W   wholesale orders welcome
E   export orders welcome

Properties, London.
C/W/E. Credit cards.
Mon-Fri 9.30am-5.30pm.

WOODWORK & PICTURE FRAMING
Lark Crafts
(Richard & Libby Kogut)
206 High Street, Batheaston BA1 7QZ
☎ 0225 852143
Bookends, mobiles, giftware,
decorative mirrors. C/W
Visit by appointment only.

CERAMICS
Lorna Hobbs
Spring Farm, Carlingcott
Peasedown St John BA2 8AP
☎ 0761 432139
Studio pottery, ceramic tiles, relief
designs. C/W
Visit by appointment.

HURDLEMAKING & TRADITIONAL
WOODLAND SKILLS
Gary Rowlands
Spring Farm, Carlingcott
Peasedown St John BA2 8AP
☎ 0225 446265
Coppice work and preparation of
wood for making traditional
country chairs. Tuitition given. C
Visit by appointment.

Bristol

STAINED GLASS
Michael G Lassen
12 Ambrose Road
Clifton Wood
Bristol BS8 4RJ
☎ 0272 277191
Design and making new windows.
Also copper foil work. C/R&R
Mon-Fri 8am-5pm, Sat-Sun 8am-3pm,
lunch-time closing.

FORGEWORK
Almondsbury Forge Works Ltd
(M J Mills)
Sundays Hill, Almondsbury BS12 4DS
☎ 0454 613315
Architectural and ornamental
ironwork. Showroom, nine employ-
ees, specialists in restoration of
ironwork. C/R&R. Credit cards.
Mon-Fri 8am-6pm, Sat 8am-4pm.

LEATHER WORK
Saddlery & Sports Shop
(G A A & S J Veater)
Station Road, Clutton BS18 4RD
☎ 0761 452460
Leather work and repairs to sports
equipment. C/R&R
Mon-Fri (closed Wed) 9.30am-6pm,
Sat 9.30am-4.30pm.

FORGEWORK
Pilning Forge
(H N Sims)
White House Lane, Pilning B12 3LS
☎ 045 45 2876
Established forge producing fireside
dogs, fire baskets, canopies and
other work to order.
C/R&R/W/E. Credit cards.
Mon-Wed 8am-5pm, Thur-Fri 8am-
4pm, Sat 9am-12 noon.

TRADITIONAL HANDMADE
BABYWEAR
Patrician Designs Ltd
(Mr and Mrs Jones)
Church House, 74 Long Ashton Road
Long Ashton BS18 9LE
☎ 0275 392278
Silk christening wear, gowns, suits
hand embroidered and babywear.
Selling in Harrods and worldwide.
Awards and exhibitions. Demon-
strations arranged. C/W/E
Mon-Fri 9am-5pm, visit by appointment.

FURNITURE
A R Designs
(A Ross)
66 Providence Lane
Long Ashton BS18 9DN
☎ *0275 392440*
Experienced designer craftsman producing high quality furniture in choice of timber. Stools supplied to leading craft shops. C/W
Mon-Sat 10am-1pm, Sun 10am-5pm, lunchtime closing
Directions: beside Scout HQ (Providence Lane runs between British Legion Club & Clevedon Road B3128).

DECOY DUCKS
Des Peters
60 Ridgeway Road
Long Ashton BS18 9ES
☎ *0275 393297*
Decoy ducks and shore bird carvings, only commissioned work. Illustrated talks and instruction given. C/R&R/E.
Visit by appointment only.

CABINET MAKING
R & R Starling
The Old School, Lansdown Lane
Upton Cheyney BS15 6ND
☎ *0272 323603*
Established business, furniture restoration, trophy bases and presentation cases. C/R&R/W/E
Mon-Fri 8am-5pm, Sat-Sun by appointment.

Clevedon

GLASS ENGRAVING
Judy's Glass
(Judy Whiting)
Studio 14, Clevedon Craft Centre,
Moor Lane, Clevedon BS21 6TD
☎ *0275 879519*

Engraving and sand blasting of commemorative glassware, sports trophies and company logos. C
Tues-Sat 10am-5pm, lunchtime closing
Directions: see under Craft Centres.

GOLD & SILVER JEWELLERY
D A Stear & J J Bright
Studio 1, Clevedon Craft Centre,
Moor Lane, Clevedon BS21 6TD
☎ *0275 872149*
Jewellery including precious/semi-precious stones and silver wares. C. Credit cards.
Mon-Sat 9am-5pm, lunchtime closing, Sun 2pm-5pm.
Directions: see under Craft Centres.

Farrington Gurney

DECOY DUCKS
Decoy Art Studio
(Mrs Sophie Ridges)
Farrington Gurney BS18 5TX
☎ *0761 52075*
Decorative ducks, geese and shore birds made both in stylised (decoy) forms and feather-textured realistic forms. Decoy books and carving supplies. Britain's only school of decorative decoy carving. C/E
Mon-Fri 9am-5.30pm, lunchtime closing. Sat 9am-12noon. Other times by appointment.

Weston-super-Mare

SILVER & WOOD MINIATURES
Silver Lining (Miniatures)
(J C Walker)
34 Holm Road, Hutton
Weston-super-Mare BS24 9RF
☎ *0934 814892*
Individually constructed miniature tableware in hallmarked sterling silver, miniature woodturning (eg

lace bobbins). C/R&R/E
Visitors welcome by appointment at any reasonable time.

MODEL COPPER BOILERS
Western Steam
(R H T Walker)
Ivy House Farm Craft Workshops,
Wolvershill Road, Banwell BS24 6LB
☎ *0934 822203*
Boilers for model locomotives and traction engines manufactured to customers' requirements. Repairs of household wares. C/R&R/E
Mon-Fri 9am-5.30pm, Sat-Sun by arrangement
Directions: see under Craft Centres.

UPHOLSTERY
Brian's Upholstery
(Mr and Mrs Pirie)
West End
Ivy House Farm Workshops
Wolvershill Road, Banwell BS24 6LB
☎ *0934 820955*
Upholstery of new and antique furniture. C/R&R/W
Mon-Fri 9am-4pm, Sat 9am-12noon
Directions: see under Craft Centres.

UPHOLSTERED FURNITURE
Lounge & Lounge Ltd
(B Powell and R Tanner)
19 Coker Road
Worle Industrial Centre
Weston-super-Mare BS20 OBX
☎ *0934 517215*
Upholstered sofas and chairs traditionally crafted to suit exact requirements. 13 employees. C/R&R/E. Credit cards.
Mon-Fri 9am-6pm, Sat-Sun 9am-6pm (Sun-viewing only)
Directions: from M5 Junc 21 follow A370 to Weston, right off roundabout, opposite Sainsbury's.

BEDFORDSHIRE CRAFT WORKSHOPS

Bedford

FURNITURE MAKING &
RESTORATION
Thomas Hudson
The Barn, 117 High Street
Odell MK43 7AS
☎ *0234 721133*
A small country workshop where
fine furniture of character is
designed and handmade in the best
English tradition. Also incised
lettering, wall panelling, church
furniture, antique furniture
restoration. C/R&R/E
Mon-Fri 8am-6pm, Sat 8am-12 noon.
Visit by appointment.

CERAMIC FIGURES
Laura Dunn Ceramics
Willow Cottage, Keeley Lane
Wootton MK43 9HR
☎ *0234 852750*
Victorian and Edwardian painted
ceramic figures about 7in tall.
C/W. Credit cards.
Visitors welcomed by appointment only.

Leighton Buzzard

FORGEWORK
Mentmore Smithy
(Roger Mildred)
Stag Hill, Mentmore LU7 0QE
☎ *0296 661760*
Five employed in commissioned
decorative ironwork of all descriptions and practical fabrications.
Large gates, fire-grates and canopies. Restoration of ornamental
ironwork and general and agricultural repairs. C/R&R/W/E
Mon-Fri 8.30am-6pm, lunchtime
closing. Visit by appointment.

Wing

FURNITURE
Simon Taylor Furniture
Old Airfield, Cublington Road
Wing LU7 0LB
☎ *0296 688905*
Quality handmade fitted furniture
for kitchens, bedrooms, bathrooms
etc. Also window frames.
C/R&R/W/E. Credit cards.
Mon-Fri 9am-5pm, Sat-Sun by
appointment.

BERKSHIRE CRAFT WORKSHOPS

Bradfield

FURNITURE MAKING
Jim Crockatt Furniture
Riverside Studios, West Mills
Bradfield RG7 6HX
☎ *0734 744728 (answerphone)*
Individual furniture designed around the grain, pattern and texture of the wood, exhibited at Liberty's, Regent Street. (The workshop is one of several; others include sculpture, violin making, photography and woodworking).
C/W/E
Mon-Sat 9am-5pm. Closed August.

Burghfield

ANTIQUE FURNITURE RESTORATION
Corwell
(S Corbyn and L Newell)
Unit 6, Amners Farm
Burghfield RG3 3UE
☎ *0734 833404*
Two qualified cabinet makers and experienced French polisher restoring and making furniture to order. C/R&R. Credit cards.
Mon-Fri 8.30am-4.30pm, Sat 8.30am-12.30pm.
Directions: from M4 Junc 12 take A4 to Reading. After 2 miles turn right to Burghfield, after 2 miles turn left to Crazeley & Three Mile Cross. After 60yd entrance on to farm right.

Great Shefford

SADDLERY
The Ilsley Saddlery
(Ian Thomson)
Northfield Farm,Wantage Road
Great Shefford RG16 7DQ

☎ *0488 398947*
Saddlery workshop employing three people making and repairing saddles, horse and donkey rugs, all types of leather and canvas work.
C/R&R/E
Mon-Fri 8.30am-5.30pm.

Hungerford

SADDLERY
Roxton Sporting Ltd
(Jon Roots)
10/11 Bridge Street
Hungerford RG17 0EH
☎ *0488 682885*
Saddlery and gunsmith workshops run by established master craftsmen, fieldsports shop.
C/R&R/E. Credit cards.
Mon-Fri 9am-5.30pm, Sat 9am-5pm
Directions: in main street between Bear Hotel and the canal bridge.

POTTER & TILE MAKER
Diana Barraclough
Croftside, 25 The Croft
Hungerford RG17 0HY
☎ *0488 682292*
Wide range of domestic and decorative stoneware and Raku pottery. Individually rolled single tiles and tile panels depicting country scenes. Many commissions undertaken, some exported. C/W/E
Visit by appointment only.

THATCHING
R J Boulton
8 Cherry Grove
Hungerford RG17 0JX
☎ *0488 683000*
Member of Oxon, Berks & Bucks Master Thatchers' Association.

Newbury

CERAMICS
Ecchinswell Pottery
(Geoffrey Eastop)
Ecchinswell RG15 8TT
☎ 0635 298220
Individually made pieces mainly
for galleries and exhibitions. Pieces
in public and private collections. C/W
Mon-Sat, visit by appointment.

CERAMICS
Cherry Tree Pottery
(Harriet Coleridge)
Common Road, Headley RG15 8LS
☎ 0635 268812
Wide range of tin-glaze earthen-
ware domestic pottery decorated in
bold colours. C/W
*Visitors welcomed at any time by
appointment.*

SADDLERY
E J Wicks
(M J A Bentick, J A Nolan, S G Nolan)
1 Newbury Street
Lambourn RG16 7PB
☎ 0488 71766
Saddle making and leatherwork,
rug and horseclothing manufactured. Royal warrant (saddlers to
HM The Queen Mother), supplying
to top racehorse trainers and
jockeys, exporting to most coun-
tries. C/R&R/E. Credit cards.
Mon-Fri 8am-5.30pm, Sat 8am-12.30pm.

GARDEN FURNITURE
J C Woodcraft
(John F Clarke)
54 Maple Crescent
Shaw RG13 1LR
☎ 0635 44042
Garden seats, tables and picnic
benches designed and built to suit
customers' individual requirements.
Also childrens' playhouses,
greenhouse staging, nesting boxes,
hutches, kennels. C
Daily 10am-5pm, lunchtime closing.
Closed for holiday last week in Aug.

CONTEMPORARY FURNITURE
Robert Kilvington
Tamarisk, Pudding Lane
Brightwalton RG16 0BY
☎ 0488 2344
Studio and workshop making
quality modern furniture by
graduate from John Makepeace
School for Craftsman in Wood. Also
handpainted silk ties.
C/R&R/W/E
Credit cards.
*Daily 9am-5pm (Sat-Sun telephone
for appointment)*
Directions: from Newbury on B4494,
take 1st left to Brightwalton, 1st right at
top of hill into Pudding Lane, 2nd
house on left.

Pangbourne

FURNITURE & WOODWORK
Bryant Crafts
(Mike Bryant)
13a Pangbourne Hill
Pangbourne RG8 7AS
☎ 0734 844309
Expanding business, making,
renovating and repairing furniture
including cane, seagrass and
upholstery seating. Also clocks,
bookcases, dolls' houses, engraving,
trophies supplied and made to
order. C/R&R/W/E. Credit cards.
*Mon-Sat 10am-6pm, lunchtime
closing*
Directions: in Pangbourne village
opposite turning to Riverview Road.

Reading

GLASS ENGRAVING
Mrs Miriam Williams
3 Downs Way, Tilehurst RG3 6SL
☎ 0734 422966
Hand engraved glass. C
Tues, Fri & Sat. Visit by appointment.

FURNITURE
John Nixon Fine Furniture
15 The Street, Aldermaston RG7 4LN
☎ 0734 713875
High quality contemporary fine
furniture with an understated,
elegant and classical feel. C/E
*Mon-Fri 9am-5.30pm, lunchtime
closing. Sat-Sun 9am-12.30pm.
Visitors welcomed by appointment only.*

CERAMICS
Aldermaston Pottery
(Alan Caiger-Smith)
Aldermaston RG7 4LW
☎ 0734 713359
Thrown and press-moulded pottery
and tiles; painted tin-glaze earthen-
ware and lustre.. C/W/E
Mon-Sat 8am-5pm.

WOODWORK & JOINERY
Mayo Furniture
(Neil Layfield and Andy Laker)
*Copyhold Farm, Lady Grove
Goring Heath RG8 7RT*
☎ 0734 845466
Woodworking business making
bars, screens, counters, bookcases.
Replacement window sashes.
Laminated and heated veneer-press
work undertaken. C/R&R/W
Mon-Fri 8am-5pm, Sat 9am-5pm.

MARBLE FIREPLACES
Theale Fireplaces Reading Ltd
(J B Woosnam)
*Mile House Farm, Bath Road
Theale RG7 5HJ*
☎ 0734 302232
Eight craftsmen designing and
manufacturing mantlepieces,
panelled rooms, furniture and all
associated marble work.
C/R&R/W/E. Credit cards.
Mon-Fri 8am-5pm, Sat 10.30am-4.30pm.

Slough

CERAMICS
House of Prayer Pottery
(Jacqueline Norris)
Britwell Road, Burnham SL1 8DQ
☎ 06286 60984
Wheel and han- built semi porcelain
ceramics. C/W/E. Credit cards.
*Mon-Sat 9am-5pm, Sat-Sun 12 noon-
6pm, visit by appointment only.*

Streatley

SPINNING & WEAVING
Daffodil
(Mary Kinipple)
*4 Warren Farm Cottages, Rectory Rd
Streatley RG8 9QG*
☎ 0491 872148
Hand weaving, spinning, dyeing
mainly rugs and tapestries, experienced in Tibetan rug making. Other
textiles also woven and colour
matched. Tuition given. C
*Tues 10.30am-4pm, other days by
appointment.*

Wokingham

THATCHING
E E Sharp & Son
18 Commons Rd, Wokingham RG11 1JG
☎ 0734 781716/732805
Long established thatching contractors and consultants.

BUCKINGHAMSHIRE CRAFT WORKSHOPS

Aylesbury

CERAMICS
Speen Pottery
(Tessa Rubbra)
Valley Cottage, Highwood Bottom
Speen HP17 0PY
☎ *0494 488206*
Tableware decorated in a distinctive
style. Commemorative plates and
bowls a speciality; christening mugs,
house plaques. C/E. Credit cards.
Mon-Sat 10am-6pm
Directions: from High Wycombe on
A4128 to Speen; through village, down
hill. On sharp bend, cross over onto
unmade track following signs.

FURNITURE
Rowland Gadsden
The Furniture Workshop
Clearfields Farm
Wotton Underwood HP18 0RS
☎ *0844 23841*
Handmade furniture and furniture
restoration. C/R&R/W/E
Visitors welcomed by appointment only.

SADDLERY, HARNESS MAKING
Tedman Harness
(Mrs Fiona Tedman)
Rookery Farm, Marsh Road
Shabbington HP18 9HF
☎ *0844 201856*
Driving harnesses for carriage
horses. C/R&R/E. Credit cards.
Mon-Fri 9.30am-5pm, lunchtime closing
Directions: 4 miles from Junc 8 on
M40. Shabbington is just off the A418.

WOODWORK & CABINET MAKING
Ivor Newton & Son
(Roger Newton)
Aston Road, Haddenham HP17 8AF
☎ *0844 291461*

Business, established in 1830, now
employing three craftsmen carrying
out commissioned work on anything made in timber. C/R&R/E.
Mon-Fri 8am-5pm.

FURNITURE
Quainton Cottage Furniture
(Jeff and Sue Curtis)
83 Station Road, Quainton HP22 4BY
Handmade English hardwood
furniture, garden furniture. Joinery,
kitchens etc. Prize winners at
county shows. C/R&R/W/E
Mon-Fri 8am-5pm, Sat 8am-4pm,
advisable to telephone first.

TRADITIONAL ALE BREWERY
The Chiltern Brewery
(Mr and Mrs R G Jenkinson)
Nash Lee Road, Terrick HP17 0TQ
☎ *0296 613647*
Brewery, founded in 1980, specialising in traditional beers with local
flavour using only quality English
ingredients. Four ales, draught and
bottled. Organised tours. Shop
selling beer-related products, eg
mustards, cheeses, malt marmalade
etc. Commemorative brews. C/W
Mon-Sat 9am-5pm
Directions: on a farm on the Lower
Icknield Way B4009 2 miles from
Wendover.

LEADED & STAINED GLASS
Saxon Abbey
(B D Seaton)
21 Main Street
Weston Turville HP22 5RR
☎ *0296 612674*
Two craftsmen employed in
workshop making and repairing
leaded and stained glass windows.

C/R&R/W/E
Mon-Fri 8am-6pm, Sat-Sun 10am-4pm.

THATCHING
Philip Craker Master Thatcher
Laceys Cottage, Upper Icknield Way
Whiteleaf HP17 0LL
☎ 0844 47929
Family thatching business, dating back to 1700, using Devonshire combed wheat reed. C/R&R

Beaconsfield

WALKING STICKS & CANES
Theo Fossel Stickmaker
(Theo and Linda Fossel)
119 Station Rd, Beaconsfield HP9 1LG
☎ 0494 672349
Established business making walking sticks, crooks and canes with outlets in USA. Components supplied. C/R&R/W/E
Visitors welcome by appointment.

Bourne End

GOLD & SILVERSMITHING
Bourne End Jewellers
(Richard and Julie Bull)
57 The Parade, Bourne End SL8 5SD
☎ 0628 527022
Making and repairing quality traditional jewellery and silverware, specialising in silver cased carriage and other clocks.
C/R&R/W/E. Credit cards.
Mon-Sat 9.30am-5.30pm.

Buckingham

WOOD CARVING
Jeremy Turner Woodcarver
Radclive Manor, Radclive
☎ 0280 815984 *(evenings)*

Hand carved bowls, dishes, house signs, wallplaques, mirror frames, sculpture. C/R&R/W/E
Open daily 10am-6pm, but appointment should be made the day before visiting. Closed 1-2 weeks August.
Directions: A421 to Bicester, at crossroads right to Radclive. Down hill, over river, up hill, take next turn left to Manor and church. Through white gates (please close), workshop on right.

Chesham

PAINTING & GILDING
Jane Conroy Knott
Little Grove, Grove Lane
Orchard Leigh HP5 3QQ
☎ 0494 778080
A painting workshop which using tradition egg tempera painting and gilding. Tuition. C/R&R
Open weekends, visitors are welcomed by appointment.

JEWELLERY & FINE ART
Bob Drake
Littlegrove, Grove Lane
Orchard Leigh HP5 3QQ
☎ 0494 77420l/778080
Unique jewellery handcrafted in carved natural roundwoods. Also abstract/symbolic mixed media paintings and drawings incorporating natural materials. C/E
Mon-Fri 9am-5pm, lunchtime closing.
Weekends by appointment
Directions:2 miles from Chesham towards Hemel Hemstead on B4505 (Rushmere Lane). Turn left at Orchard Leigh into Grove Lane. House on left.

RESTORATION WORK
Michael Wallis
Norfolk Cottage
Harridge Common HP5 2UH
☎ 0240 298172

Restoration of all kinds; woodwork, glassware, brass locks, inlaid brass work, piano actions. Polishing, metal polishing, cane and rush work. Challenging jobs welcome. C/R&R
Mon-Fri 8am-6pm, Sat-Sun 9am-6pm by appointment.

Great Missenden

PAINTED CHINA & PORCELAIN
Basia
(Basia Watson-Gandy)
Squirrel Court, Hare Lane
Little Kingshill HP16 OEF
☎ *02406 5441*
Finest china, porcelain and miniatures hand-painted by internationally-known porcelain artist. 'One-offs' a speciality. Most work done to commission. C/W/E
Visitors welcomed by appointment only. Closed in January.

ANTIQUE FURNITURE
RESTORATION
Cony Crafts
(S T Naler)
Hale Acre Workshops, Watchet Lane
Great Missenden HP16 0DR
☎ *0240 65668*
Antique and modern furniture restoration and period upholstery. Polishing, cabinet work, cane and rush seating. C/R&R. Credit cards.
Mon-Fri 9am-5.30pm.

High Wycombe

FURNITURE
Stuart Groves Furniture
Kitchener Works, Kitchener Road
High Wycombe HP11 2SJ
☎ *0494 446460*
Any type and style of furniture designed and made. C/R&R/W/E
Mon-Fri 9am-6pm, Sat 9am-12noon.

ANTIQUE FURNITURE
RESORATION
Charles K Hole
Derwent, 1 Green Lane
Radnage HP14 4DJ
☎ *0494 483263*
Restoration of antique furniture, including buhl, marquetry, carved and gilded pieces, lacquered pieces. Illustrated adult education lectures.
Visitors welcome by appointment.

GLASS BLOWING & STAINED GLASS
The Glass Market
(Gerald Paxton)
Broad Lane, Wooburn Green HP10 0LL
☎ *0494 671033*
Stained glass and glass blowing. Unique decorative glass centre in 300-year old barn. Suppliers of stained glass for Euro Disney, Paris. C/R&R/E. Credit cards.
Mon-Sat 9.30am-5pm.

Long Crendon

CONTEMPORARY FURNITURE
Ralph J Martin Furniture Maker
Workshop Number Ten, Notley Farm
Long Crendon HP18 9ER
☎ *0844 291258/201779*
Designing and making fine contemporary furniture on commission and for sale at craft fairs. C/R&R
Mon-Fri 8am-5pm.

Marlow

UPHOLSTERY
David James Traditional Upholsterer
22 Bisham Village, Marlow SL7 1RR
☎ *0628 483366*
Experienced upholsterer, restoration work. Author of book on upholstery published 1990. R&R

Tues, Thur, Fri 11am-4pm, please telephone in advance.

Milton Keynes

DRESSMAKING
Snap Dragon UK Ltd
(Mr Z Lucek and Mrs P Barnard)
The Glebe, Nash Road
Whaddon MK17 0NQ
☎ 0908 501928
High-quality, distinctive children's clothing, ladies' dresses and knitwear made to original designs by Maggie Petch. Reversible frocks and smocks a speciality. Everything hand-finished. W/E. Credit cards.
Mon-Fri 9am-5pm, Sat-Sun by appointment.

North Marston

STENCILLED & PAINTED FABRICS
Church Street Designs
(M Gwynne-Jones and J Stevens)
lst Floor, Shakespeare House
North Marston MK18 3PD
☎ 0296 67540
Hand-decorated fabrics and lampshades in a craft centre/interior design business with local watercolours, pottery and furniture. C/R&R/W/E
Tues/Wed/Fri 10am-3pm, Sat 10am-5pm.

Old Amersham

CONTEMPORARY FURNITURE
Richard Williams
5 The Maltings, School Lane
Old Amersham HP7 0ES
☎ 0494 729026
Fine furniture designed specifically for customers made by prizewinning woodworker. C
Open daily 8.30am-6pm, by appointment.

Olney

CERAMICS
Olney Pottery
(Deborah Hopson-Wolpe)
Holes Lane, Olney MK46 4BX
☎ 0234 712306
Thrown and hand-built stoneware. C/W/E
Visitors welcomed by appointment
Directions: Holes Lane is a turning off the High Street at the north end of Olney.

Preston Bissett

DRIED FLOWERS
Littlebrook Farm Dried Flowers
(Mr and Mrs David Smith)
Littlebrook Farm, Pound Lane
Preston Bissett MK18 4LX
☎ 0280 848646
Family business growing and drying flowers, making up arrangements using only natural colours. C/W. Credit cards.
Open Aug-Dec only 9am-5.30pm, lunchtime closing. Please telephone during growing season.

Princes Risborough

CABINET MAKING
Peter Wilder
Lydebrook, North Mill Road
Bledlow HP27 9QP
☎ 08444 2213
High-quality furniture designed to suit customers' requirements, based on 18th and early 19th century designs; bureaux, chairs, glass-fronted bookcases, dining tables. Unusual work undertaken. Producer of lute & cittern kits. C/W/E
Mon-Fri 8am-8pm, Sat-Sun 8am-1pm by appointment.

CAMBRIDGESHIRE CRAFT WORKSHOPS

Abbotsley

FURNITURE
Ashwell Cabinets
(Stephen R Sissens)
New Barns, St Neots Road
Abbotsley PE19 4UU
☎ 0767 77922
Restoration and specialist joinery.
Fitted pine kitchens. Free standing
furniture a speciality, including
Japanese style furniture. C/R&R/W/E
Mon-Fri 8am-5pm, Sat 8am-1pm.

Cambridge

THATCHING
William F R Pepper
Wheatfields, 222 Wimpole Road
Barton CB3 7AE
☎ 0223 264856
Longstraw/water reed thatching.
Re-thatching roofs, repairs, new
extensions, surveys.
Mon-Fri 7.30am-4.30pm.

CERAMICS
Anna McArthur Ceramic Artist
Skachbow, 95 Green End
Landbeach CB4 4ED
☎ 0223 860629
Sculptural forms and vases. Inlaid
porcelain bowls and earrings.
Interlocking stoneware boxes. C/W
Mon-Sat any reasonable time, by
appointment.

MUSICAL INSTRUMENT MAKING
Richard Wilson
The Workshop, 70a Hartington Grove
Cambridge CB1 4UB
☎ 0223 411071
Bows for string instruments. C/R&R/E

Mon-Fri 8.30am-6pm, lunchtime
closing. Sat 9am-12.30pm.
Directions: From city centre take Hills
Road A1307. Pass Homerton College
on right, turn left into Hartington
Grove. 70a is beyond crossroads, on
right; car park opposite workshop.

MUSICAL INSTRUMENT MAKING
Cambridge Violin Makers
(Juliet Barker and Christopher Beament)
The Workshop, 70a Hartington Grove
Cambridge CB1 4UB
☎ 0223 411071
Violin making. C/R&R
Mon-Fri 9am-6pm, Sat 9am-1pm.
Directions: see above entry.

BOOKBINDING
Brignell Bookbinders
(Barry Brignell)
2 Cobble Yard, Napier Street
Cambridge CB1 1HP
☎ 0223 321280
Specialists in fine bindings and
restoration of old books and
manuscripts. Also boxes and
writing boxes, gold lettering.
Magazine/journal binding. Leather
table-tops. C/R&R/W/E
Mon-Fri 9am-5pm, Sat 9am-12noon
by appointment.

ARCHITECTURAL CARVING &
LETTERING
Keith Bailey's Studio
63 Eden Street, Cambridge CB1 1EL
☎ 0223 311870
Memorial tablets, headstones. Also
heraldry and sculpture. C/R&R
Visitors welcome by appointment.

LEATHERWORK
Momo The Cobbler
(Aid Mohamad Iraninejad)
Craft Centre Unit 8, Cobble Yard,
Napier Street, Cambridge CB1 1HR
☎ 0223 358209
Established specialist shoemaker frequently featured in newspaper articles. Specialist in surgical shoes and unusual fittings and sizes. Also bags, belts and other leather goods. C/R&R. Credit cards.
Mon-Sat 9am-5pm.

BRASS RUBBING
The Cambridge Mediaeval Brass Rubbing Centre
(Kristin Randall)
St Giles Church, Castle Street
Cambridge CB2 5QD
☎ 0223 464767/811621
Over 110 brasses with information sheets available for brass rubbing; charge (£2+) includes materials. Groups catered for, worksheets available for schools etc. C/W/E
May-Sept Mon-Sat 10am-6pm, winter months 1-4 pm. Other times by appointment.

CERAMICS
Angela Mellor Ceramics
5 Stulpfield Road
Grantchester CB3 9NL
☎ 0223 840528
Hand-built porcelain and stoneware. C
By appointment only.

SCULPTURE
Laurence Broderick ARBS NDD
Thane Studios, Waresley SG19 3DA
☎ 0767 50444
Sculpture in stone and bronze; figurative and abstract. Wildlife, nudes, busts. Annual exhibitions at studio. Commissions for public places, notably *The Leaping Salmon* (Chester Business Park). C/E
Open at any time, by appointment.
Please ring for directions.

Elsworth

FURNITURE
ASP Cabinet Maker
(Adrian Parfitt)
The Brick Barn, Dears Farm
Elsworth CB3 8HX
☎ 0954 7312
Contemporary country furniture combining traditional craftsmanship with modern simplicity. Boardroom tables, dining suites, cabinets and desks for home or office. Traditional joints are used, often exposed with ebony wedges as part of the design. C
Mon-Sat 9am-5pm, lunchtime closing.

Ely

ANGORA GARMENTS
Tifrum Fibre Farm
(Mrs Shona Murfitt and Mrs Mitta Rowberry)
St Owens, 7 Third Drove
Little Downham CB6 2UE
☎ 0353 698060
Five employed breeding angora rabbits, clipping and spinning their hair, then knitting into garments. C/R&R. Credit cards.
Visitors welcome by appointment.

BRASS & BRONZE CASTING
Louis Lejeune Limited
(Sir David Hughes)
The Berristead, Station Road
Wilburton CB6 3RP
☎ 0353 740444
Long established business making

original small sculpture in bronze carrying their distinctive mark. Extensive selection of car mascots and trophies. Sculpture and casting, high quality metal finishing and chasing. C/R&R/W/E
Visitors welcome by appointment.

CERAMICS & WOODWORK
Twentypence Pottery
(Peter Dodge and Chris Arnold)
Twentypence Road
Wilburton CB6 3RA
☎ 0353 741353
Two person partnership; hand thrown pottery and woodwork (turned items and furniture). C/W
Open daily 9am-6.30pm
Directions: on A1123/B1049 junction.

CABINET MAKING
New From Old
(Richard William Forward)
The Engine House, Little Ouse CB7 4TG
☎ 0353 76227
Business designing and making handcrafted domestic furniture using antique reclaimed and reconditioned timbers. C/R&R/W/E
Mon-Fri (closed Thurs) 8am-12noon, Sat-Sun by appointment.

CERAMICS
Prickwillow Pottery
(Margot and Derek Andrews)
52 Main Street, Prickwillow CB7 4UN
☎ 0353 88316
Established business in typical fenland village 4 miles E of Ely. Hand-thrown and cast pottery in stoneware and traditional slipware; sculpture. C/E. Credit cards.
Mon-Sat 9am-5.30pm.
Directions: pottery is between the Drainage Engine Museum and the Baptist church.

Harston

GLASS ENGRAVING
Roger Phillippo Glass Engraving
The Bakehouse, Church Street
Harston CB2 5NP
☎ 0223 870277
Engraved glassware, exhibited in England and abroad. Also design glassware for industry. C/R&R/W/E
Mon-Fri (closed Tues) 9am-6pm, Sat 9am-1pm.

Huntingdon

FURNITURE RESTORATION
D J Dawson
1 Safefield Farm Cottage
Alconbury Weston
Alconbury Hill PE17 5JW
☎ 0487 3376
Restoration, French polishing, wood turning, picture framing. C/R&R
Mon-Thur 8.30am-5pm.

PICTURE FRAMING
& SOFT FURNISHINGS
Taggart Gallery
(Jane Ann Taggart)
Robin Hood Cottage
Great Staughton PE19 4BB
☎ 0480 860314
Picture framing workshop, decorative cushions and covered furniture. C/R&R/W/E. Credit cards.
Wed-Sun 10am-5pm.

CERAMICS
Clover Pottery
(Desmond Clover)
5 Oldhurst Road, Pidley PE17 3BY
☎ 0487 841026
Stoneware decorated with distinctive designs and colours. C/W. Credit cards.
Mon-Fri 9am-7pm, Sat-Sun 10.30am-4pm.

FURNITURE
Mel Heald Handmade Furniture
8 Harding Way, St Ives PE17 4WR
☎ *0480 67559*
Six cabinet/furniture makers
producing all types of furniture
including copies of period pieces.
Fire surrounds. Carving. C/R&R/W
Mon-Fri 9am-5pm, Sat 9.30am-
4.30pm, lunchtime closing.

GARDEN FURNITURE
Cromwell Woodcraft
(Bryan Moore)
College Farm, Great Paxton PE19 4RN
☎ *0480 72139*
Garden benches, tables, picnic
benches and seats made in
softwood, West African iroko and
oak. C/R&R/W/E
Mon-Fri 8.30am-5pm, Sat-Sun by
appointment.
Directions: in Great Paxton, take
Adams Lane leading to College Farm.

GLASS ENGRAVING
Britton Crystal
(John and Lyn Britton)
Unit 33, Earith Business Park,
Meadow Lane, Earith PE17 3QF
☎ *0487 740374*
Hand-engraved glassware and
crystal for every occasion; gifts,
trophies, prizes etc. to order. Over
200 lines to choose from, each item
individually designed. Customers'
own glass can be engraved, if
suitable.
C/R&R/W/E. Credit cards.
Mon-Fri 8.30am-5pm, Sat 8.30am-
12noon by appointment. Closed Bank
Holidays.

Linton

BASKETWORK
English Willow Baskets
(Clare Neville)
14 Hill Way, Linton CB1 6JE
☎ *0223 893251*
Basic collection of baskets available.
Basketwork undertaken to custom-
ers' requirements. C/R&R
Open at any reasonable time,
advisable to telephone first.

Peterborough

BESPOKE FURNITURE
Thorpe Hall Carpenter's Shop
(Renato Antonelli)
Thorpe Hall, Thorpe Road
Peterborough PE3 6LW
☎ *0733 263389*
Furniture repaired, handmade and
made-to-measure with environmental bias. Pyrography and
toymaking; artefacts by local
craftspeople sold in shop. C/R&R
Mon-Fri 9.30am-6pm (lunchtime
closing) by appointment. Sat-Sun
11am-4pm showroom only.
Directions: from railway station in
Peterborough, follow Thorpe Rd over
railway, past hospital on right, over
roundabout. Thorpe Hall is on left.

CERAMIC GARDEN SCULPTURE
David Warren
50 St Pega's Road, Peakirk PE6 7NF
☎ *0733 253212*
Garden sculpture; sundials, bird
baths, planters, figures. Also some
pots in ceramic. Work in collections
nationwide. C/W/E
Open daily 10am-5pm
Directions: Peakirk is on B1443, just
off the A15 Peterborough to Lincoln
road, signposted to Wildfowl Trust.

FURNITURE
R Taylor — The Woodworker
Eastgate Cottage
Deeping St James PE6 8HH
☎ 0778 343381
Solid hardwood furniture made to customers' requirements. All types of woodwork undertaken including production and site work. Guild member. C/R&R/W/E
Open any time, advisable to telephone first
Directions: 7 miles north of Peterborough.

Royston

FURNITURE
Country Woodcrafts
(W A G Ward)
Whitehall Farm Road, Ermine Way
Arrington SG8 0AG

☎ 0223 207951
Handmade pine furniture, kitchens, general woodwork. Commissioned work. Showroom. C/R&R/W/E
Mon-Fri by appointment, Sat 9.30am-4.30pm.

Somersham

CERAMICS
Somersham Pottery
(Stuart and Jacky Marsden)
3 & 4 West Newlands
Somersham PE17 3EB
☎ 0487 841823
Three employed making handmade ceramics specialising in house signs and one-off commissions. Clients throughout the world.
C/R&R/W/E
Mon-Sat 9am-5pm, Sun 1-5pm.

CHESHIRE CRAFT WORKSHOP CENTRES

Barn Studios
(Iain Robertson)
Top Farm Centre
High Street
Farndon CH3 6PT
Several craft workshops; designer carpenter/furniture maker, textile artist, picture framing, contemporary artist, photography.
Mon-Fri 9am-5.30pm, advisable to telephone first.

Craftsmen at Arley
Arley Hall
Nr Northwich CW9 6NA
☎ 0565 777353
Independent craftsmen with workshops including: fine bone china pottery, garden furniture, cabinet making and stone masonry.
Workshops open throughout the year; see individual entries for opening
times. Arley Hall and Gardens open March to October 2-6pm.
Directions: 5 miles from M6 and M56.

Dagfields Craft Centre
(Ian Bennion)
Dagfields Farm
Walgherton CW5 7LG
☎ 0270 841336/882597
See craftsmen at work in picturesque Victorian barns set in beautiful Cheshire countryside. Crafts include: dried flowers, leatherwork, toys and woodcrafts, model farm buildings, woollens, traditional oak garden furniture, fabrics/soft furnishings. Farm animals and tea rooms.
Open daily, evenings arranged.
Parking and admission free.
Directions: on B5071 (off A51 between Nantwich and Woore)

CHESHIRE CRAFT WORKSHOPS

Aldford

CABINET MAKING
Silver Lining Workshops
(Mark Boddington and Adrian Foote)
Bank Farm, Chester Road
Aldford CH3 6HJ
☎ 0244 620200
Workshop of five craftsmen specialising in furniture using rare woods and silver. Pieces to be found in many important private and public collections in UK and abroad. C/W/E. Credit cards.
Mon-Fri 8.30am-5.30pm, Sat 9am-12.30pm.

Chester

CANDLEMAKING
Cheshire Candle Workshops
(Richard James)
Cheshire Workshops
Burwardsley CH3 9PF
☎ 0829 70401
Visitors welcomed to watch candles being made and sculpted. Also glass artist making figurines. Craft shop, restaurant, children's play area. Groups/parties especially welcomed. C/W/E. Credit cards.
Open daily 9am-5pm
Directions: 30 minutes from Chester; follow Tourist Board signs from Tarporley or Tattenhall.

BOOKBINDING
Delrue Bookbinders
(Paul Delrue)
Studio 2, 1 City Walls
Chester CH1 2JG
☎ *0244 345106*
Hand bookbinding and restoration work carried out by Fellow of Designer Bookbinders and assistant. C/R&R/E. Credit cards.
Mon-Sat 9.0am-5.30pm.

CERAMICS
Top Farm Pottery
(Willie Carter)
High Street
Farndon CH3 6PT
☎ *0978 364812*
Functional stoneware in the domestic ware tradition allowing for modern tastes in interior design. C/W/E. Credit cards.
Mon-Fri 9am-5pm, Sat 9am-1pm. Advisable to telephone if making a special journey.

BOOK & PAPER RESTORATION
Conway (Conservation)
(L D Conway, BSc, MBIM)
1 The Mill, Walkers Lane
Farndon CH3 6QY
☎ *0829 270202*
All types of repairs and restoration of books, maps and documents. Member of the Society of Bookbinders. C/repairs, restoration
Mon-Fri 9am-4.45pm, Sat-Sun by appointment.

BAROMETER RESTORATION
Derek Rayment Antiques
Orchard House, Barton Road
Barton, Nr Farndon SY14 7HT
☎ *0829 270429*
Specialist dealers and restorers of antique mercury and aneroid barometers and associated instruments. Cabinet work also undertaken.C/R&R/W/E
Open at any time by appointment.

Helsby

STAINED GLASS
Alvanley Stained Glass Studio
(Miss S L Worthington)
Ivy Dene, Alvanley Road
Alvanley WA6 9BL
☎ *0928 724871*
Design and production of hand-painted stained glass work on commission. C/R&R/E
Daily 9am-5pm by appointment.

Knutsford

FORGEWORK
DC & S Broadbent Engineering
Barnshaw Smithy, Pepper Street
Mobberley WA16 6JH
☎ *0565 872938*
Five craftsmen working in metal on traditional and contemporary designs including balustrades, dividing screens, hand rails, brackets. Full restoration on antique gates, balustrades, overthrows etc. New ironwork in Classic English style. C/R&R/W/E
Tues-Sat 10am-4pm.

Malpas

SADDLERY
J W Wycherley & Son
Church Street
Malpas SY14 8NU
☎ *0948 860316*
Master saddlers, makers of high quality riding saddles and equipment. All types of repairs under-

taken. R&R/E. Credit cards.
Open daily 8.15am-5.30pm (Weds 8.15am-1pm), lunchtime closing.

PATCHWORK & APPLIQUE WORK
Country Scene
(Pam Nicholls)
7 Church Street
Malpas SY14 8NU
☎ *0948 860031*
Quilts, cushions, wall hangings etc. Workshops with retail shop and mail order supplying needlework packs. Stencils, paints and glazes. Tuition: 'Make a Quilt a Day' and 'Creative Experimentation in Patchwork & Quilting'.
C/R&R/W/E
Tues, Thur, Fri & Sat 10am-5pm, lunchtime closing.

ANTIQUE FURNITURE RESTORATION
Marcus Moore Antiques & Restorations
(M G J Moore and M P Moore)
Holly Cottage
No Mans Heath SY14 8DY
☎ *0948 85500*
Four craftsmen restoring, repairing and repolishing all types of furniture and architectural timberwork. C/R&R/W/E
Mon-Fri 8.30am-6pm.
Directions: opposite Wheatsheaf Inn, No Mans Heath.

Middlewich

FORGEWORK
Byley Smithy
(J Clark & Son)
Byley CW10 9NF
☎ *0565 653677*
Family run village blacksmiths

making and repairing wrought ironwork, tractor repairs etc.
C/R&R/W
Mon-Fri 8am-6pm, Sat 8am-12noon.

Nantwich

KNITWEAR
Woolly Jumper
(Ian Bennion)
Dagfields Craft Centre
Walgherton CW5 7LG
☎ *0270 841336*
Knitwear and woollen goods for all the family. Large selection of patterns for custom-knit sweaters and cushions. Credit cards.
Open daily 11am-5pm
Directions: see under Craft Centres.

LEATHERWORK
Alan Kinsey
Dagfields Craft Centre
Walgherton CW5 7LG
☎ *0270 626054*
Leather handbags and belts made to measure.
C/R&R/W/E. Credit cards.
Open daily 10am-5.30pm
Directions: see under Craft Centres.

DRIED FLOWERS
The Dried Flower Workshop
(Phillip and Stephanie Wiles)
Dagfields Craft Centre
Walgherton CW5 7LG
☎ *0270 882597*
Dried flower arrangements, also selling basketware and other crafts. C/R&R/W. Credit cards.
Open daily 9.30am-6pm
Directions: see under Craft Centres.

WOODEN TOYS
Timbertoys
(Peter Cookson and Ros Fair)
Dagfields Craft Centre
Walgherton CW5 7LG
☎ *0270 842040*
Model farm and equestrian buildings with complimentary accessories; dolls' houses, forts and castles, garages etc. Also a large selection of templates, pencil trays, keyrings and badges all made on the premises. C/R&R/E. Credit cards.
Open daily 11am-5pm
Directions: see under Craft Centres.

WOODEN TOYS
The Woodcraft Workshop
(Colin Hurst)
Dagfields Craft Centre
Walgherton CW5 7LG
☎ *0270 842000*
Wooden toys, clocks, dolls' houses and other wood crafts.
C/R&R/W. Credit cards.
Open daily 9.30am-5.30pm
Directions: see under Craft Centres.

GARDEN FURNITURE
Stevenson Woodcraft
The Woodcraft Workshop
(David Stevenson)
Dagfields Craft Centre
Walgherton CW5 7LG
☎ *0270 842065*
All types of garden furniture made from coppiced oak; seats, tables, bird tables. Also archways, fencing and pergolas.
C/R&R/W/E. Credit cards.
Open daily 9am-5.30pm
Directions: see under Craft Centres.

KNITTING
Metropolitan Machine Knitting Centre
(Mark and Carol Hocknell)
The Pinfold
Poole CW5 6AL
☎ *0270 628414*
Machine knitting workshop and school with tuition for all levels. Large groups welcomed. Catering. W/E. Credit cards.
Mon-Sat 10am-4pm.

CERAMICS
Firs Pottery
(Joy and Ken Wild)
Whitchurch Road
Aston CW5 8DE
☎ *0270 780345*
Large selection of functional and decorative hand-built and thrown stoneware for sale. Tuition: Pottery-making workshops for adults and children — telephone for details. Part of South Cheshire Festival of Gardening, 1992 with workshops/ pottery trail. C/W/E. Credit cards.
Open most days all year round.
Directions: 4 miles from Nantwich on A530 to Whitchurch.

Northwich

FORGEWORK
Bartington Forge
(David Wilson)
Bartington CW8 4QU
☎ *0606 851553*
Small rural blacksmiths specialising in architectural and decorative work. Canalside setting opposite boatyard. C/R&R/W/E
Mon-Fri 8am-5pm, Sat 9.30am-12.30pm.
Directions: situated on main A49 Warrington to Whitchurch road on N side of Trent & Mersey canal.

CERAMICS
Arley Pottery
(Robert Geoffrey Dawson)
Craftsmen at Arley, Arley Hall
Northwich CW9 6NA
☎ *0565 777416*
Sculpture and hand-thrown
stoneware. Bone china figurines and
giftware. Regular exhibitions at
craft museums and galleries. Work
commissioned by businesses.
C/W/E. Credit cards.
Open daily 2pm-5pm, March to
October. (Arley Hall and Gardens
open March to October 2-6pm).
Directions: see under Craft Centres.

GARDEN FURNITURE
Arley Garden Furniture
(Peter Holt)
Craftsmen at Arley, Arley Hall
Northwich CW9 6NA
☎ *0925 601131*
Hardwood garden/patio furniture
C/W/E
Open daily 2-6pm throughout the
year.
(Arley Hall and Gardens open March
to October 2-6pm).
Directions: see under Craft Centres.

Sandbach

CERAMICS
The Potters Barn
(Andrew Pollard and Steve Marr)
Roughwood Lane
Betchton
Domestic reduction fired stoneware
and terracotta garden pots. Also
giftware. C/W/E. Credit cards.
Open daily 10am-5pm
Directions: from M6 junc 17 head
towards Sandbach on bypass. Turn
left at lights onto A533, turn right at
New Inn to Hassell Green. Turn left at
Romping Donkey over canal; pottery
on left.

LEATHERWORK
A E Collars
(Mr and Mrs A Elphick)
Studio 9, The Mill
Warmingham CW11 9QW
☎ *0270 668329*
Decorative collars, leads and
harnesses, specialising in Stafford-
shire bull terrier show equipment.
Brass stud decorated belts etc.
C/R&R/W/E
Due to show commitments,
answerphone lists open times per day.
Directions: from M6 follow to
Middlewich town centre. Follow signs
for Sandbach or Crewe, then follow
signs for Warmingham.

Tarporley

ROCKINGHORSES
W R Greenwood
15a Park Road
Tarporley CW6 0AN
☎ *0829 732006*
Two craftsmen making
rockinghorses selling in London
and New York. C/R&R/W/E
Sat 9am-5pm.

CORNWALL CRAFT WORKSHOP CENTRES

Land's End Craft Workshops
(Land's End Ltd)
Land's End TR19 7AA
☎ *0736 871501*
Six craft workshops including glass
engraving, leather work, pottery,
wood carving, jewellery. Workshops situated at Greeb Cottage,
typical Cornish croft. Many other
attractions at Land's End. Refreshments available.
Open throughout the year from 10am.
Charges for entry (free for the
disabled)
Directions: 12 miles from Penzance.

Sloop Craft Market
Capel Court, St Ives TR26 1LS
☎ *0736 796051*
Various craft workshops in covered
market
Individual workshop opening times.

The Cornwall Donkey & Pony
   Sanctuary
(TJA and AC Belton)
Lower Maidenland
St Kew, Nr Bodmin PL30 3HA
☎ *0208 84242*
Craft workshops (dried flowers and
ceramic portraits) attached to
sanctuary. Various other attractions.

Coffee shop & restaurant.
Easter-end Oct open daily 10am-5pm.
Nov-Easter open Sun only.
No charge to Craft Workshops
Directions: follow signs from St Kew
Highway on A39.

The Guild of Ten
19 Old Bridge Street, Truro TR1 2AH
☎ *0872 74681*
A co-operative shop formed by a
group of craftsmen, manned on a
rota basis. Crafts include ceramics,
jewellery, clothes, hand painted silk
& textiles, leather goods, wood
turning, wood carving, knitwear.
Mon-Sat 9.30am-5.30pm.

Tregreenwell Farm Craft Centre
St Teath, Bodmin PL30 3JJ
☎ *0208 851171*
Workshops converted from old
buildings set around a medieval
cobbled courtyard. Crafts include
ceramics, knitwear, bridal wear and
silk painting. Also antiques, crafts
gallery and tea room.
Open daily 10am-5.30pm in summer,
11am-5pm from Oct.
Directions: Situated in the Allen
Valley, just off A39 between
Wadebridge and Camelford.

CORNWALL CRAFT WORKSHOPS

Bodmin

CERAMICS
Helland Bridge Pottery
(Paul Jackson and Rosie Jenson)
Helland Bridge PL30 4QR
☎ *0208 75240*

Hand-thrown brightly decorated
earthenware to which a range of
techniques is applied (sponging,
spraying, resist etc). C/W/E
Mon-Fri 10am-4pm, lunchtime
closing.

CERAMICS
Wenford Bridge Pottery
(Seth and Ara Cardew)
St Breward PL30 3PN
☎ *0208 850471*
Established family pottery. Traditional rustic functional stoneware. Situated on the River Camel with showroom and museum, tours of workshops by arrangement. C/W/E
Mon-Fri & Sun 8am-6pm.

TEXTILE PAINTING
Images on Silk
(Jenni Milne)
Tregreenwell Farm Craft Centre
St Teath PL30 3JJ
☎ *0208 851171*
Painting on silk, greeting cards, framed pictures, waistcoats, lampshades, scarves etc. One-man shows throughout the south west. C/W/E. Credit cards.
Open daily 10am-5.30pm Easter-Oct, Wed-Sun 11am-5pm winter.
Directions: see under Craft Centres.

ANTIQUE FURNITURE
RESTORATION
Acorn Antiques
(Paul Smith)
Tregreenwell Farm Craft Centre
St Teath PL30 3JJ
☎ *0208 851171*
Repairs & restoration. Credit cards.
Wed-Sat 11am-5pm
Directions: see under Craft Centres.

KNITWEAR
Clever Ewe
(J Smith)
Tregreenwell Farm Craft Centre
St Teath PL30 3JJ
☎ *0208 851171*
A small flock of expert knitters and designers producing individual hand and machine knitted garments, some embroidered.
C/E. Credit cards.
Open daily 10am-5.30pm Easter-Oct, Wed-Sun 11am-5pm winter.
Directions: see under Craft Centres.

ARTIST
Allan Stone Watercolours
Tregreenwell Farm Craft Centre
St Teath PL30 3JJ
☎ *0208 851171*
Watercolours of landscapes and local scenes and paintings in mixed media (ink, pastel, gouache etc).
C. Credit cards.
Open daily 10am-5.30pm Easter-Oct, Wed-Sun 11am-5pm winter.
Directions: see under Craft Centres.

WATERCOLOURS
& DRIED FLOWERS
Mulberry Studio
(Jennifer Lilley)
Tregreenwell Farm Craft Centre
St Teath PL30 3JJ
☎ *0208 851171*
Watercolours and dried flower arrangements. C/E. Credit cards.
Open daily 10am-5.30pm Easter-Oct, Wed-Sun 11am-5pm winter.
Directions: see under Craft Centres.

CERAMICS
Blisland Porcelain
(Mary-Jane Hill)
Tregreenwell Farm Craft Centre
St Teath PL30 3JJ
☎ *0208 851171*
Porcelain sculptures and pots.
C/W/E. Credit cards.
Open daily 10am-5.30pm Easter-Oct, Wed-Sun 11am-5pm winter.
Directions: see under Craft Centres.

Boscastle

LEATHERWORK
The Leather Shop
(Rob & Teresa Lloyd)
The Old Mill, Boscastle PL35 0AR
☎ *0840 250515*
All kinds of leather objects made on the premises, specialising in the unusual, eg clocks, barometers and flowers. Tankards and water bottles. Leather *cuir bouilli* armoury & shields made to order.
C/R&R/W. Credit cards.
Open daily, 10am-9pm in summer, 11am-5pm in winter.

Bude

PAINTED WOODEN FIGURES
Lyn Muir
Aboukir House
Marhamchurch EX23 0HB
☎ *0288 361561*
Original wooden figures, animals and cards all hand-painted, supplying galleries and craft shops throughout the country. C/W/E
Wed 2-8pm, other times by arrangement.
Directions: just off A39 holiday route opposite village pub.

CERAMICS & JEWELLERY
Marylyn Hyde Pottery
& Richard Hyde Jewellery
Lower Stursdon
Morwenstowe EX23 9HU
☎ *0288 83412*
Individual pots fired to stoneware, based on figurative themes, for plants etc. and smaller pieces. Silver jewellery and body decoration.
C/W/E. Credit cards.
Open Wed/Sat 10am-5pm, or at any time by appointment.

Hayle

GLASSBLOWING
Norman Stuart Clarke Glass
The Glass Gallery
St Erth TR27 6HT
☎ *0736 756577*
Distinctive style of art glass vases etc. Working at the furnace, glass is decorated and blown to final form. Each piece individually made and signed. Exhibited worldwide.
W/E. Credit cards.
Mon-Fri 10am-5pm, lunchtime closing. Sat 10am-1pm.

Helston

CERAMICS
Heather Swain Pottery
The Old Chapel, Helston PL33 9EY
Thrown and hand-built tableware, slip dipped and hand-painted. Steady production supplying shops and galleries wholesale. Also mail order. C/W/E
Mon-Fri 10am-5pm, lunchtime closing. Sat-Sun 2-5pm. Advisable to telephone first.

CERAMIC SCULPTURE
Kennack Pottery
(Michael Hatfield)
Kennack Sands TR12 7LX
☎ *0326 290592*
Detailed models of animals, figures, boots etc, miniatures a speciality. Facilities for clay modelling for up to 60 people. Groups welcomed out of season. Shop and tea room.
Mon-Sat 9am-5pm (9pm in summer), Sun 10am-1pm (7pm in summer).
Directions: take A3083 to Lizard from Helston. Left onto B3293, past Goonhilly Satellite Station, right at crossroads. Past Coachhouse Café at Kuggar, left at T-junc for Kennack Sands.

CERAMICS
Trelowarren Pottery
(Nic and Jackie Harrison)
Trelowarren, Mawgan TR12 6AS
☎ 0326 22583
A wide range of oven-to-table ware
and domestic stoneware is hand-
thrown in the studio where visitors
are always welcome. Showroom
with many individually designed
pieces and a selection of hand-
woven rugs. C. Credit cards.
Open daily 10am-6pm (Sat-Sun closed
during winter).
Directions: take A3083 from Helston,
turn left onto B3293 then left at Garras
and follow drive to Trelowarren House.

Land's End

GLASS ENGRAVING
Bill Davenport
Greeb Cottage Workshops
Land's End TR19 7AA
☎ 0736 871501
Assoc Fellow of Guild of Glass
Engravers; workshop. C
Open daily 10am-5.30pm
Entry charges to Land's End
Directions: see under Craft Centres.

LEATHER WORK
Greeb Cottage
Greeb Cottage Workshops
Land's End TR19 7AA
Belts, bags, purses and other leather
goods made. C/E
Open daily 10.30am-6.30pm (closed
Jan/Feb). .Entry charges to Land's End
Directions: see under Craft Centres.

JEWELLERY
Edward Williams
Greeb Cottage Workshops
Land's End TR19 7AA
☎ 0736 50503

Jewellery designed and made in
silver and other materials.
C/R&R/E
Mon-Fri 10am-5.30pm, Sun 10am-
5.30pm (summer time only). Lunch-
time closing. Closed Dec-Mar.
Entry charges to Land's End
Directions: see under Craft Centres.

WOODWORK
Tom Brooke Woodcrafts
Greeb Cottage Workshops
Land's End TR19 7AA
☎ 0736 787143
Fretwork/woodcarving. Unusual
decorative signs (house and
personal name) made to order, also
clocks, bookends etc. C/W/E
Open daily 10am-6pm (closed Dec-
Mar). Entry charges to Land's End
Directions: see under Craft Centres.

CERAMICS
Doug Francis
Greeb Cottage Workshops
Land's End TR19 7AA
☎ 0736 763295
Handmade pottery and thrown
ceramics, pottery house nameplates.
'Throw Your Own Pot'. C/R&R/W/E
Open daily 10am-5.30pm (closed Jan-
Feb). Entry charges to Land's End
Directions: see under Craft Centres.

Launceston

TEXTILE WALL PANELS
Wendy Stedman Textiles
East Trevadlock Farmhouse
Trevadlock
Lewannick PL15 7PW
☎ 0566 82795
Knitted and woven wall panels
produced by freelance textile artist. C
Open most days by prior arrange-
ment.

CERAMICS
Trehane Pottery
(Mrs Suzannah Stack)
Trehane, Troswell
North Petherwin PL15 8NA
☎ *0566 85243*
Hand-thrown earthenware vessels
and pots for domestic use. C
Open daily 10am-6pm by appointment.
Directions: from Brazacott Cross,
pottery will be found on right on
entering hamlet of Troswell.

Liskeard

SADDLERY
Blisland Harness Makers
(Jane Talbot-Smith)
Higher Harrowbridge
Bolventor PL14 6SD
☎ *0579 20593*
Master saddler with experience in
carriage building making the full
range of saddlery and harnesses for
all animals with a complete repair
service. Leather goods include
cartridge bags and belts, bellows
and sailing equipment. C/R&R/E
*Mon-Fri 8.30am-5pm (excepting show
days), Sat-Sun by appointment.*

Mullion

WOODEN TOYS
Shoogly Lums Folk Toys
(Anthony and Judy Peduzzi)
Southernwood, Predannack Wartha
Mullion TR12 7HA
☎ *0326 240273*
Unusual and traditional moving
folk toys, magical illusions and jig-
dancing dolls made in brightly
painted wood. Suitable for all ages.
Author's book *Making Moving
Wooden Toys* available. C
Open all hours all year.

CERAMIC FIGURES
Norman Underhill
Collector's Corner
Trecarne, Meaver Road
Mullion TR12 7DN
☎ *0326 240667*
Original and humorous character
studies in earthenware. Range of
about 40 hand-modelled and signed
pieces. C/E. Credit cards.
*Open daily 10am-10pm summer,
10am-5pm winter time.*

Newquay

WOODEN TOYS
It's Childsplay
(C W Sharpe and P W Sharpe)
Treworthal Barn, Treworthal TR8 5PJ
☎ *0637 830896*
Family run business making
traditional toys, games and nursery
furniture. C/W/E. Credit cards.
*Mon-Fri 8.30am-5pm, Sat-Sun 10.30-
3.30pm*

HAND-PRINTED TEXTILES
Lorna Wiles Textiles
(Lorna Wiles and Bob Cann)
Wesley Yard, Newquay TR7 1LB
☎ *063787 6840*
Attractive and unusual co-
ordinated clothing, table linen,
cushions and scarves all made from
hand-printed textiles. C
Mon-Fri 8am-5pm.

Penzance

CERAMICS
Delan Cookson
Lissadell
St Buryan TR19 6HP
☎ *0736 810347*
Thrown porcelain decorative
pottery.

C/W/E. Credit cards.
Visitors welcome by appointment
Directions: in St Buryan, turn behind
church leading to A30 Lands End;
Lissadell on right.

Redruth

MODEL MAKING
The Dragons Lair
(Mrs Lynne Smith)
The Cottage, Piece
Nr Carnkie TR16 6SF
☎ 0209 216902
Dragons individually modelled in
brightly coloured, oven hardened
modelling clay. Dragon jewellery,
display shelves, garden sculpture.
C/W/E
Mon-Fri 10am-5pm, Sat-Sun 10am-
4pm, advisable to telephone first.
Directions: take Pool exit off A30,
follow signs to Fourlanes.

Saltash

FORGEWORK
The Forge, St Germans
(Lt Cdr Jack Prior RN CEng MIMarE)
Newport Road, St Germans PL12 5NS
☎ 0503 30411
All kinds of decorative and func-
tional metalwork, shot blasting,
welding, repair work. Commissions
undertaken for parks, magazine
publications, council authorities.
C/R&R/W
Mon-Fri 9am-6pm, lunchtime closing.
Sat 9am-1pm.

St Austell

CERAMICS
Rashleigh Pottery
(David Carew)
Wheal Martyn China Clay Museum
Carthew

☎ 0726 64014 *out of working hrs.*
Hand-thrown stoneware pottery
with freehand brushwork decoration and some reduced magnesium
glaze and sgraffito decoration. Also
terracotta garden pots.. C/W/E
Open daily 9am-6pm.

UPHOLSTERY
St Austell Upholstery
(Mr and Mrs B Beaumont)
36 Carlyon Road
St Austell PL25 4LN
☎ 0726 74626
Family business; traditional
upholstery and soft furnishings by
award winning craftsmen.
C/R&R/E
Mon-Fri 9am-5.30pm, Sat 9am-5pm.
Lunchtime closing.

St Ives

KNITWEAR
Corinne Carr Designs
Bolenna
Towednack TR26 3AP
☎ 0736 796176
Small workshop with design studio.
Production of knitted garments
(jackets, pullovers, hats). Also
jewellery and knitting wool oiled on
cone for sale. C/W/E. Credit cards.
Visitors welcome by appointment.

CERAMICS
Coldharbour Pottery
(Bob Berry)
Coldharbour Studios
Towednack TR26 3AU
☎ 0736 798316
Raku ceramics; visitors are welcome
to attend Raku firings by prior
arrangement, (charge £2). Work
exhibited throughout the south
west. C/W/E

Mon-Fri (closed Weds) 10am-4pm,
Sat-Sun by arrangement.

CERAMICS
Leach Pottery
(Janet Leach)
The Stennack
St Ives TR26 1RT
☎ *0736 796398*
Handmade studio pottery.
Credit cards.
Mon-Fri 10am-5pm, also Sat in
season.

ENAMELLING
Floyd & Gregory Enamels
Sloop Craft Market
St Ives TR26 1LS
☎ *0736 796051*
Enamelling, jewellery.
W/E. Credit cards.
Mon-Sat 10am-5pm.

WOOD CARVING
D J Oldbury
Sloop Craft Market
St IvesTR26 1LS
☎ *0736 796051*
Figurative carving in the round and
relief panels with oil/wax finish or
polychromed. C/E
Mon-Fri 10am-5pm, lunchtime
closing. Sat 10am-2pm.

WALKING STICKS
Roy Harrison
Sloop Craft Market
St Ives TR26 1LS
☎ *0736 796051*
Walking sticks and wood carvings
in English hardwoods.
R&R/W/E. Credit cards.
Mon-Fri 10am-5pm.

CERAMICS
Ginnie Bamford Pottery
Sloop Craft Market
St Ives TR26 1LS
☎ *0736 796051*
Domestic range of earthenware
pottery with sponged, slipped or
hand-painted decoration.
C/W/E. Credit cards.
Mon-Fri 10am-5pm, Sat-Sun by
appointment.

CERAMICS
Cornish Ceramics
(Graham Blow)
Sloop Craft Market
St Ives TR26 1LS
☎ *0736 796051*
Bone china hand-painted and
personalised for any occasion; free
samples on request. C/W/E
Mon-Fri 10am-4pm, lunchtime
closing.

GOLD & SILVER JEWELLERY
J E and S R Simpson
Sloop Craft Market
St Ives TR26 1LS
☎ *0736 796051*
Jewellery designed and made in
gold, silver and gemstones.
C. Credit cards.
Mon-Fri 11am-5pm, lunchtime
closing. (Closed Nov-Mar).

LEATHERWORK
Sloop Leather Crafts
(John Grey)
Sloop Craft Market
St Ives TR26 1LS
☎ *0736 796051/740321*
Leather handbags, belts etc.
C/R&R/W

Truro

CERAMICS
Carnon Downs Pottery
(Hugh West and Christine Kent)
Carnon Downs Garden Centre,
Quenchwell Road
Carnon Downs TR3 4LN
☎ *0872 863374*
Hand-thrown hand decorated
ceramics in porcelain, stoneware
and earthenware. Four potters
undertaking large and small
commissions and contracts and
for export. C/W/E. Credit cards.
Mon-Sat 9am-5pm, Sun 10am-5pm.
Directions: A39 Truro to Falmouth
Road, turn right at Carnon Downs
roundabout, 1st right into Quenchwell Road, Garden Centre 400yd on
left.

CERAMICS
John Davidson
The Pottery, New Mills
Ladock TR2 4NN
☎ *0726 882209*
Tour charge includes wine
tasting. Refreshments available.
C/W/E
Mon-Fri 10am-6pm, Sat 10am-1pm.

Wadebridge

KITE MAKING
Highflyers
(The Kite Site Ltd)
The Mowhay, Trebetherick PL27 6SE
☎ *0208 862567*
Kites, windsocks and banners
manufactured and sold in old barn.
Ultra-light, high-technology stunt
kites. Repairs undertaken. Export
worldwide. Ripstop nylon, carbon
spars and other materials for sale to
public. C/R&R/W/E. Credit cards.
Open daily 9.30-6pm.

Isles of Scilly

CERAMIC SCULPTURE
Bourdeaux Pottery
(John Bourdeaux)
The Barn, Old Town, St Mary's
☎ *0720 22025*
Ceramic sculptor working in
porcelain and stoneware, specialis-
ing in bird studies. Visitors wel-
come in workshop and gallery.
Work exhibited and held in private
collections worldwide. Winner of
Richards Scholarship (USA) 1990.
W/E. Credit cards.
Mon-Fri 10am-5pm, Sat 10am-
12.30pm (closed during Nov).

CUMBRIA CRAFT WORKSHOP CENTRES

Brougham Hall
Brougham
Penrith CA10 2DE
A 19th century residential castle undergoing restoration. Stables and servants' quarters converted to craft workshops including goldsmiths, soft furnishing & upholstery, woodturning, art metalwork. Gift shop, art gallery, museum. Tea room.
Directions: take A6 south from Penrith, take 1st left through Eamont Bridge. Approx 2 miles from Penrith.

CUMBRIA CRAFT WORKSHOPS

Alston

DESIGNER KNITWEAR
Chameleon
(Mrs Julia Neubauer)
Front Street, Alston CA9 3SE
☎ 0434 382097
Workshop producing individual knitwear in natural yarn and shop selling range of knitwear clothing, textiles and crafts.
C/W/E. Credit cards.
Mon-Sat 10am-5pm (Tues 10am-1.30pm). Closed Feb.

Ambleside

HORN, SADDLERY & LEATHER GOODS
Hide-Horn
(Peter Hodgson)
Dixons Court, 101 Lake Road
Ambleside LA22 0DB
☎ 05394 33052
Wide variety of horn work. Also saddlery and leather goods and comprehensive repair service for suitcases, locks, straps, zips etc.
C/R&R/W/E
Mon-Sat (closed Thur) 9am-5pm
Directions: next to Lakeland Sheepskin Centre. Under arch, 2nd workshop down yard.

GLASS MAKING
Adrian Sankey
Rydal Road, Ambleside LA22 9AN
☎ 05394 33039
An open workshop where visitors can watch the making of traditional and contemporary lead crystal studio glass. C/W/E. Credit cards.
Mon-Sat 8.30am-5.30pm, Sun 9.30am-5.30pm.

UPHOLSTERY/RUSH & CANEWORK
The Chair Lady
(Jennifer S Borer)
Old Stamp House Yard
Ambleside LA22 0AD
☎ 05394 32641
Master upholsterers; traditional upholstery, cane and rush seating, bespoke soft furnishings. C/R&R/E
Mon-Fri 9am-5pm, Sat 9am-12noon by appointment.

SPINNING & WEAVING
Fibrecrafts
(Meg and Martin Riley)
Barnhowe, Elterwater CA22 9HW
☎ 09667 346
Handspinning, weaving. dyeing. Introductory and advanced tuition given; residential accommodation

for long courses. Textile craft supplies and mail order. W/E. Credit cards accepted
Mon-Sat 10am-5pm Easter-Nov. Winter visiting and Sun by appointment.
Directions: Elterwater is on B5343 to Skelwith Bridge/Ambleside.

Askham-in-Furness

WOODWORK
Greenside Woodcraft
(John and Wendy Gott)
Greenside House, Duke Street
Askham-in-Furness LA16 7AD
☎ 0229 63716
Woodturning and carving, cabinet-making, antique restoration and traditional upholstery. Furniture made to traditional designs in many types of wood. C/R&R/W/E
Visitors welcome by appointment.

Brampton

WOODWORK
Peter Lloyd
The Old School House, Hallbankgate
Brampton CA8 2NW
☎ 06977 46698
Boxes made in English hardwoods individually designed around the wood, lined with sycamore, velvet, silk or leather. C/W/E. Credit cards.
View by appointment.

ANIMAL SCULPTURE
Art Ducko
(Miss Kirsty Armstrong)
12 Market Place, Brampton CA8 1RW
☎ 06977 41012
Small animal sculpture, especially ducks. Also 'chair-people'; made-to-order caricatures of people in chairs created from photographs. Gallery sells other craft work. C/R&R/W

Mon-Sat (Thur half day) 10am-5pm. lunchtime closing.

Carlisle

CERAMICS
Jim Malone Pottery
(Jim and Audrey Malone)
Hagget House, Towngate
Ainstable CA4 9RE
☎ 0768 86444
High fired stoneware pottery. Functional domestic ware, vases, bowls, jugs etc fired in traditional oriental kiln. Exhibited widely in UK and abroad. C/W/E
Open daily 8.30am-6pm.

WOVEN AND KNITTED TEXTILES
Eden Valley Woollen Mill
(Stephen Wilson)
Front Street, Armathwaite CA4 9PB
☎ 0699 2457
Wool, cotton, mohair and other natural fibres made into rugs, scarves, shawls, ties and garments. Knitting wools also for sale. Regular hand-weaving courses, telephone for details. C/W/E. Credit cards.
Easter-end Oct: open daily 9.30am-5.30pm, Nov/Dec: Mon-Sat, Jan-Easter: Sat and other times by appointment.

BASKETMAKING
Hedgerow baskets
(S J and S P Fuller)
Daffiestown Rigg, Longtown CA6 5NN
☎ 0228 791187
Strong English willow baskets for logs, shopping, pets, bicycle, linen, washing, cribs etc. Some repair work undertaken. C/R&R/wholesale.
Open to visitors at weekends only.
Directions: Take 3rd turning off A71 mile after crossing river at Longtown.

Carnforth

CHESS SETS & NURSERY FURNITURE
Robin & Nell Dale
Bank House Farm, Holme Mills
Holme LA6 1RE
☎ 0524 781646
Original handmade and hand-painted chess sets (classical book character themes, eg Alice in Wonderland). Unusual commissions welcome. Nursery furniture and original paintings. C/R&R/W/E
Mon-Fri 10am-4pm, Sat-Sun by appointment.

Cleator

FURNITURE
Michael Harbron — Crftsman in Wood
Longlands, Kinniside CA23 3AQ
☎ 0946 861662
Woodturning; high quality furniture from English hardwoods. C/R&R
Mon-Fri 9am-6pm, Sat-Sun by arrangement.

Dalton-in-Furness

SILK PAINTING & NEEDLECRAFT
Sew Special
(Jane Brierley)
Whitriggs, Tytup LA15 8JU
☎ 0229 65986
Hand-painting on silk, quilting, appliqué and embroidery.
C/W/E. Credit cards.
Mon-Sat 9am-5pm, lunchtime closing 12-1pm.

CERAMICS
Dalton Pottery
(Sue Thwaites)
8 Nelson Street
Dalton-in-Furness LA15 8AF
☎ 0229 65313
Terracotta plant pots of all sizes. Often a range of seconds available. C/W. Credit cards.
Open daily 9am-dusk. (Sat-Sun sometimes attending shows so advisable to telephone first)
Directions: from Market Street, turn into Nelson Street (Duke of Wellington on corner). Pottery is down 2nd turning on right.

Gosforth

CERAMICS
Gosforth Pottery
(Richard and Barbara Wright)
Gosforth CA20 1AH
☎ 09467 25296
Visitors are welcome to see work in progress at pottery. Throwing/ 'Have a Go' on Thurs and Sun summer months only.
C/W/E. Credit cards.
Open daily 10am-5.30pm (closed Mon Jan-Easter).
Directions: Gosforth is just off A595 West Coast Road.

Ireby

MODEL MAKING/PLASTER & RESIN CASTING
Old Barn Studios
(Kevin O'Connor and Ruth Charlton)
Ruthwaite CA5 1HG
☎ 09657 690
Giftware; miniatures, wall plaques, brooches, hand-painted depicting climbers, walkers, campers, skiers and rural scenes. Also pyrography and ceramics.
C/W/E. Credit cards.
Mon-Sat 10am-5pm.

Kendal

CABINET MAKING
Peter Hall & Son
Danes Road, Staveley LA8 9PL
☎ *0539 821633*
Family business designing and
making fine furniture. Also antique
furniture restoration, traditional
upholstery and woodturning.
C/R&R/E. Credit cards.
*Mon-Fri 9am-5pm, Sat showroom
only 10am-4pm.*

Keswick

GATES & GARDEN FURNITURE
Robin Hood Sawmill
(T & M Beattie)
Bassenthwaite CA12 4RL
☎ *07687 76437*
Small sawmill specialising in fencing,
larch gates and quality garden furniture. Visitors welcomed. C/R&R/W/E
*Mon-Fri 8am-4.30pm, other times by
appointment.*
Directions: from A591 at Castle Inn
crossroads, Bassenthwaite, take road
signed to Ireby, Uldale, Caldbeck.
Sawmill half mile on right.

UPHOLSTERY
Simon P Ireland
Unit 3, County Hotel
Southey Street, Keswick CA12 4HR
One-man upholstery business.
C/R&R/W/E
Open daily 8.30am-5.30pm.

Kirkby Stephen

GOLD & SILVER JEWELLERY
Clifford House Crafts & A J Designs
(Jane Chantler and Ailsa McKenzie)
Clifford House, Main Street
Brough CA17 4AX
☎ *07683 41296*
Wide range of contemporary gold
and silver jewellery designed and
made on the premises. Craft shop in
restored barn displaying many local
and other crafts. Coffee shop with
home baking on premises.
C/R&R/W/E. Credit cards..
*Open daily 10am-5.30pm (coffee shop
East-Oct only).*

FURNITURE MAKING
Beckside Gallery & Workshop
(J Kendall)
Scar Cottage
Ravenstonedale CA17 4NQ
☎ *05396 23259*
Fine furniture traditionally made by
hand in all types of hardwood.
Designs are based on traditional
English and period styles. Special
commissions welcome. All forms of
antique furniture restoration
undertaken and pine stripping.
C/R&R/W/E. Credit cards.
*Open daily 9am-6pm in summer,
10am-5pm in winter.*

CERAMICS
Langrigg Pottery
(Mrs J R Cookson)
Winton CA17 4HL
☎ *07683 71542*
Pots for everyday practical use that
are also attractive and pleasant to
handle. Tableware, plant pots, lamp
bases etc. Small selection of purely
ornamental plates made. Lettering
and specific designs undertaken. C
*Open on most days 9am-5pm,
advisable to telephone first.*
Directions: 1 mile from Kirkby
Stephen, just off A685 to Brough.

Milnthorpe

PAINTED TILES
Maggie Angus Berkowitz Tiles
21/23 Park Road, Milnthorpe CA7 7AD
☎ *05395 63970*
Pictures painted on industrially
produced tiles by artist with world-
wide training. Designs are unique
and site specific, working with
architect/client. C/E
Visitors welcome by appointment.

Penrith

DRYSTONE WALLING
S T & S E Allen Stonewalling
(Steven and Susan Allen)
Hillcroft, Mount Pleasant
Tebay CA10 3TH
☎ *05874 536*
Master craftsman full time profes-
sional dry stone waller; holder of
Best Waller in Britain DSWA Grand
Prix trophy. C/R&R

PAINTING & PRINT MAKING
Alan Stones
Blencarn, Penrith CA10 1TX
☎ *0768 88688*
Paintings and prints. Visitors are
welcome to visit the workshop and
gallery. C/W/E. Credit cards.
Open Wed 1-6pm, other times by
appointment.

DRIED FLOWERS
Blooms Dried Flower Shop
(Miss Carole Halliday)
39 Great Dockray
Penrith CA11 7BN
☎ *0768 63466*
Original dried flower arrangements.
C
Mon-Sat (Wed closing) 9am-5pm.

WOODTURNING
Walter Gundrey
Brougham Hall, Brougham CA10 2DE
☎ *0434 381563*
Sole craftsman producing original
work at a range of prices. Bowls,
vessels and platters. Also traditional
items, eg milking stools. C/R&R/W
Thur-Sun 11am-6pm
Directions: see under Craft Centres.

GOLD & SILVERSMITHING
The Goldsmiths Workshop
(Mark Heeley-Creed)
Brougham Hall, Brougham CA10 2DE
☎ *0768 66363*
A wide range of jewellery and
objets d'art in gold, silver and
precious stones. Awards from
Goldsmiths' Hall, London. C/R&R/W
Tues-Sun 11am-4pm
Directions: see under Craft Centres.

METALWORK
John Harrison Art Metalwork
Brougham Hall, Brougham CA10 2DE
☎ *0768 890558*
Dishes, bowls, vases, candle-
holders, brooches etc. in copper,
brass, stainless steel, gilding metal.
Beaten metalwork, antique repairs,
trophies, repoussé work.
C/R&R. Credit cards.
Open daily 9am-5pm, closed Xmas-
New Year
Directions: see under Craft Centres.

UPHOLSTERY & SOFT FURNISHING
Country Furnishings
(Dennis and Hazel Walker)
Brougham Hall, Brougham CA10 2DE
☎ *0768 890144*
Upholstery of period and modern
furniture, loose covers and curtains.
Interior design workshop and
showroom with other crafts on sale.

C/R&R/W
Mon-Fri 9am-5.30pm, Sat-Sun 1pm-5.30pm.
Directions: see under Craft Centres.

PICTURE FRAMING
Fellside Framing
(John and Pauline Macdonald)
Beck Mill, Langwathby CA10 1NU
☎ 0768 881371
Bespoke picture framing and
associated services; tapestries and
needlework, heat sealing and dry
mounting. Hand-painted line-
washed mounts. C
Open daily 9am-6pm.
Directions: from Penrith take A686 to
Langwathby, under railway, turn
right at Kirkland signpost. After 1
mile, turn left at public footpath sign
to Briggle Bridge, than half a mile to
workshop.

DOLLS HOUSE MINIATURES
Carol Black Miniatures
Sun Hill, Great Strickland CA10 3DF
☎ 0931 2330
Miniatures (one-twelfth scale) for
collectors. Wide selection in glass,
brass, copper, wood, ceramics etc
made by various craftsmen.
Miniature patchwork quilt kits.
C/E. Credit cards.
Visitors welcome by appointment only.

CERAMICS
Wetheriggs Country Pottery
(Peter Strong)
Clifton Dykes CA10 2DH
☎ 0768 62946
Terracotta and traditional English
slipware. 19th century steam
pottery housing a workshop with
museum. Tea rooms, restaurant,
shop, garden centre and caravan
park. C/R&R/W/E. Credit cards.

Open daily 10am-6pm.
Directions: 4 miles south of Penrith off A6.

Sedbergh

ANTIQUE FURNITURE
RESTORATION
Merlin Restorations
(R Udall)
Merlin Cragg, Flowergill LA10 5HU
☎ 05396 20719
Family business restoring period
furniture. C/R&R/W/E
*Mon-Fri 9.30am-5pm, lunchtime
closing.*

FURNITURE
Little Oak Furniture
(Avril and Michael Priestley)
*Bridge End Cottage, Denthead
Dent LA10 5QX*
☎ 05875 330
Hand-crafted traditionally made
oak and pine furniture. Also wood
carvings in limewood, turnings in
burr woods. Workshops and
showroom. C
*Open daily 9am-5pm, lunchtime
closing.*

WOOLLEN MILL
Pennine Tweeds
(Bryan and Carol Ann Hinton)
Farfield Mill, Sedbergh LA10 5LW
☎ 05396 20558
Working Victorian woollen mill on
the banks of River Clough. Weaving
using traditional Dobcross looms
from 1930s; warping, winding,
weaving, mending. Tweeds, cloth,
wool, travel rugs, scarves & shawls.
Retail showroom. C/W. Credit cards.
*Open daily 9.30am-5.30pm, (closed
Sun Nov-Apr).*
Directions: 1 mile from Sedbergh on
A684 Hawes road.

CABINET MAKING
Colin Gardner
Stone House , Cowgill
Dent LA10 5RL
☎ 0587 5380
Specialist in gun-cabinets tradition-
ally made in solid hardwood; any
specific requirement met. Also
Welsh dressers, tables, chairs etc.
and commissioned items. C/R&R
Mon-Fri 9am-5pm, lunchtime closing.
Directions: 5 miles east of Dent
towards B6255.

EARLY MUSICAL INSTRUMENTS
Dennis Wooley
Tubhole Barn, Dent LA10 5RE
☎ 0587 5361
Harpsichords, spinets, virginals
made to order in a variety of sizes
and styles based on classic originals
from Europe. Also forte pianos and
early grand pianos. C/R&R/E
Visitors welcome by appointment.

Ulverston

LEAD CRYSTAL
Cumbria Crystal Ltd
Lightburn Road, Ulverston LA12 0DA
☎ 0229 54400
High quality hand-crafted crystal
tableware and giftware; engraved
items and trophies. 36 employees
who can be viewed glass making
and hand-cutting. Engraving
service, retail shop.
C/R&R/W/E. Credit cards.
*Factory: Mon-Thur 9am-4pm, Fri
9am-3pm. Shop: Mon-Fri 9am-5pm,
Sat 10am-4pm (June-Sept 12noon-
4pm)*
*Charges: adults 90p. Children & OAPs
40p. Family ticket £2. Party bookings
(15+) 25p each.*

DOLLS' HOUSES
Dolls' House Man
(Anthony and Julie Irving)
Furness Galleries, Theatre Street
Ulverston LA12 7AQ
☎ 0229 57657
Workshop where dolls' house
making can be viewed. Showroom
with houses and miniatures on sale.
Also wooden toys, model farms.
Commissions from abroad; televi-
sion appearances. Coffee room.
C/R&R/W/E. Credit cards.
*Mon-Fri (closed Wed) 10am-5pm, Sat
9.30am-5pm.*

Wigton

GOLD & SILVERSMITHING
Jewellery by Michael King
Oakleigh, Todd Close
Curthwaite CA7 8BE
☎ 0228 710756
Exclusively designed hallmarked
jewellery set with precious and
semi-precious stones. Has designed
for many notable people, including
members of the Royal Family.
C/R&R/E
Tues-Sat 9.30am-5.30pm
Directions: Under 1 mile from
Thursby off A595.

WOODTURNING
Maurice Mullins
Brickhouse Cottage
Hesket Newmarket CA7 8HY
☎ 06998 645
Renowned woodturner producing a
wide range of work including
highly precisioned boxes and art
forms, large decorative bowls and
semi-sculptural forms. Work sold
nation-wide and abroad. Tuition at
all levels. Award winner and TV
appearances.C/W/E. Credit cards.

Mon-Fri 9am-6pm, Sat-Sun 9am-12.30pm, visitors welcome by appointment only; please telephone before making the journey.

FURNITURE
Ian Laval — Cabinet Maker
Meadow Bank Farm
Curthwaite CA7 8BG
☎ *0228 710409*
Well-made furniture distinguished by the extensive use of inlays and highly figured veneers which are sawn in the workshop. Both contemporary and traditional designs using hardwoods felled by Ian Laval. C. Credit cards.
Mon-Fri 9am-5pm, telephone at weekends.
Directions: Under 1 mile from Thursby off A595

CERAMICS
Mike Dodd Pottery
Wellrash, Bolton Gate
Nr Wigton CA5 1DH
☎ *0965 7615*
Comprehensive range of hand-thrown pots using many natural materials from the Lake District. Widely exhibited in UK and abroad. C/W/E
Visitors welcomed by appointment only.
Directions: approx midway between Carlisle and Cockermouth on A595 sign to Weary Hall/Thornthwaite just east of Mealsgate. Take turn, continue for just over 2 miles until pottery sign on right.

DERBYSHIRE CRAFT WORKSHOP CENTRES

Caudwell's Mill Craft Centre
(Richard and Patricia Priestley)
Bakewell Road, Rowsley DE4 2EB
☎ 0629 734374
Craft workshops including glass blowing, ceramics, woodturning, furniture making and restoration. Shop, gallery, restaurant. Working flour mill with guided tours for visitors, school parties and adult groups. Nature trail.
Open daily 10am-6pm summer time, 10am-4.30pm winter time. (Jan-Feb weekends only).
Directions: situated on A6 between Matlock and Bakewell where Rivers Wye and Derwent meet. From Bakewell take 1st right after mill entrance. From Matlock, take left turn opposite Peacock Hotel.

Derbyshire Craft Centre
Calver Bridge, Nr Baslow
☎ 0433 31231

Lathkill Dale Craft Centre
Manor Farm, Over Haddon DE4 1JE
☎ 0629 812390
Award winning craft workshop centre; ceramics, furniture, clocks and barometers, needlework, stained glass, fine art. Gift shop and tea room. Mostly on ground floor level with facilities for the disabled. Also English Nature offices.
Open daily 10am-5pm.
Directions: situated at Over Haddon, 2 miles south west of Bakewell. From Bakewell take B5055 to Monyash; signpost to craft centre after 1 mile.

Melbourne Hall Craft Centre
Melbourne DE7 1EN
☎ 0332 862502
Once the home of the Victorian Prime Minister, William Lamb, 2nd Viscount Melbourne, the house and grounds are open to the public. Craft centre, shop and tea room open most days throughout year.
Directions: 9 miles south of Derby on B587.

Ridgeway Cottage Industry Centre
Kent House Farm
Ridgeway S12 3XR
☎ 0246 231111
Conversion of old farm providing 14 craft workshops near Eckington. Includes gift shop and tea room.
Tues-Sun 10.30am-5pm, also Bank Holidays.
Directions: On B6054 from Sheffield at Ridgeway turn right into Main Road and left opposite Swan Inn.

DERBYSHIRE CRAFT WORKSHOPS

Ashbourne

TRADITIONAL CHAIRS
M and G Tolley
The Old Post Office
Bradley DE6 1PG
☎ 0335 370112
Specialists in handmade Windsor country chairs to traditional designs using locally grown timbers where possible. All chairs made to customers' specification. C/R&R/E.
Mon-Fri 9am-5.30pm, Sat 10am-5pm, Sun by arrangement.

FURNITURE
Neil A Clarke
High Croft, Chapel Lane
Kniveton DE6 1JP
☎ *0335 43901/43604*
Contemporary and traditional
furniture. Also small wooden boxes.
C/R&R/W/E.
Mon-Fri (closed Tues) 9am-5pm, Sat
10am-12noon.

ANTIQUE FURNITURE
RESTORATION
Spencer Crafts
(Peter R Spencer)
The Riddings Farm, Kirk Ireton DE4 4LB
☎ *0335 370331*
Cabinet maker and upholsterer
repairing and restoring all types of
furniture; woodturning, carving,
veneering, polishing. Commissions
undertaken for internal quality
woodwork. C/R&R
Mon-Fri 9am-5pm, Sat-Sun 10am-5pm.
Directions: 2 miles from Hulland Ward
on A517 Ashbourne to Belper road. 2
miles from B5035 Ashbourne to
Wirksworth road off Hognaston by-pass.

GLASS BLOWING
Derwent Crystal Craft Centre
(Keith Doggett)
off Shaw Croft, Ashbourne DE6 1GH
☎ *0335 45219*
Crystal glass made; blowing and
cutting/decorating. Engraving and
sandblasting servic, re-grinding/
repairs. Established business
employing 16. Commissioned work
includes pieces for the Royal
Family. C/R&R/E. Credit cards.
Factory shop: Mon-Sat 9am-5pm
including Bank Holidays. Workshop
and glassblowing demonstrations:
please telephone for details.

Bakewell

ANTIQUE FURNITURE
RESTORATION
Stephen Simmons & Helen Miles
Caudwell's Mill Craft Centre
Rowsley DE4 2EB
☎ *0629 732227*
Partners working only to commis-
sion restoring antique furniture and
French polishing using traditional
methods and materials. Short
courses available. C/R&R
Open daily by appointment, 10am-4pm.
Directions: see under Craft Centres.

CERAMICS
Susan Mulroy Ceramics
Caudwell's Mill Craft Centre
Rowsley DE4 2EB
☎ *0629 733185*
Thrown stoneware specialising in
ash glazes. C/W/E. Credit cards.
Mon-Fri 10am-6pm summer time,
10am-5pm winter time. Sat-Sun
10am-6pm.
Directions: see under Craft Centres.

CERAMICS
Lori Cannon Ceramics
Caudwell's Mill Craft Centre
Rowsley DE4 2EB
☎ *0742 757828*
Multi-influenced domestic and
decorative ceramics in high fired
stoneware. Each piece is individu-
ally handmade and unique in
character. C/W/E. Credit cards.
Open daily 10am-6pm.
Directions: see under Craft Centres.

GLASS MAKING
Greenhalgh Glass
(Darrell and Joy Greenhalgh, BAHons)
Caudwell's Mill Craft Centre
Rowsley DE4 2EB

☎ 0773 520671

Visitors welcome to watch making of studio glass. Rich layers of colour given to molten lead crystal by adding gold and cobalt, silver and other metals, using traditional skills. Work widely exhibited.
C. Credit cards.
Open daily 10am-5pm.
Directions: see under Craft Centres.

FURNITURE
Dan Shotton Handmade Furniture
The Old Sawmill Yard
Rowsley DE4 2EB
☎ 0629 733935
Furniture making of all types and specialised joinery. Pine traditional styles and oak and mahogany reproductions. C/R&R. Credit cards.
Open daily 9am-5.30pm, lunchtime closing 12.30-1.30pm.

CERAMIC SCULPTURE
Kate Thorpe
Studio 5, Lathkill Dale Craft Centre,
Manor Farm, Over Haddon DE4 1JE
☎ 0629 812851
Distinctive fine ceramics in stoneware and porcelain made by trained artist/Royal Doulton designer. Work widely exhibited and becoming well known. Also murals, *trompe l'oeil* and other modern designer techniques. C/E
Open daily (closed Tues) 10am-5pm, lunchtime closing.
Directions: see under Craft Centres.

ARTIST
Romey T Brough NDD ATD
Studio 4, Lathkill Dale Craft Centre
Over Haddon DE4 1JE
☎ 0629 814337
Oil paintings and works in mixed media including greeting cards.

C/W/E. Credit cards.
Open daily (closed Mons) 10am-5pm, lunchtime closing.
Directions: see under Craft Centres.

STAINED GLASS
Glasslights
(Rosalind Ann Jones)
Studio 2, Lathkill Dale Craft Centre
Over Haddon DE4 1JE
☎ 0629 815112
Stained glass work, mainly windows made to order. Also lampshades, mirrors, terrariums, jewellery. C/R&R. Credit cards.
Open daily (except Wed) 10am-5pm.
Directions: see under Craft Centres.

CHILDRENS' DRESSMAKING
Heirs & Graces
(Ann Esders)
Lathkill Dale Craft Centre
Over Haddon DE4 1JE
☎ 0629 814335
Christening robes and traditional childrens' clothes. Cot layettes and crib dressings. Also ladies' lingerie and nightwear. C/R&R/E
Open daily 10.30am-5.30pm.
Directions: see under Craft Centres.

CLOCKS & BAROMETERS
Green Dale Crafts
(Joseph Hardy)
Studio 8, Lathkill Dale Craft Centre
Over Haddon DE4 1JE
☎ 0629 813923
Clocks and barometers old and new. Also repairs and restoration by member of British Horology Institute and instrument maker. C/R&R
Mon-Wed-Fri 10am-5pm, Sat-Sun 10.30am-5.30pm.
Directions: see under Craft Centres.

FURNITURE & ROCKING HORSES
M D Smith
Lathkill Dale Craft Centre
Over Haddon DE4 1JE
☎ *0629 812797*
Furniture designed and made to
customers' requirements. Clock
cases, rocking horses, settles, small
boxes etc. Solid timber construction
a speciality. C/R&R
Open daily except Thurs 10am-5pm.
Directions: see under Craft Centres.

Brailsford

MODEL MAKING
Cannon Craft
(P Torry and A Fletcher)
Sundial Farm, Brailsford DE6 3DA
☎ *0335 60480*
High quality engineering work.
Products include models of 19th-
century cannons and field guns
made from brass and oak, brass
sundials and armillary spheres,
rolling ball clocks. Engraving of
most metals. C/R&R
Mon-Fri 9am-5pm, lunchtime closing
(advisable to telephone first).

GOLD & SILVERSMITHING
Peter J George Goldsmiths
Saracens Head Coaching House Yard
Brailsford DE6 3DA
☎ *0335 60994*
Specialist in handmade gold
jewellery; handmade chains,
bracelets, rings etc. Commissioned
work for people of note. C/R&R
Mon-Sat 9am-5pm.
Directions: on the main road, Brailsford.

WOODTURNING
Reg Slack
Saracens Head, Coaching House Yard
Brailsford DE6 3AS

☎ *0335 60829*
Woodturning tuition, woodturners'
supplies, turned items. Courses run
by widely experienced wood
machinist vary in length (meals
provided). Local accommodation
can be arranged. C/R&R
Mon-Wed, Sat-Sun 9am-5pm.

Buxton

FURNITURE MAKING
The Wood Factor
(Andrew Heywood)
The Old Smithy, Church Street
Buxton SK17 6HD
☎ *0298 71673*
Woodturning and furniture, much
of it influenced by 19th century Arts
and Crafts movement. Workshop
and showroom (also displaying
paintings by local artists) in one of
oldest buildings in Buxton behind
St Anne's church. C. Credit cards.
Open daily (except Wed) 10am-5pm,
lunchtime closing.

Chesterfield

DRY STONE WALLING
Joseph Hill
90 Holymoor Road
Holymoorside S42 7DX
☎ *0246 569433*
All aspects of dry stone walling
undertaken by qualified instructor
and examiner. Special garden
features (wishing wells, stiles, steps,
curved walls). C/R&R

DRY STONE WALLING
Andy Craig
90 Holymoor Road
Holymoorside SH2 7DX
☎ *0246 569433*
All aspects of dry stone walling

undertaken by qualified instructor and examiner. C/R&R
Mon-Sat 8am-5pm.

Coxbench

CERAMICS
Glyn Ware Reproductions
(Glyn Colledge)
Hilltop Cottage, Horsley Lane
Coxbench DE2 5BH
☎ *0332 880705*
Decorative stoneware; vases, bowls, lamps, jugs, mugs and tankards. Reproductions of Denby pottery 'Glynware' made in the 1950s, using same materials and techniques. C/E
Open daily, visitors welcome by appointment.

Crich

CERAMICS
Crich Pottery
(Diana and David Worthy)
Market Place, Crich DE4 5DD
☎ *0773 853171*
Prize-winning pottery with six craftsmen making high quality stoneware on the potters' wheel using a unique method of glazing. Sold throughout England and abroad. Seconds' shop. C/W/E
Mon-Fri 9am-6.30pm, Sat 9am-6pm, Sun by appointment.

FURNITURE
Ram Furniture
(Keith Fretwell)
Cliffside Workshop
Crich DE4 5DP
☎ *0773 857092*
Makers of traditional furniture in solid English oak. C/R&R/E.
Mon-Fri 8.30am-5pm, Sat 9am-12noon.

Denby

CERAMICS
Denby Pottery Visitors Centre
Derby Road, Denby DE5 8NX
☎ *0773 743644*
Large manufacturers of stoneware pottery made from Derbyshire clay. Factory tours, shop, restaurant, museum. Potting can also be seen in craftroom. C/W/E. Credit cards.
Open daily (Sat-Sun shop/Craftroom only) 9.30am-4.30pm. Factory tours weekdays only, last tour 3.30pm. Pre-booked tour charge: £1.75 adults, £1 OAPs & children. Craftroom charges: £1.25 adults, £1 OAPs & children.
Directions: 2 miles south of Ripley, 8 miles north of Derby on B6179.

Derby

PRESSED FLOWER ARTWORK
Consider The Lilies/Little Trees Studio
(Mrs Angela Thomas)
5 Highfield Road , Little Eaton DE2 5AG
☎ *0332 833492*
Commemorative pictures, greeting cards and other artwork. C
Tues, Wed & 1st Sat of month 11am-4pm. Closed 15 Dec-31 Jan.
Directions: 4 miles north of Derby off B6179. Near village school turn into New Street and into Highfield Road.

CANE & RUSH WORK
Joan Gilbert
50 Ashbourne Road, Derby DE3 3AD
☎ *0332 44363*
Experienced caneworker restoring chairs, settees, bedheads etc. All patterns of cane, restoration of furniture upholstered instead of caned. Also seagrass work. Recommended by *The Times*, 1991. C/R&R
Mon-Fri 10am-4pm, (away August).

FURNITURE
Matthew Morris
Markeaton Park Craft Village
Derby DE3 3BG
☎ 0332 298460
Furniture handmade to classical
and modern designs in English and
exotic hardwoods by well qualified
craftsman. C/R&R/W/E
Mon-Fri 8am-6pm, Sat 9am-4pm,
lunchtime closing.

Eckington

PICTURE FRAMING
Picturesque
(John M Hogan)
Gallery & Upper Court
Ridgeway Cottage Industry Centre
Kenthouse Farm, Ridgeway S12 3XR
☎ 0742 477238
Picture framing and decorative
mountwork. Art conservation.
R&R. Credit cards.
Tues-Sun 10.30am-5.15pm.
Directions: see under Craft Centres.

SADDLERY
R Birdsall Saddler
West View
Ridgeway Cottage Industry Centre
Main Road, Kenthouse Farm
Ridgeway S12 3XR
☎ 0246 230286
City & Guilds qualified saddler.
C/R&R/W. Credit cards.
Open daily 10.30am-5pm.
Directions: see under Craft Centres.

STAINED GLASS
Tiffany Land Stained Glass
(Sylvia Allen)
The Garden Room
Ridgeway Cottage Industry Centre
Kenthouse Farm, Ridgeway S12 3XR
☎ 0742 477104

Stained glassware; gifts, terrarium,
lampshades, lamp bases etc.
C/R&R/E. Credit cards.
Tues-Sun 10.30am-5pm and Bank
Holidays.
Directions: see under Craft Centres.

MACRAMÉ & DRIED FLOWERS
Marilyn's Macramé & Flowers
(Marilyn Cartwright)
Ridgeway Cottage Industry Centre
Kenthouse Farm, Ridgeway, S12 3XR
☎ 0742 476953
Designs in macramé, also kits and
materials. Dried flower arrange-
ments in variety of containers.
C/W. Credit cards.
Tues-Sun 10.30am-5pm and Bank
Holidays.
Directions: see under Craft Centres.

GOLD & SILVERSMITHING
Colante (Sheffield) Ltd
(A H, M A and S R Bisby)
Ridgeway Cottage Industry Centre
Kenthouse Farm, Ridgeway S12 3XR
☎ 0742 477028
Three craftsmen producing a large
range of jewellery in gold and silver
with precious stones.
C/W/E. Credit cards.
Tues-Sat 9.30am-5.30pm, Sun 2.30-
5.30pm.
Directions: see under Craft Centres.

Grindleford

FURNITURE
Andrew Lawton
Goatscliffe Workshops
Grindleford S30 1HG
☎ 0433 631754
Fine furniture made mainly to
commission. Small showroom with
pieces on sale. C/E
Mon-Fri 9am-5pm, Sat 10am-4.30pm.

Hartington

CERAMICS
Rookes Pottery
(David and Catherine Rooke)
Mill Lane, Hartington SK17 OAN
☎ *0298 84650*
Visitors welcome to watch four craftsmen producing terracotta garden pots. Commemorative items undertaken. Also mail order. C/W/E. Credit cards.
Mon-Fri 9am-5pm, Sat 10am-5pm, Sun 11am-5pm.

Hulland Ward

CABINET MAKING
W K Morley
Main Road, Hulland Ward DE6 3EX
☎ *0335 370175*
Furniture made by award-winning cabinet maker. C/E
Mon-Sat 8.30am-6.15pm, lunchtime closing. Advisable to telephone for appointment.

MUSICAL INSTRUMENT MAKING
Northworthy Musical Instruments
(Alan Marshall & Warwick Stevenson)
Above Hulland Motors, Main Road Hulland Ward DE6 3EA
☎ *0335 370806*
Two craftsmen making acoustic and electric guitars, basses, mandolins, mandolas, dulcimers. C/R&R/W/E
Mon-Fri 9.30am-5.30pm, Sat 9.30am-1pm (hours flexible).

Matlock

GLASS BLOWING
Hothouse
(Paul Barcroft and Anthony Wassell)
7 Lumsdale Mill
Lower Lumsdale DE4 5EX
☎ *0629 580821*
Handmade studio glass; perfume bottles, vases, bowls, paperweights. One of Britain's leading contemporary glass studios set in a picturesque conservation area. Visitors are welcome to watch glassblowing. Wheelchair access at rear. C. Credit cards.
Mon-Fri 10am-4pm, Sun (seasonal) 11am-5pm, lunchtime closing.
Directions: Lumsdale is signed from A615 Matlock to Alfreton/Nottingham road.

KNIFE MAKING
Harry Boden Hand Crafted Knives
Via Gellia Mill, Bonsall DE4 2AJ
☎ *0629 825176*
Knife making, leather belts and pouches. Repair of golf clubs. C/R&R
Mon-Fri 8.30am-5pm (lunchtime closing 1.30-2.30pm). Sat 9am-12noon.
Directions: under 1 mile from Cromford on A5102 Buxton road.

RURAL CRAFTS
Dethick Crafts
(Harold Groom/Joy Stephenson)
Manor Farm, Dethick DE4 5GG
☎ *0629 534207/534246*
Signwriting, hedgelaying, hurdle making, spinning, sheep fleeces, knitting wools. Accommodation. C/E
Open daily 10am-4pm
Directions: from A6 take road to Lea until sign for Dethick/Riber. From M1 junc 28 follow road for Matlock, turn left at sign for Dethick.

STAINED GLASS
Stained Glass Designs
R M and Y M Severn
Unit 205, Via Gellia Mills
Bonsall DE4 2AJ
☎ *0629 825422*
Repair, re-leading and construction

of all leaded and stained glazing. Coloured leaded units for double glazing. Reproduction Victorian glass. Sealed units. C/R&R/W/E. Credit cards.
Mon-Fri 9am-5pm, Sat 9am-12.30pm.
Directions: under 1 mile from Cromford on A5102 Buxton road.

WOODTURNING
Chapman Woodturning
(David Chapman)
Unit 307 Via Gellia Mill
Bonsall DE4 2AJ
☎ *0629 636068*
Woodturning, mainly copy lathe work for furniture/building trades. Also one-off items and restoration work. C/R&R/W/E
Mon-Fri 9am-5pm, Sat-Sun various.
Directions: under 1 mile from Cromford on A5102 Buxton road.

FURNITURE
Nigel Griffiths
The Old Cheese Factory
Grangemill DE4 4HU
☎ *0629 650720*
Furniture handmade in finest quality quarter-sawn English oak. Each signed piece produced by one craftsman. C/E
Mon-Sat 8.30am-5pm.
Directions: Grangemill is on crossroads of A5012/B5056 west of Matlock.

CERAMICS
Pottery Workshop
(Josie Walter)
Via Gellia Mill, Bonsall DE4 2AJ
☎ *0629 825178*
Individual handmade earthenware pottery, mostly domestic ware decorated with coloured slips. Designs based on fruit and vegetables, flowers and fish. C/W/E

Mon-Sat 10am-5pm (closed 15 Dec-3 Jan). Advisable to telephone first.
Directions: under 1 mile from Cromford on A5102 Buxton road.

GLASS ENGRAVING
Glass Scribe
(John David Balcombe)
Yew Tree Cottage, Willersley Lane
Willersley DE4 5JG
☎ *0629 57917*
Glass etching, engraving, sand blasting. Windows/door panels, shower screens, tableware, mirrors. Window hanging plates (leaded). C/R&R/W/E
Mon-Fri 8am-5pm, Sat 8am-12noon.

Melbourne

BASKETS, RUSH & CANEWORK
B T & M Sturgess
Melbourne Hall Craft Centre
Melbourne DE7 1EN
☎ *0332 864663*
Makers of all kinds of English willow baskets. Also restoration and chair re-seating in rush and cane. Also a ntique furniture restoration.
C/R&R/W/E. Credit cards.
Tues-Sat 10am-4.30pm.
Directions: see under Craft Centres.

FURNITURE RESTORATION
Melbourne Hall Furniture Restorers
(Neil Collumbell)
Melbourne Hall Craft Centre
Melbourne DE7 1EN
☎ *0332 864131*
C/R&R
Mon-Fri by appointment. Sat-Sun 1.30-4.30pm.
Directions: see under Craft Centres.

PICTURE FRAMING
Townsend Picture Framers
(Albert and Julia Townsend)
Melbourne Hall Craft Centre
Melbourne DE7 1EN
☎ *0332 862461*
Picture frames, mounts, etc. Oils,
watercolour pictures and prints.
C/R&R/W/E. Credit cards.
Tues-Sat 9am-5pm, Sun 1-5pm.
Directions: see under Craft Centres.

LACE WORK
Nosmit Lacecraft
(I Timson)
Melbourne Hall Craft Centre
Melbourne DE7 1EN
☎ *0332 863655*
Household linen and lingerie; all
machine lace work, particularly
Nottingham lace. Handmade lace
from abroad. Items made or altered
while visitors browse around craft
centre. R&R
Wed, Sat & Sun 11am-5pm.
Directions: see under Craft Centres.

Shardlow

HEDGELAYING/STICK-MAKING/
HORNWORK
Coppice Crafts
(R J McConnell)
The Coppice, Ambaston Lane
Shardlow DE7 2GU
☎ *0332 792739*
Award-winning craftsman produc-
ing horn and antler work, walking
sticks and shepherds crooks,
wildlife wood carvings. Jan-Mar
(Mon-Wed) hedgelaying. C/R&R/E.
Jan-Mar Thurs-Fri 9am-6pm, April-Dec
Mon-Fri 9am-6pm. Sat-Sun 9am-6pm.
Directions: from M1 Junc 24 take A6
west. Coppice crafts on western end
of Shardlow beyond B6540 junc.

Tideswell

GEM & MINERAL WORK
Tideswell Dale Rock Shop
(Don Edwards)
Commercial Road, Tideswell SK17 8NU
☎ *0298 871606*
Pietra Dirc inlay in Ashford Black
marble; jewellery in Blue John
stone. Leading dealers in mineral
and fossil specimens for collectors.
Exhibitors in USA. E. Credit cards.
Thur-Sun 10am-5.30pm, lunchtime closing.

West Hallam

CERAMICS
Bottle Kiln Pottery & Gallery
(Charles and Celia Stone)
High Lane West
West Hallam DE7 6HP
☎ *0602 329442*
Unusual and distinctive salt-glazed
ceramic sculpture based on fantasy
and poetic themes. Fine-art gallery
and café and a conifer tree nursery
on the premises. Credit cards.
Tues-Sun 10am-5pm.
Directions: 8 miles from Derby on A609.

Wirksworth

ANTIQUE FURNITURE
RESTORATION
Frank Pratt
(T C Jones)
Old Grammar School, Church Walk
Wirksworth DE4 4DP
☎ *0629 822828*
Established business employing
three craftsmen restoring antique
furniture, also handmade and hand-
carved furniture. C/R&R/E
Mon-Fri 8.30am-5pm, lunchtime closing.
Sat 9am-1pm or by appointment.

DEVON CRAFT WORKSHOP CENTRES

Bickleigh Mill Craft Centre & Farms
(W V Shields)
Nr Tiverton EX16 8RJ
☎ *0884 855419*
Many craft skills demonstrated in and around the old watermill, craftwork displayed in large shop.
Open daily Easter-Dec 10am-6pm, Jan-March weekends only. Admission charge.

Devon Rural Skills Trust, Cockington Court
Cockington Village, Torquay TQ2 6XA
☎ *0803 605377*
Visitors to the Devon Rural Skills Centre can see craftspeople working at their many skills, eg stained glass, etching, patchwork, rush & cane seating, chair bodging, wheelwrighting etc. Periodic demonstrations of various traditional rural skills (telephone for further information). Organic garden where help and advice is available. Licenced café and craft shop. Country park with lakes, church and horse riding.
Open daily Apr-Oct 10.30am-5pm. Admission charge: £1.00 adults, 50p children.

Otterton Mill Centre
(Director: Desna Greenhow)
Budleigh Salterton EX9 7HG
☎ *0395 68521*
Restored working water-powered cornmill in the lower Otter valley. Studio workshops include pottery, woodturning, stained glass, fine art and printing, glass-blowing, sign-making. Gallery (special exhibitions) museum and display of Devon lace. Bakery and restaurant.

Open daily 10.30am-5.30pm Easter-Jan, 11.30am-4.30pm Jan-Easter. Free admission to workshops. Mill tour charge £1.25 adults, 60p children (less for groups).
Directions: north of Budleigh Salterton, off A376.

Sorley Tunnel Farm & Craft Workshop Centre
(James Noye Balsdon)
Sorley Farm, Kingsbridge TQ7 4BP
☎ *0548 857711*
Leisure and educational working farm with craft workshops; wood-turning, doll making, tie-dyeing, knitwear, pottery. Farm and nature trail walks, childrens' play area, pony trekking. Tearoom, picnic area.
Craft workshops open all year 10am onwards, Farm Centre 10am-6pm Easter-Oct.
Free admission to workshops. Charges for farm and leisure areas £2 adults, £1.40 children (free under three years).
Directions: north of Kingsbridge between B3196 and A381.

The Round House Craft Centre
(David & Angelene Perryman)
Buckland-in-the-Moor
Ashburton TQ13 7HN
☎ *0364 53234*
Various craft workshops in an idyllic picturesque village setting of thatched cottages.
Open daily 9.30am-5.30pm.

The Wheel Craft Centre
(John and Anne Buckley)
Clifford Street
Chudleigh TQ13 0LH
☎ *0626 853712*
Craft workshops and studios

housed in historic mill building; water mill with one of the largest working backshot water wheels in UK. British Toymakers' Guild permanent exhibition in old mill building. Gift shops, licensed restaurant and tea rooms.
Directions: situated just off A38 south of Exeter (follow ETB signs). From A380 follow Ugbrooke House / Chudleigh signs.

ALSO CRAFT WORKSHOPS / CRAFT CENTRES AT:

Axminster Carpets
Axminster
☎ 0297 32244

Colyton Tannery
Colyton
☎ 0297 52282

Buckfast Butterfly Farm
Buckfastleigh
☎ 0364 42916

Dartington Glass Ltd
Great Torrington
☎ 0805 23797

DEVON CRAFT WORKSHOPS

Axminster

DOLLS' HOUSES
Dolphin Miniatures
(J B Nickolls)
Myrtle Cottage, Greendown
Membury EX13 7TB
☎ 04048 8459
Sole craftsman making dolls' houses and miniature furnishings in one-twelfth scale. Award winner and author of *Making Dolls' Houses* (1991). C/R&R/E
Visitors welcome by appointment only.
Visiting charge: £1 each (half donated to Save The Children fund).
Directions: 3.5 miles north of Axminster.

Barnstaple

CERAMICS
Newport Pottery
Denis & Wendy Fowler
72 Newport Rd, Barnstaple EX32 9BG
☎ 0271 72103

Glazed earthenware specialising in house nameplates, giftware and commemorative pieces. C
Mon-Sat 10am-6pm (winter) -8pm (summer).

Beaworthy

CERAMICS
Duckpool Cottage Pottery
(Svend Bayer)
Duckpool Cottage
Sheepwash EX21 5PW
☎ 0409 23282
Wood-fired stoneware garden and domestic pots. Awards won. C/W/E.
Open daily 9am-6pm, lunchtime closing.
Directions: Pottery lies over 1 mile north of Sheepwash. At Okehampton traffic light turn right, take A386 to Hatherleigh, turn left towards Holsworthy on A3072. At Highampton turn right to Sheepwash.

CERAMICS
Shebbear Pottery
(Clive Bowen)
Shebbear EX21 5QZ
☎ 040928 271
Traditional small workshop
producing range of wood-fired
earthenware garden and domestic
pots and larger individual pieces.
Widely exhibited in UK and
abroad. C/W/E.
Mon-Sat 9am-5pm.
Directions: Shebbear lies east of
A388 between Holsworthy and Stibb
Cross junc with B3227.

Bideford

GLASS ENGRAVING
John Beard Engraved Glass
South Yeo, Yeo Vale EX39 5ES
☎ 0237 451218
Stipple and flexible drill engrav-
ing of glass. C/E
*Mon-Fri 10am-5pm, Sat 10am-
12.30pm. Advisable to telephone first.*

FURNITURE
David Savage
21 Westcombe
Bideford EX39 3JQ
☎ 0237 479202
Ten craftsmen making fine
English hallmarked furniture,
specialising in contemporary
pieces made to specific requirements. C/E. Credit cards.
*Mon-Fri 9.30am-4.30pm, lunchtime
closing. Sat-Sun by appointment.*
Directions: from N Devon link road
cross Torridge Bridge to roundabout, take 1st turn to Bideford.
Down hill, turn right immediately
after Raleigh Garage on right, past
PO on right, school on left, coal yard
on right; workshop sign beyond.

CERAMICS
Springfield Pottery
(Frannie and Philip Leach)
Hartland EX39 6BG
☎ 0237 441506
Handmade oven-to-table ware,
small garden pots. Also handmade,
and hand screenprinted tiles.
C. Credit cards.
Mon-Sat 9am-6pm.

TRADITIONAL CHAIRS
Millthorne Chairs
(Bob and Sue Seymour)
10 Fore Street, Hartland EX39 6BD
☎ 0237 441590
Small workshop making childrens'
and full size Windsor chairs in
beech, ash and elm; wide range of
chair splats available. Also stools. C
*Mon-Fri 8.30am-6pm, weekends
advisable to telephone first.*

TOY MAKING
Hilary Bix Puzzles
(Jerry and Hilary Bix)
36 Lower Gunstone, Bideford EX39 2DE
☎ 0237 470792
Award winning makers of hand-
painted wooden jig-saw puzzles,
name plaques, nursery accessories,
badges. Puzzle repair service.
C/R&R/W/E. Credit cards.
*Mon-Fri 9.30am-5pm (lunchtime
closing 12.30-2.30pm).*
Directions: from Bideford Quay, up
High Street 40yd, turn right into Mill
Street (pedestrian way), then 1st left
into Lower Gunstone. Workshop in
private house 150yd up on right.

CERAMICS
Hartland Pottery
(Clive C Pearson)
North Street, Hartland EX39 6DE
☎ 0237 431693

Thrown stoneware mainly in oatmeal and soft blue and rusty colouring. Coffee, wine and tea sets; oil lamps. C/W/E
Mon-Sat 10am-6pm, lunchtime closing.
Directions: from Bideford take A39 for Bude and 1st right after Clovelly Cross roundabout. Follow road signs to Hartland (approx 3 miles).

FURNITURE MAKING & TUITION
David Charlesworth
Harton Manor Workshop
Hartland EX39 6BL
☎ 0237 441288
Fine furniture-making courses to exhibition standards. Individual tuition, 3-12 months; introductory weeks available. C
Mon-Fri 9am-5pm, lunchtime closing.

Bovey Tracey

CERAMICS
Lowerdown Pottery
(Tim Andrews)
Bovey Tracey TQ13 9LE
☎ 0626 833076
Hand-thrown stoneware, decorated porcelain and Raku pottery; work particularly noted for brush decoration. C/W/E
Mon-Fri 10am-5.30pm, Sat in summer or by appointment.
Directions: from Bovey Tracey take B3344 westwards, turn 1st left. Pottery on crossroads.

GLASS BLOWING
House of Marbles & Teign Valley Glass
The Old Pottery, Pottery Road
Bovey Tracey TQ13 9DS
☎ 0626 835358
A wide range of decorative glass and marbles. Also traditional games and puzzles using marbles. Show-room and seconds' shop. Museum of glass and marble history. W/E. Credit cards.
Mon-Sat 9am-5pm, Sun Easter-Sept.
Directions: from A38 Plymouth-Exeter road take A382 Bovey Tracey (2 miles). At roundabout turn left into Pottery Road. Factory on right.

Chittlehampton

CERAMICS
Chittlehampton Pottery
(Roger Cockram)
Victoria House
Chittlehampton EX37 9PX
☎ 0769 540420
Studio making individual ceramics, domestic stoneware and decorative plant containers; work influenced by training and experience in zoology and marine biology. Also paintings and prints.
C/W/E. Credit cards.
Open daily 9am-6pm (Sat-Sun advisable to telephone first).

Chumleigh

THATCHING
Tristan Johnson
The Old Rectory
Wembworthy EX18 7RZ
☎ 0837 83694
All thatching and restoration work; 15 employed. Also materials supplied. C/R&R/wholesale.

Crediton

BASKET MAKING
Sue Reece
19 The Green, Crediton EX17 2BD
☎ 0363 773442
Traditional willow baskets. Com-

missions welcome. Chair re-seating. C/R&R/W/E
Mon-Sat 9am-5pm (Tues & Thurs 2-5pm), lunchtime closing. Advisable to telephone first.

Cullompton

FURNITURE
Alan Peters Furniture
Aller Studios, Kentisbeare EX15 2BU
☎ *0884 6251*
Specialist workshop making high quality modern craft furniture. Single items to complete room schemes to order. Showroom. C/E
Tue-Fri 9am-6pm, advisable to telephone first. Sat by appointment.

Dartmouth

FORGEWORK
John Churchill
The New Forge, Capton Workshops Capton TQ6 0JE
☎ *0804 21535*
General blacksmithing, mainly door and window furniture, traditional and modern. Stove design and construction. Restoration of antique ironwork; reproductions. C/R&R/E. Credit cards.
Mon-Fri 9am-6pm, Sat 9am-1pm, lunchtime closing.
Directions: from Dartmouth (4 miles) on B3207, turn right at opposite Sportsmans Arms (Dittisham road). Left turn for Capton, Forge at end of village.

Dawlish

TEXTILES
The Vivian Gallery/Vivian Individual Designs
(Sarah Vivian)
2 Queen Street, Dawlish EX7 9HB
☎ *0626 867254*
Patchwork cord jackets, painted silk tops, made-to-measure of all kinds. Patternweave tweed for coats, shawls, scarves, bags etc. Gallery selling work of many crafts people; handmade textiles, accessories and jewellery. C/R&R/W. Credit cards.
Mon-Sat 9am-6pm.
Directions: Gallery is on corner of The Strand and Queen Street opposite Barclays Bank and Car Park.

MARBLEWORK
Marble Craft
(W M Carew)
Shutterton Industrial Estate Exeter Road, Dawlish EX7 0LA
☎ *0626 864856/865748*
Variety of marble gifts; clocks, barometers, table lighters, pen holders, desk calendars etc. Special commissions welcome. C/R&R/W/E
Mon-Fri 8am-5pm, lunchtime closing. Sat 9am-12noon.

Hatherleigh

CERAMICS
Elizabeth Aylmer Pottery
Widgery House, 20 Market Street Hatherleigh EX20 3JP
☎ *0837 810624*
Studio employing two potters making oven-to-table ware influenced by African handcraft. Also garden pots. C/W/E. Credit cards.
Mon-Fri 10am-5pm, Sat-Sun by appointment.
Directions: take A386 to Hatherleigh; pottery is on main street up hill beyond Tally Ho pub.

Holsworthy

CERAMIC SCULPTURE
John and Jan Mullin
September House, Parnacott EX22 7JD

☎ 0409 253589
Small ceramic studio and show-room. Mostly humorous animal sculptures, especially birds. C/W/E. Credit cards.
Mon-Fri and most weekends 9am-6pm or by appointment.

POTTERY & TEXTILES
Haytown Pottery
(David and Margaret Cleverly)
Haytown EX22 7UW
☎ 0409 261476
Hand-thrown domestic ware with modelled painted animal decoration mouse pottery and individual animals. Also embroidery of animals, landscapes and batik. C/W/E. Credit cards.
Easter-October Mon-Fri 11am-5pm; at other times advisable to telephone first.

Honiton

CERAMICS
The Pottery Shop (Honiton) Ltd
(Paul Redvers)
30-34 High Street, Honiton EX14 8PU
☎ 0404 42106
Pottery and crafts workshop specialising in hand-painted work. C/W/E. Credit cards.
Mon-Sat 9am-5pm.

Kingsbridge

BOOKBINDING
Bookbuild
(Judith Lamb and Colin Roberts)
The Gatehouse, West Charleton TQ7 2AL
☎ 0548 531294
Bookbinding and antique book refurbishment by specialists in book repair and conservation. Equipped to cover a wide range of hand bookbinding requirements. C/R&R

Mon-Fri 9am-6.30pm, Sat-Sun any time. Advisable to telephone first.
Directions: from Kingsbridge, West Charleton is 1st village on A379 going east to Dartmouth. Workshop is last but one house on right.

CERAMICS
Fuchsia Ceramics
(Paul H Metcalf)
Cutlands Pottery, Chillington TQ7 2HS
☎ 0548 580390
Hand-thrown stoneware pottery decorated with unique fuchsia design; domestic ware, vases etc. Also other crafts and gifts. C/W/E
Mon-Fri 9am-5.45pm, lunchtime closing. Sat 9am-1pm summer.

LEATHERWORK
Sleepy Hollow
(Andrew & Francis Brown)
Avon Mill, Woodleigh Road
Loddiswell TQ7 4DD
☎ 0548 550210
Small workshop employing three craftsmen making quality leather shoes, hats, slippers and fashion accessories.
C/R&R/W/E. Credit cards.
Mon-Sat 10.30am-5pm. Closed Feb.

Lydford

WOODCARVING
Lydford Gorge Woodcarver
(Rodney Smith)
Larrick Cottage, Lydford EX2 4BJ
☎ 0822 82288
Deep relief wood carvings, pictures or free-standing tableaux depicting birds and animals in natural surroundings. Also carvings of customers' pets or homes. Winner of the Henry Taylor Award for woodcarving and six gold medals.

Selected for membership of Society of Wildlife Artists 1990. C/E
Open during normal working hours, advisable to telephone.
Directions: Larrick Cottage is just under 1 mile south of Lydford in lane west of Gorge Road.

Modbury

WOODTURNING
Woodturners (South Devon) Craft Centre
(Mr and Mrs J and Miss H Trippas)
New Road, Modbury PL21 02H
☎ 0548 830405
Visitors are welcome to watch the craftsmen and women making items from wood ranging from egg cups to furniture. Showroom also has other craftwork for sale. C/R&R
Mon-Sat 10.30am-6pm. Closed Xmas-1st week, February.

Moreton Hampstead

CERAMICS
Moorland Pottery
(Catherine Engelsen)
44a Court Street
Moreton Hampstead TQ13 8LB
☎ 0647 40708
Hand-crafted mugs, jugs, coffee and teapots, tableware and vases glazed in rich moorland colours. C/W/E
Mon-Sat 10.30-5.30pm, advisable to telephone first.
Directions: workshop is situated on Princetown Road next to free car park.

Newton Abbot

HAND PAINTED SILK
Pom Stanley
7 Whitehill Road, Highweek TQ12 1QD
☎ 0626 51835
Individually designed hand-painted

silk scarves and silk paintings. Also hand weaving. C/W/E
Visitors welcome by appointment only.

UPHOLSTERY
Peter James Hatt Upholsterers & Furnishers
69 Fore Street, Chudleigh TQ13 0HT
☎ 0626 852656
Traditional and modern upholstery, curtain making, loose covers. Also small craft showroom.
C/R&R/W. Credit cards.
Tues-Sat 9am-5pm, lunchtime closing.

UPHOLSTERY
Chivers Upholstery
(S P Chivers)
Kings Mews Warehouse
Jetty Marsh Road
Newton Abbot TQ12 2SL
☎ 0626 335446/0803 875263
Upholstery, antique furniture restoration; polishing, woodwork repairs. Contract work undertaken.
C/R&R/W/E
Mon-Fri 8.30am-5.30pm. Closed Bank Holidays.

CERAMICS
Robert Tinnyunt Ceramics
96 Exeter Rd, Kingsteignton TQ12 3LU
☎ 0626 61011
Local clay used to make stoneware. Work includes domestic and individual pieces, hand-thrown and decorated with brushwork. C/W/E
Mon-Sat 9.30am-5.30pm. Sun by appointment.

North Tawton

CERAMIC SCULPTURE
Gil Tregunna
5 The Square, North Tawton EX 20 2ER
☎ 0837 82513

Renowned ceramic sculptor; work much influenced by the flora, fauna and landscapes of the south west. Work widely exhibited in UK and abroad. TV appearances and subject of a documentary film. C/W/E
Mon-Sat 11am-5pm.

Okehampton

SUNDIALS
Solar Time
(Hugh Franklin)
4 Harveys in Town
Sampford Courtenay EX20 2SX
☎ *0837 82848*
Sundials; hand-engraved commissions in slate, especially for anniversaries. C/E
Visitors welcome by appointment.

Paignton

CALLIGRAPHY & BOOKBINDING
Trinity Studios
(Roy Collier)
2B Roundham Road
Paignton TQ4 6EZ
☎ *0803 522679*
Work mainly to commission; calligraphy and bookbinding by experienced teacher of the craft. C/R&R/E. Credit cards.
Open daily 9am-5pm.

Plymouth

SADDLERY
Ermington Mill Saddlery
(Bruce Crawford)
National Shire Horse Centre
Yealmpton PL8 2EL
☎ *0752 881133*
Master saddler undertaking any kind of leatherwork. Bellows

made and refurbished. C/R&R/W/E
Mon-Fri 8.30am-5pm, Sat-Sun 9am-5pm. Lunchtime closing.

Seaton

CERAMICS
Branscombe Pottery
(Paul and Linda Wilson)
The Bulstone, Seaton
☎ *0395 577844*
Hand-thrown domestic stoneware; practically designed tableware (teapots guaranteed non-drip) and lamp bases, vases etc, all carrying pottery mark.
Opening hours vary — advisable to telephone first.
Directions: pottery is signposted from A3052 Branscombe Cross.

South Brent

CERAMICS
Rose Cottage Pottery
(Audrey Price)
Harbourneford TQ10 9DT
☎ *0364 72550*
Earthenware pottery; a wide range of domestic and decorative pots. C
Mon-Sat 9.30am-5pm
Directions: 2 miles from South Brent. Turn off A385 signed Dartington/Totnes; head north to Harbourneford.

South Molton

SADDLERY
Acorn Saddlery
(Frank Edwards)
3A East Street
South Molton EX36 4BY
☎ *076957 3847*
Saddles and allied leather goods made and repaired. C/R&R/W/E
Mon-Fri 8.30am-5.30pm, Sat 8.30am-12.30pm.

CERAMICS
Sandy Brown Pots
(Sandy Brown)
38 East Street, South Molton EX36 3DF
☎ 0769 572829
Potter producing her own distinctive, highly-individual pots. The work is exuberant and colourful with abstract designs. C/W/E
Visitors welcome by appointment.

Tavistock

CERAMICS
Wren Pottery
(Tim Farmer)
*Village Arcade, 10 Brook Street
Tavistock PL19 0HD*
☎ 0822 616896
Small pottery producing hand-thrown tableware and novelty items. C/W/E. Credit cards.
Mon-Sat 9am-5pm.
Directions: pottery is at eastern end of Tavistock Main Street.

FORGEWORK
Redferns Smithy
(John R Allen)
Brentnor PL19 0LR
☎ 0822 810496
Fireside furniture, canopies, ornamental ironwork, wind-vanes. Old keys copied. General forgework: repair work to farm and domestic implements; knife, scissor, chisels, picks, mower blades sharpened. Brass and copper platework. C/R&R
Mon-Sat 9am-12.30pm, afternoons by appointment.

Tiverton

FURNITURE
Nicholas Chandler
Woodpeckers, Rackenford EX16 8ER
☎ 0884 88380
Contemporary furniture designed and manufactured by third generation cabinet maker; woodturning, lettercutting. C/R&R/W/E.
Mon-Sat 9am-6pm, lunchtime closing. Closed August.
Directions: From M5 junc 27 Tiverton, take A361 to roundabout. Continue 7 miles, take 4th turn on right Bickham/Woodburn for 400yd. Turn into drive at white signboard to Woodpeckers; after 350yd, 5-bar gate on right.

FURNITURE
Michael S Scott
Popes, Shillingford EX16 9BP
☎ 0398 331999
Commissioned designed furniture made by graduate of John Makepeace School for Craftsmen in Wood. C/W/E
Mon-Fri 9am-5pm, Sat 9am-12noon.

Topsham

ANTIQUE FURNITURE
RESTORATION
Tony Vernon
15 Follett Road, Topsham EX3 0JP
☎ 0392 874635
Restoration and conservation of all periods of antique furniture; also cabinet making. On the register of Conservation Unit, Museums & Galleries Commission. C/R&R/E.
Mon-Fri 8.30am-5.30pm.

Torquay

WHEELWRIGHTING &
WOODWORK
The Wheelwright Shop
*Devon Rural Skills Trust,
Cockington Court
Cockington Village TQ2 6XA*
☎ 0803 605377/690094

Wheelwrighting, gates, wheelbarrows. Hurdle making and general woodwork. C/R&R/W/E
Apr-Oct open daily except Sat 10am-5pm, lunchtime closing. Winter by appointment. Charges: see Craft Centres.

STAINED GLASS
Paul Edwardes
Devon Rural Skills Trust, Cockington Court Cockington Village TQ2 6LA
☎ *0803 605377/690094*
Leaded lights and stained glass manufacture and repairs. C/R&R
Apr-Oct open daily except Sat 10am-5pm, lunchtime closing. Winter by appointment. Charges: see Craft Centres.

ETCHING
Sarah Ringrose
Devon Rural Skills Trust, Cockington Court Cockington Village TQ2 6XA
☎ *0803 605377/690094*
Limited edition etchings and illustration. C/W/E. Credit cards.
Apr-Oct open daily except Sat 10am-5pm, lunchtime closing. Winter by appointment. Charges: see Craft Centres.

PATCHWORK
Jackie Wills
Devon Rural Skills Trust, Cockington Court Cockington Village TQ2 6XA
☎ *0803 605377/690094*
or 295548 (evenings)
Internationally known creator of Fibonacci patchwork. Specialist in hand-sewn hexagon mosaic work. Wall hangings, waistcoats, cushions etc. C/R&R/W/E. Credit cards.
Apr-Oct open daily except Sat 10am-5pm, lunchtime closing. Winter by appointment. Charges: see Craft Centres.

RUSH & CANE WORK
Cane Corner
(Brigitte L Stone)
Devon Rural Skills Trust, Cockington Court Cockington Village TQ2 6XA
☎ *0803 605377/690094*
Antique and modern furniture professionally reseated with split cane and rush. C/R&R
Apr-Oct open daily except Sat 10am-5pm, lunchtime closing. Winter by appointment. Charges: see Craft Centres.

TRADITIONAL CHAIRS
Living Wood Chairs
(Mike Abbott and Jason Griffiths)
Devon Rural Skills Trust, Cockington Court Cockington Village TQ2 6XA
☎ *0803 605377/690094*
Chair bodging and chair making using locally grown hardwoods and traditional methods to produce quality Windsor chairs. Also courses run in local woodlands. C/E. Credit cards.
Apr-Oct open daily except Sat 10am-5pm, lunchtime closing. Winter by appointment. Charges: see Craft Centres.

WOODTURNING
Nick Agar Woodturner
Devon Rural Skills Trust, Cockington Court Cockington Village TQ2 6XA
☎ *0803 605377/690094*
General woodturning. C/R&R/W/E
Apr-Oct open daily except Sat 10am-5pm, lunchtime closing. Winter by appointment. Charges: see Craft Centres.

Totnes

CERAMICS
Lotus Pottery
(Michael Skipwith)
Old Stoke Farm, Stoke Gabriel TQ9 6SL
☎ *0804 28303*
A wide variety of wood-fired
stoneware and porcelain, especially
pots for plants. Also stockists of
pottery materials. C/W/E
*Mon-Fri 8.30am-5.30pm, lunchtime
closing. Sat 8.30am-1pm. (Annual
holiday mid-summer).*
Directions: 3 miles from either Paignton
or Totnes; pottery signs from Paignton.

TOYS
David Plagerson
28 Bridgetown, Totnes TQ9 5AD
☎ *0803 866786*
Toy maker specialising in wooden
Noah's Ark set, painted or natural
wood versions. C/R&R/W/E
Visitors welcome by appointment.

CERAMICS
Colin Kellam
*The Lion Brewery, South Street
Totnes TQ9 5DZ*
☎ *0803 863158*
Several potters producing general
stoneware pottery with floral
designs. C/R&R/W/E. Credit cards.
Mon-Fri 9am-5pm, lunchtime closing.

CERAMICS
Dartington Pottery
(S Course and P Cook)
Shinners Bridge, Dartington TQ9 6JE
☎ *0803 864163*
Pottery producing Janice Tchalenko
award winning contemporary
designs. Tableware and bowls, jugs,
vases. Also patterns to clients'
requirements. C/W/E. Credit cards.

*Mon-Fri 9am-6pm, Sat-Sun 10am-
6pm (closed Sun Jan-Feb).*

WOODTURNING
Rendle Crang (Woodcraft)
20 Burke Road, Totnes TQ9 5JA
☎ *0803 865447*
Woodturner specialising in wooden
bowls from English and foreign
hardwoods. Also stools, lamps and
kitchenware. Newel posts and
bannisters made to customers'
requirements. C/R&R/W/E
*Mon-Fri 9am-5pm, Sat-Sun by
appointment.*

CABINET MAKING
Ashridge Workshops
(Christopher Faulkner)
*Ashridge Barn
Tigley, Nr Dartington TQ9 6EW*
☎ *0803 862861*
Large workshop making traditional
and contemporary furniture;
bureaux, tables, chairs, chests-of-
drawers etc. Commissioned and
designed pieces. Tuition given; one-
year courses. Showroom exhibiting
work of ten cabinet makers. C
*Mon-Fri 10am-4pm, lunchtime closing.
Visit by appointment only. Closed Aug.*
Directions: 3 miles west of Totnes off A385.

Winkleigh

SADDLERY
P H Saddlers
(Pippa Hutchinson)
Park Farm, Winkleigh EX19 8LE
☎ *0837 83757*
Two experienced saddlers making
new tack and leatherwork to order.
Repairs on all saddlery, leatherwork
and canvas work. C/R&R/W/E
*Mon-Fri 8am-5pm, Sat-Sun by appoint-
ment.*

DORSET CRAFT WORKSHOP CENTRES

Broadwindsor Craft Centre
(Ian, Kate and Ruth Guilor)
Broadwindsor, Beaminster DT8 3PX
☎ 0308 68362
Ten workshops in converted farm buildings with gallery and crafts shops. Crafts include woodwork, ceramics, crystal and gemstones, millinery, upholstery, woodturning. Restaurant. Large car park.
Open daily 10am-5pm, closed Xmas fortnight.
Directions: situated on edge of Broadwindsor on Beaminster road.

Luccombe Farm Business & Craft Centre
Milton Abbas DT11 0BD
Converted Georgian and Victorian farm buildings housing craft workshops, studios and offices. Crafts include bespoke joinery/kitchen furniture, antique furniture restoration, sculpture and stonework, picture framing, printing.
Individual opening times.
Directions: 1.5 miles north of A354, off road between Milton Abbas and Winterborne Whitechurch.

Poole Pottery Craft Centre
Poole Pottery Ltd
The Quay, Poole BH15 1RF
☎ 0202 668681

Three floors of independent crafts people housed at pottery, demonstrating and selling their products. Crafts include glass blowing, enamel jewellery, pottery, wood turning, macramé, découpage, fine art. Museum, pottery showroom and tea room.
Crafts centre and museum open daily Mar-Xmas 10am-4pm. Showroom and tea room open daily all year 10am-5pm, also evenings in summer.
Directions: follow signs for Poole Quay and Pottery. Car parking close by and on Quay.

Walford Mill Craft Centre
Stone Lane
Wimborne Minster BH21 1NL
☎ 0202 841400
A former flour mill converted to house workshops, craft shop, exhibition hall, education centre and commissioning gallery. Resident silk weaver and painter/printmaker giving lithographic printing demonstrations. Courtyard stalls for hire in summer. Courses, workshops and seminars throughout the year. Childrens' courses during holidays. Restaurant.
Open daily 10am-5pm
Directions: on outskirts of Wimborne, signposted from B3078 Cranborne road.

DORSET CRAFT WORKSHOPS

Abbotsbury

CERAMICS
Chapel Yard Pottery
(Richard Wilson)
Abbotsbury DT3 4LF
☎ 0305 871663

Terracotta garden pots. Hand-decorated domestic stoneware and earthenware; egg cups to casserole dishes, tea services etc. Specialists i marbling decorations. C/W
Tues-Sat 9am-6pm, lunchtime closing

Beaminster

CERAMICS
Chedington Pottery
(Beresford Pealing)
Manor Farm House
Chedington DT8 3HY
☎ *0935 891482*
Good, functional kitchen and tableware hand-thrown in stoneware. Also more unusual ceramic objects (range of Dorset clocks with quartz movements). C/W/E
Open most daylight hours.

ANTIQUE FURNITURE RESTORATION
Christopher Booth
Holeacre Farm, Mapperton DT8 3NR
☎ *0308 862585*
Restoration of French and English 18th century furniture by trained craftsmen. C/R&R
Open daily, normal working hours.

CERAMICS
Eeles Family Pottery
(David, Benjamin and Simon Eeles)
Mosterton DT8 3HN
☎ *0308 68257*
Large variety of wood-fired stoneware and porcelain made by three craftsmen (also sold at The Pot Shop, Lyme Regis). Work very widely exhibited and sold. David Eeles has lectured abroad and was founder member of the Craftsmen Potters Association. C/W/E
Visitors welcome by appointment only. Directions: on A3066 between Beaminster and Crewkerne.

FORGEWORK
Hill Forge
(Michael Henderson)
Mount Pleasant, Seaborough DT8 3QY
☎ *0308 68781*
Rural forge workshop/studio specialising in 'hot forge' work in contemporary and traditional styles. Sculptures in steel, architectural and ornamental ironwork, cutting and welding. C/R&R
Tues, Wed & Fri 9am-5pm, lunchtime closing. Please telephone first.

UPHOLSTERY & SOFT FURNISHING
Country Seats
(Helena and Reg Fielder)
4a Broadwindsor Road Trading Estate Beaminster DT8 3DW
☎ *0308 862927*
Traditional re-upholstery, custom-made curtains, pelmets and soft furnishings. Extensive range of fabrics and wallpapers. Full-time interior designer and restoration craftsmen. C/R&R/E
Mon-Fri 9am-5pm, lunchtime closing. Sat 9am-12.30pm.

ANTIQUE FURNITURE RESTORATION
Croft Antiques Restorations
(Clare Ross)
The Old Post Office
Stoke Abbott DT8 3JT
☎ *0308 68202*
Restoration, polishing etc of antique furniture. R&R
Visitors welcome by appointment.

Blandford

CERAMICS
Foxdale Pottery
(Keith & Carol Burbidge)
The Old Bakery
Child Okeford DT11 8EF
☎ *0258 860039*
Handmade stoneware mugs, jugs,

bowls, vases, cheese and butter dishes. Terracotta garden patio items, wooden tubs and bric-a-brac. C/W
Open daily 10am-6pm, advisable to telephone first.

FURNITURE MAKING
Southern Crafts
(J N Hodges and L J P Baily)
The Old Brewery, Durweston DT11 0QE
☎ 0258 453987
Established workshop employing eight craftsmen designing and making high quality fitted furniture for kitchens, bedrooms and all domestic interiors. Northern hardwoods used such as oak, ash, cherry and maple. C/W/E
Mon-Fri 9am-5pm.
Directions: workshop situated in the village of Durweston on A357.

ANTIQUE FURNITURE RESTORATION
Milton Antique Restoration
(Nigel Church)
Luccombe Farm Business & Craft Centre, Milton Abbas DT11 0BD
☎ 0258 880668
Restoration of antique furniture, re-polishing etc. C/R&R/W
Mon-Fri 9am-5pm
Directions: see under Craft Centres.

SCULPTURE AND STONEWORKS
Green Man
(Leó Reynolds)
The Cartshed, Luccombe Farm Business & Craft Centre
Milton Abbas DT11 0BD
☎ 0258 880121/454161
Free-standing and relief sculpture in local limestones, alabaster/marble, plaster, ceramic/terracotta, resin, mixed media. Former cathedral stonemason experienced

in architectural stonecarving and restoration work, heraldry, ecclesiastical works. Inscriptions and memorials. C/R&R/W/E
Mon-Sat 9am-5pm. Advisable to telephone first.
Directions: see under Craft Centres.

FURNITURE/WOODTURNING & SILVERWARE
Cecil Colyer MA, FSD-C
Orchardene, Candys Lane, Shillingstone DT11 0SF
☎ 0258 860252
Furniture of all sorts and turned items such as bowls, mirrors and mazers made from rare or special hardwoods. Also church and domestic silverware; wedding, anniversary and presentation pieces are a speciality. Most work to commission.
Visitors welcome by appointment only, (closed lunchtime).
Directions: Candys Lane is off A357 near Village Hall.

UPHOLSTERY
R E Hogg
The Causeway
Milborne St Andrew DT11 0JD
☎ 0258 87392
Family business; upholstery and specialist makers of ironback chairs and sofas in the traditional way. Master upholsterers and award winners; guild members. C/R&R/W/E
Mon-Fri 9am-6pm, Sat 9am-1pm.

FURNITURE
Christopher Stubbs Country Kitchens
(Albert Andrews)
Luccombe Farm Business & Craft Centre, Milton Abbas DT11 0BD

☎ 0258 881031
Hand-finished kitchen furniture
and accessories. Solid wood doors
using old reclaimed timber or from
renewable sources. Attention given
to high quality detail. Free planning
and design service. C/R&R/E.
Mon-Sat 9am-5.30pm
Directions: see under Craft Centres.

Bournemouth

PAPER SCULPTURE
Miriam Troth
25 Seafield Road
Bournemouth BH6 3JL
☎ 0202 427296
Sculptures and works of art created
from recycled paper. Also jewellery.
Work exhibited at the Barbican and
other main galleries.
C/W/E. Credit cards.
Tues-Thur 2-5pm.

Bridport

ANTIQUE FURNITURE
RESTORATION
Richard Bolton BAFRA
Ash Tree Cottage, Whitecross
Netherbury DT6 5NH
☎ 0308 88474
Restoration work carried out by
members of British Antique
Furniture Restorers Association.
Commissioned work for antique
trade and private collectors.
C/R&R/W/E
*Mon-Fri 8.30am-6pm or by appoint-
ment. Lunchtime closing.*

CERAMICS
Mellors Garden Ceramics
(Kate Mellors and John Davis)
Rosemead
Marshwood DT6 5QB

☎ 02977 217
Variety of glazed garden ceramics;
lanterns, bird baths, water features,
planters. C/W. Credit cards.
*Mon-Sat 1pm-5pm, other times by
arrangement. Advisable to telephone
first.*

BOOKBINDING
The Old Granary Bindery
(Dr David Evans)
Marsh Barn, Burton Road
Bridport DT6 4PS
☎ 0308 421759
Bookbinding, restoration and
conservation. Gold lettering on
leather goods. C/R&R/E.
*Mon-Fri 10am-5pm. Other times by
appointment.*
Directions: On main road between
Bridport and Burton Bradstock.

CANDLEMAKING
Bridport Candlemakers
(J and T A Cottington)
10 Beaumont Avenue
Bridport DT6 3AU
☎ 0308 22176
Beeswax and paraffin wax candles.
Also beeswax polishes and
beeswax/resin fillers.
C/R&R/W
Mon-Fri 9am-5pm.

Broadwindsor

FURNITURE
Dorset Pine Craft
(Ivor and Rosa Downton)
The Old Chapel, West Street
Broadwindsor DT8 3QQ
☎ 0308 68814
Traditional furniture handmade in
kiln-dried pine and hardwoods, to
customers' designs. Sycamore
boxes, stools, trays and spoons

DORSET · BROADWINDSOR · CERNE ABBAS · DORCHESTER

etched and painted with name, nursery rhyme or design. C/R&R
Open during working hours, advisable to telephone first.

CHAIR RESTORATION
Broadwindsor Chairs & Crafts
(Graham Wilkinson)
Old Exchange Workshop
Drimpton Road
Broadwindsor DT8 3RS
☎ 0308 68909/68784
Chair repair work, rush and cane seating. Restored stools and chairs for sale. Also batik pictures and embroidered samplers. C/R&R
Mon-Fri 9.30am-5.30pm, lunchtime closing. Sat-Sun by arrangement.

UPHOLSTERY
Malcolm Duncan
Unit 2, Broadwindsor Craft Centre
Broadwindsor DT8 3PX
☎ 0308 68306
Upholstery and antique furniture restoration; French polishing, cabinet making etc. C/R&R
Open daily 10am-5pm, lunchtime closing. Closed Jan-Feb.
Directions: see under Craft Centres.

Cerne Abbas

FORGEWORK
Cerne Valley Forge
(Mr Eiles Clarke)
Mill Lane
Cerne Abbas DT2 7LA
☎ 0300 341298
Ornamental hand-forged metal-work including fire-baskets, fire-irons, fire-screens, gates, railings, garden furniture and various gift items. C/R&R/W
Visitors welcome by appointment.

CERAMICS
Abbey Pottery
(Paul Green)
Cerne Abbas DT2 7JQ
☎ 0300 341865
Handmade stoneware and porce-lain pottery, mainly functional oven and tableware in a variety of glazes. Some purely decorative pieces. Workshop and showroom. C/W. Credit cards.
Tues-Sun 9.30am-6pm
Directions: situated close to Cerne Giant and Cerne Abbey.

Dorchester

CERAMICS
Warwick Parker Pottery
The Dairy House
Station Road
Maiden Newton
Dorchester DT2 0AE
☎ 0300 20414
Wide range of individual and domestic ceramics. C/W/E
Mon-Fri 9am-5.30pm, Sat 9am-12noon. Advisable to telephone first.

RUSH & CANE WORK
A W Hammacott
19 Main Street
Broadmayne DT2 8EB
☎ 0305 854082
Restoration and repairs of all cane, rush and seed grass furniture. Courses. R&R/W. Credit cards.
Tues-Sat 9.30am-5.30pm.

SPINNING & WEAVING
Woolly Shepherd Spinning Emporium
(A & C Hammacott)
19 Main Street
Broadmayne DT2 8EB

☎ *0305 854082*
All forms of hand spinning and weavers of a wide range of wool/mohair goods, especially individual floor rugs from rare breed fleeces. Spinning and dying courses. Equipment supplied.
C/W/E. Credit cards.
Tues-Sat 9.30am-5.30pm.

ANTIQUE FURNITURE RESTORATION
Michael Barrington Restorers
The Old Rectory
Warmwell DT2 8HQ
☎ *0305 852104*
This workshop has restored the casework and pipe gilding of the organ in Lulworth Castle chapel, possibly one of the most important organ restorations this century, and has made one of the world's largest traditional mahogany dining tables. Primarily restoration of antique furniture, polishing, gilding, upholstering and making furniture to order. Metalwork restoration and mechanical toys (notably steam) and small mechanical bits for full size engines and vehicles. C/R&R
Mon-Fri 8.30am-5pm, Sat 8.30am-1pm.

FURNITURE RESTORATION
The Old Pine Workshop
(Martin Elsegood and Christine Haskett)
Unit A, Duck Farm
Bockhampton DT2 8QL
☎ *0305 848656*
Pine furniture restoration and reproduction. C
Mon-Sat 10am-5pm.
Directions: from Puddletown turn onto Rhododendron Mile. At T junc turn right, Duck Farm on left.

Evershot

GILDING, FURNITURE CONSERVATION & WOODWORK.
Claire Timings Gilding
Unit 4, Westhill Barn
Westhill DT2 0LD
☎ *0935 83267*
Small family business specialising in the design and production, to commission, of decorative architectural woodwork, fine mirror and picture frames, and furniture. Three full-time qualified staff with up to 20 years' experience in architectural design, conservation, carving, gilding, joinery and cabinet making including work for churches and large country houses. C/R&R/E
Open daily 10am-4pm by appointment.
Directions: 9 miles south of Yeovil off A37. in converted barn at western end of Evershot, entrance 150yd on right past West Hill Farmhouse.

Ferndown

ANTIQUE FURNITURE RESTORATION
G A Matthews
The Cottage Workshop
Dudsbury GG Camp
Ferndown BH22 8SS
☎ 0202 572665
Experienced, qualified cabinet maker restoring antique furniture.
C/R&R
Mon-Fri 9am-5pm.

Gillingham

LEATHERWORK
Blackmore Vale Saddlery
(Susan W Harvey)
Four Winds, West Bourton SP8 5PE
☎ *0747 840741*

Saddler also making bellows, belts and bags and small leather accessories. C/R&R. Credit cards.
Mon-Fri 8.30am-6pm, advisable to telephone first. Sat-Sun by apointment.
Directions: on B3081 one mile off A303.

THATCHING
George East
1 Church Close, Gillingham SP8 4DR
☎ *0747 822693*
Thatching contractor. C/R&R/W/E

Lyme Regis

BASKETMAKING
And Beautiful Baskets
(Ivor and Joy Uglow)
1 Drake's Way, Lyme Regis DT7 3QP
☎ *0297 444714*
Basket-making workshop. Unusual and colourful baskets of Somerset willow, hand-dyed and woven on the premises. Shop selling other crafts and gifts. C/W. Credit cards.
Open daily May-Jan 10am-5pm, lunchtime closing. Fri-Sat Jan-May 10am-5pm. Closed February.
Directions: Drake's Way is off Broad Street in town centre.

CERAMICS
Alan Wallwork Ceramics
Whitty Down Farm
Higher Rocombe, Uplyme DT7 3RR
☎ *0297 443508*
Handmade ceramics. W/E.
Mon-Fri 10am-5pm, often open Sat-Sun.

Poole

WOODTURNING & FURNITURE
Woodway
(Wayne and Julia Marshall)
Poole Pottery, The Quay
Poole BH5 3RE

☎ *0202 666200 xt 238*
Three craftsmen; woodturning, cabinet making and pyrography. Fruitbowls, coasters, lace bobbins. Rocking horses. Solid wood furniture made to commission. C/W/E. Credit cards.
Mon-Sat 10am-4.30pm. Closed Jan-Mar.
Directions: follow signs for Poole Quay and Pottery. Car parking close by and on Quay.

Puddletown

ANTIQUE FURNITURE RESTORATION
Tolpuddle Antique Restorers
(Raymond Robertson)
The Stables
Southover House
Tolpuddle DT2 7HF
☎ *0305 848739*
Specialist conservation and restoration of antique furniture. Prizes and awards won; work commissioned by V & A museum.
C/R&R. Credit cards.
Mon-Fri 9am-5pm, Sat 2-5pm. Please telephone first.

Shaftesbury

GOLD & SILVER SMITHING
Craft Studio
(Chris Morphy)
4 Gold Hill Parade, Shaftsbury SP7 8LY
☎ *0747 54067*
Designing and making jewellery in precious metals using traditional techniques. Precious and semi-precious stones used.
C/R&R/W. Credit cards.
Tues-Sat 9am-5pm, lunchtime closing.

CABINET MAKING
J R Twist
Millbush Farm
East Orchard SP7 0LH
☎ 0747 811712
Fine replica furniture mading and
furniture restoration by four
craftsmen. C/R&R/W/E
Mon-Fri 8.30am-5pm.

FORGEWORK & FARRIERY
Wing & Staples
(J Wing and L Staples)
Motcombe Forge, The Street
Motcombe SP7 9PE
☎ 0747 53104
Farriery and decorative and
practical hand-forged ironwork in
old forge. Winners of many awards.
Work undertaken for National
Trust, English Heritage, Dept of
Environment. C/R&R
Mon-Fri 9am-4pm, lunchtime closing.
Sat 9am-12noon.
Directions: 2 miles from Shaftesbury,
forge lies between B3081 and A350.

CERAMICS
Studio Pottery
(Anne Chase BA)
15 Bell Street, Shaftesbury SP7 8AR
☎ 0747 52198
Unusual one-off porcelain and
stoneware pieces thrown and hand-
built (no moulds). Anne Chase has
a wide reputation for her individual
and different style of work.
Summer opening: Mon-Sat incl Bank
Holidays 10am-5pm, Wed-Sat.
Autumn opening: 10am-4pm. Closed
Jan-Apr.

CAR RESTORATION
The Old Forge Car Restoration
(Tim Kerridge)
The Old Forge, Compton Abbas SP7 0NQ
☎ 0747 811881
Restoration and repair of pre- and
post-war cars, classic cars and
sports cars. R&R
Mon-Fri 9am-5.30pm, lunchtime
closing. Other times by appointment.
Directions: 3 miles south of
Shaftesbury on A350 with B & B sign.

Sherborne

FORGEWORK
Trent Smithy
(Michael Malleson)
42 Rigg Lane, Trent DT9 4SS
☎ 0935 850957
Forged ironwork for gardens,
homes, hotels, restaurants, churches
etc. Emphasis on original designs
for contemporary settings, but any
style undertaken. C/R&R
Mon-Fri 9am-5pm but advisable to
telephone first. Other times by
appointment.

SADDLERY
Godden Saddlery
(Richard H Godden)
41 Cheap Street
Sherborne DT9 3PU
☎ 0935 817350
Master saddlers making side
saddles, bridlework, fancy leather
goods. C/R&R/W/E
Mon-Fri 9am-5pm.

Sturminster Newton

FORGEWORK
Newton Forge
(Ian John Ring)
Stalbridge Lane
Sturminster Newton DT10 2JQ
☎ 0258 72407
Wrought ironwork including gates,
railings, fire baskets, fire screens,

gifts, weather vanes etc made by well-qualified blacksmiths. C/R&R/E. Credit cards.
Mon-Sat 9am-5pm, lunchtime closing.

SCULPTURE & HERALDY
Antony Denning
Frith Farm Workshops
Stalbridge DT10 2SD
☎ 0963 251433
Sculpture, wood carving, stone carving, welded metal. Mostly commissioned work which has included coats of arms (for Fishmongers Co, London), crests (Order of the Bath for Westminster Abbey) and many other notable commissions. C/R&R/E.
Visitors welcome by appointment only.

Swanage

CERAMICS
The Owl Pottery
(Leslie Gibbons ATD)
108 High Street
Swanage BH19 2NY
☎ 0929 425850
Finely decorated handmade earthenware and artwork. Artist/potter's interest in owls and wildlife is shown in detailed paintings on dishes etc. Also many intricate abstract patterns. Work in collections worldwide. Talks and demonstrations given to schools and colleges.
Open at any reasonable time, lunchtime closing. Closed Sun and some Thur afternoons.
Directions: A few minutes' walk from town centre past Town Hall on right of High Street, near mill pond and parish church.

ENGRAVING
Aquarius Designs
(Simon Winch)
Whitecliff Manor Farm, Swanage
☎ 0929 423158
Various methods of engraving, hand-printed from copper plates. Work sold throughout UK and Europe. C/R&R/W/E
Visitors welcome by appointment.

Wareham

JEWELLERY
Paws Jewellery
(D Watts)
Southbrook Workshop
Bere Regis BH20 7LN
☎ 0929 471808
Jewellery made. Also lost-wax castors, platers and diamond merchants, stampers, model makers, diamond and stone setters 'all under one roof'. C/R&R/W/E
Mon-Fri 10.30am-5pm, lunchtime closing 1-2.15pm, 7pm-midnight. Sat-Sun 2.15pm-midnight.

STONEWORK
Tony Viney
Sandy Hill Workshop
Sandy Hill Lane, Corfe Castle BH20 5JF
☎ 0929 480977
Polished stone artefacts; stone bowls and plates. Also stone work-surfaces, fireplaces etc. C/R&R/E. Credit cards.
Mon-Fri 9am-6pm.

STONE CARVING
Jonathan Living-stone
(Jonathan Mark Sells)
Sandy Hill Workshop, Sandy Hill Lane
Corfe Castle BH20 5JF
☎ 0929 480044

Stone sculpture, mainly figurative; animals and people. Also stone carving, stone masonry, letter cutting, stone artefacts, fireplaces. C/R&R/E. Credit cards.
Mon-Fri 9am-6pm, Sat 10am-5pm.

PLASTER CASTING
Casting Images
(Miss Annie Forster)
21 East Street
Wareham BH20 4NN
☎ *0929 551163*
Reproduction of old carvings into wall plaques, garden ornamentalia, planters, water fountains. C/R&R/W
Mon-Fri 9am-5.30pm and other times by appointment.

THATCHING
R V Miller
Belhuish Farm
Combe Keynes BH20 5PS
☎ *0929 462465*
Thatching contractors and suppliers of reeds, straw, spars, reed fencing and coppice materials. C/R&R/W/E

Weymouth

METALWORK
Roy Kilburn
The Craft Courtyard
Brewers Quay, Weymouth
☎ *0305 889563*
All types of metalwork from wrought-iron to silver jewellery; copper, brass and other metals. Etching. Gates, interior and exterior furniture, decorative objects. C/R&R/W/E. Credit cards.
Open daily (closed Wed) 10am-5pm.

Wimborne

FORGEWORK
Saxon Forge
(Mr R D Haigh and Mrs Z A Coakes)
Verwood Road
Three Legged Cross BH21 6RR
☎ *0202 826375*
Ornamental ironwork, traditionally forged iron; lanterns, door furniture, mangers, companion sets, fire canopies, gates, wall screens etc. Traditional sheet-metal work on jugs, skillets etc undertaken. C/R&R/W/E
Mon-Sat 8am-6pm, lunchtime closing.

CERAMICS
The Edmondsham Pottery
(Timothy Paul Dancey)
Smallbridge Farmhouse
Edmondsham
Cranborne BH21 5RH
☎ *07254 251*
Holder of world pot-throwing record. Range of pottery; vases, ovenware and ornaments. C/W/E
Open daily 9am-6pm.

CERAMICS
Hare Lane Pottery
(Jonathan Garratt)
Cranborne BH21 5QT
☎ *0725 4700*
Handmade terracotta garden pots with quiet decoration; frost-proof, wood-fired to give distincitve dark colour.
Thurs-Fri 9am-5pm, Sun 10am-5pm, lunchtime closing. Other times by appointment.
Directions: 2 miles from Cranborne on Alderholt road.

DURHAM, CLEVELAND, TYNE & WEAR CRAFT WORKSHOPS

Darlington

FORGEWORK
Little Newsham Forge
(Brian Russell AWCB DIP AD)
Winston, Co Durham DL2 3QN
☎ 0833 60547
Traditionally made wrought-ironwork forged to unique designs. Mostly to commission. C/R&R/E
Mon-Fri 8.30am-6pm, Sat 8.30am-4pm, visit by appointment.

Newcastle-upon-Tyne

GLASSWORK
Woolhouse & Kinsella
(Sue Woolhouse MA (RCA), BA (Hons) and Stephen Kinsella BA (Hons))
Unit 3, Level 1 The Ouseburn Warehouse Workshops
36 Lime Street, Byker NE1 2PN
☎ 091 438 4313
Architectural and cast glass; sculptural forms — colours, shapes, textural qualities. Lost-wax method used. Two highly qualified artists in glass whose work is widely exhibited. C/W/E
Mon-Fri 9am-5pm, Sat-Sun 10am-3pm.

North Tyneside

CERAMICS
Northumbrian Craft Pottery
(A M Harding and M Palmer)
Unit 7, Backworth Workshops,
Station Road, Backworth NE27 0RD
☎ 091 216 0820
Reduction-fired stoneware and terracotta hand-thrown pottery. Gifts, domestic ware, promotional and commemorative ceramics for

places of interest etc. C/W
Mon-Fri 8am-4pm, Sun 10.30am-3pm.

Stockton-on-Tees

FORGEWORK & FARRIERY
Peat Oberon, Blacksmith
Preston Hall Museum, Yarm Road
Stockton-on-Tees TS18 3RH
☎ 0642 785543
Prizewinning blacksmiths' workshop employed in forgework and farriery. Architectural work. C/R&R
Mon-Fri 9.30am-5pm, Sat 10.30am-4pm.

CERAMICS
Sedgefield Pottery
(W R Todd)
The Old Smithy, Cross Street
Sedgefield, Cleveland TS21 3AH
☎ 0740 21998
Set in Sedgefield's conservation area near the church and Green, the workshop produces studio pottery ; hand-thrown and decorated stoneware and terracotta; fettling and pierced decoration. C/W/E
Mon-Sat (Wed half-day closing) 9am-5.30pm.

Weardale

DRIED FLOWERS
Wearside Cottage Crafts
(Mrs Patricia Whitehouse)
45 The Causeway
Wolsingham, Co Durham DL13 3AZ
☎ 0388 528885
Home-grown dried flowers and arrangements. Crafts for sale. C/W
Mon-Fri (closed Wed) 9.30am-5pm, Sat 9am-12noon, Sun 12-5pm. Closed Jan-Mar.

ESSEX CRAFT WORKSHOP CENTRES

Burntwood Barn Craft Centre
(John Jones)
Stonyhills Farm, Warley Street
Great Warley, Brentwood
Craft workshops include John Jones'
pottery: functional and decorative
earthenware, (tuition given).
Jewellery workshop: Trudi Penman
making earrings, necklaces,
bracelets etc. Ornamental metalwork by Ron Bawden (☎ 04022
50090 answerphone & evenings):
mainly small items such as hanging
baskets, brackets, flower planters,
gates etc. Craft shop and tea room.
Open all year Fri-Sun 10.30am-5pm.
Directions: from Brentwood station
north towards M25, past Thatchers
Arms to Stonyhills Farm on right.

Lower Barn Farm Craft Centre
(L Argentieri)
London Road, Rayleigh SS6 9ET
☎ *0268 780991*
Newly established craft/cultural
centre. Workshops include:
fireplace makers, hand-crafted
chess sets, pot pourri, dried and
fabric flowers, ceramics, satin and
lace work, picture framing, wooden
garden and outdoor artefacts, fine
art, egg decorating and forgework.

Also farm/animal sanctuary and
falconry centre (demonstrations —
small fee charged). Renovated barn
for functions (licensed). Tea rooms.
*Open daily 10am-6pm. Free entry and
car parking.*
Directions: from A129/A130 intersec-
tion, take road to Rayleigh; Craft
centre is 1st property on left.

Mistley Quay Workshops
Swan Basin, Mistley
Manningtree CO11 1HB
☎ *0206 393884*
Established craft workshop group;
makers of harpsichords and maker
of traditional stringed musical
instruments, specialised woodworking and ceramics. Showroom.
Vegetarian restaurant.
Open daily 10am-6pm.

Oakwood Arts Centre
(Oakwood House Ltd)
2 High Street, Maldon
☎ *0621 856503*
Five craft studios; spinning and
weaving, needlecraft co-operative,
pottery, jewellery, picture framing.
Refreshments available in café.
*Mon-Fri 10am-5pm (closed Wed), Sat
9am-5pm.*

ESSEX CRAFT WORKSHOPS

Braintree

THATCHING
The Thatchers
(Colin Neil McGhee)
Oak Cottage, Dunmow Road
Great Bardfield CM7 4SD
☎ *0371 810606/0860 287310*

Master Thatchers, six employed
thatching roofs and building
timber-framed garden buildings (eg
Wendy houses, kennels, poolside
bars). Thatching cup winners.
C/R&R/W/E
Visit by appointment.

Castle Hedingham

CERAMICS
Castle Hedingham Pottery
(John and Margaret West)
37 St James Street
Castle Hedingham CO9 3EW
☎ 0787 60036
Hand-made domestic ware and garden pots. Also pottery supplies. Tuition: throwers master classes, seminars and kiln courses, day and evening classes. One-day courses with home-cooked lunch. C/W/E
Tues-Fri 9am-5.30pm, Sat-Sun 10am-5.30pm.
Directions: from Halstead north on A604, after 2 miles onto B1058. Follow ETB signs to pottery.

Chelmsford

ANTIQUE FURNITURE RESTORATION
Lomas Pigeon & Co Ltd
The Workshops, rear of Beehive Lane
Chelmsford CM2 9SU
☎ 0245 353708
Antique furniture restoration (members of BAFRA), cabinet making and upholstery. Leather desk linings, French polishing, woodturning, wood carving, cane and rush seating. Picture restoration and gilding. C/R&R. Credit cards.
Mon-Fri 9am-5pm. Sat 9am-12noon.

FARRIERY & FORGEWORK
Pleshey Forge
(Harold and Ian Clements)
Back Lane, Pleshey CM3 1HL
☎ 0245 37233
Two specialist wrought-iron craftsmen making ornamental and architectural ironwork; gates, weathervanes, pub signs, lanterns, balustrades. Also farriery. C/R&R
Mon-Fri 8am-5pm, lunchtime closing. Sat 8am-1pm.

Coggeshall

CERAMICS
Coggeshall Pottery
(Peter Turner)
49 West Street, Coggeshall CO6 1NS
☎ 0376 561217
Handmade stoneware pottery, mostly domestic ware. Also terracotta garden pots. W
Thurs-Sun 10am-5pm.

Felsted

CERAMICS
Brick House Crafts
(R Batt)
Cock Green
Felsted CM6 3JE
☎ 0371 820502
Pottery making and decorating. Tuition and seminars. Pottery supplies. C/W
Mon-Fri 9am-5pm, Sat 9am-1pm.

Great Dunmow

CERAMICS
White Roding Pottery
(Deborah Baynes)
Brett's Farm
White Roding CM6 1RF
☎ 0279 876326
Salt-glazed wheel-made decorative and practical pottery. Tuition: pottery holiday (residential and non-resident) courses in 16th century timber-framed house. Also weekend courses and specialist salt-glazing courses. C/W/E
Open at any reasonable time; advisable to telephone first.

Halstead

FORGEWORK
Frank R Nice
The Forge, Church Street
Gestingthorpe CO9 3AZ
☎ *0787 60126*
Hand-forged traditional wrought
ironwork. General blacksmithing
work in 200-year-old forge. C/R&R
Mon-Fri 8.30am-5pm, Sat-Sun 9am-
12noon, by appointment.

Ingatestone

FORGEWORK
Geo Carford Ltd
(B Careless)
Ingatestone Forge, 3a High Street
Ingatestone CM4 9ED
☎ *0277 353026*
Three craftsmen making ornamen-
tal ironwork. C/R&R
Mon-Fri 8.30am-5pm.

Laindon

FORGEWORK
V and J Quelch & Son
The Forge, Dunton Road
Laindon SS15 4BU
☎ *0268 543976*
General blacksmithing, ornamental
ironwork. C/R&R
Mon-Fri 8am-5.30pm, Sat 8.30am-1pm.

Manningtree

ENGRAVING ON GLASS
Jennifer Conway
31 Oxford Road
Mistley CO11 1BW
☎ *0206 396274*
Engraving on all sizes of clear float
glass, mostly to commission.
Screens, windows and doors

decorated with figures, creatures
etc. Lead crystal goblets etc decorated with architectural subjects,
landscapes, figures, portraits. C/E
Visitors welcome by appointment only.

CERAMICS
Manningtree Pottery
(M J Goddard)
18-20 High Street
Manningtree CO11 1AG
☎ *0206 396380*
Four craftsmen making domestic
stoneware, gardenware and floor
tiles. Work exhibited throughout
England and Europe. C/W/E
Mon-Sat 9am-5.30pm.

MUSICAL INSTRUMENT MAKING
Anne & Ian Tucker
Mistley Quay Workshops, Swan Basin
Mistley CO11 1HB
☎ *0206 393884*
Established business making
harpsichords, spinets, virginals and
clavichords for British and overseas
markets. Traditional methods used
and rigorous selection of materials,
with all parts manufactured in the
workshop. C/R&R/E. Credit cards.
Visitors to workshop welcome
strictly by appointment.

WOODWORK
Graham Pearson
Mistley Quay Workshops, Swan Basin
Mistley CO11 1HB
☎ *0206 393884*
Maker of furniture and woodwork.
C/R&R/W/E. Credit cards.
Open daily 10am-6pm.

MUSICAL INSTRUMENT MAKING
Con Rendell
Mistley Quay Workshops, Swan Basin
Mistley CO11 1HB

☎ 0206 393884
Maker and repairer of lutes, violins, cellos, mandolins and guitars. C/R&R/E. Credit cards.
Mon-Fri 9am-6pm.

CERAMICS
Bennett Cooper
Mistley Quay Workshops, Swan Basin Mistley CO11 1HB
☎ 0206 393884
Potter, Master of Arts (RCA) and assistant making decorated tableware, planters and house plaques. C/W/E. Credit cards.
Open daily 10am-6pm.

Rayleigh

FIREPLACES
Essex Fireplaces
(Graham Marlow)
Lower Barn Farm Craft Centre, London Road, Rayleigh SS6 9ET
☎ 0268 783700/770892
Fireplace design and construction. Also Tudor style interiors and antique restoration; feature brickwork. C/R&R/W/E
Mon-Fri 9.30am-2.45pm, Sat-Sun 10am-6pm.
Directions: see under Craft Centres.

CHESS SETS
Hand Crafted Chess
(John W Roe)
Studio 7, Lower Barn Farm Craft Centre, London Road Rayleigh SS6 9ET
☎ 0702 616276 (evenings)
Chess sets made from marble/resin and slate/resin. Veneered or glass boards, furniture to customers' requirements. C/R&R/W. Credit cards.
Open daily 10am-5pm.
Directions: see under Craft Centres.

POT POURRI & CANDLES
Pot Pourri Fragrant Oil Burners & Wall Plaques
(Mr and Mrs R Larnder)
Studio 4, Lower Barn Farm Craft Centre, London Road Rayleigh SS6 9ET
☎ 0702 77380
Pot pourri, tile trays and stands, candles. W
Open daily 10am-5pm.
Directions: see under Craft Centres.

CERAMICS
Gill Hopson — Ceramic Artist
Lower Barn Farm Craft Centre, London Road, Rayleigh SS6 9ET
☎ 0268 780991
Hand-crafted ceramics; thrown and sculptured forms (animals, teddybears etc). Tuition given. C/W
Wed-Sun 10am-5pm.
Directions: see under Craft Centres.

DRIED FLOWERS
Flower Craft
(Sue Andrews)
Lower Barn Farm Craft Centre, London Road, Rayleigh SS6 9ET
☎ 0268 780991
Dried flowers in natural colours; arrangements to order. Tuition. C/W
Wed-Sun 10am-5pm.
Directions: see under Craft Centres.

LACEWORK
The Lacy Lady
(Debbie Hanning)
Lower Barn Farm Craft Centre, London Road, Rayleigh SS6 9ET
☎ 0268 750991
Satin and lace work. Soft furnishings. Also ornamental kitchen baskets (containing bread etc). C/W
Wed-Sun 10am-5pm.
Directions: see under Craft Centres.

PICTURE FRAMING
Archive Arts
(Martin and Mel Crocker)
Lower Barn Farm Craft Centre,
London Road, Rayleigh SS6 9ET
☎ *0268 780991*
All forms of picture framing and
découpage.
C/R&R/W. Credit cards acepted
Wed-Sun 10am-5pm.
Directions: see under Craft Centres.

WOODWORK
Mr and Mrs L Argentieri
Lower Barn Farm Craft Centre,
London Road, Rayleigh SS6 9ET
☎ *0268 780991*
Woodwork for the garden and out-
of-doors; bridges, windmills,
dovecotes, wishing wells etc.
C/W/E. Credit cards.
Wed-Sun 10am-5pm.
Directions: see under Craft Centres.

ARTIST
Frank Hyde
Lower Barn Farm Craft Centre,
London Road, Rayleigh SS6 9ET
☎ *0268 780991*
Water colour and oil portraits to
commission. Also nautical paintings
and model ships. C/R&R/E
Wed-Sun 10am-5pm.
Directions: see under Craft Centres.

CERAMICS
Petal Porcelain
(Marilyn Houson)
Lower Barn Farm Craft Centre,
London Road, Rayleigh SS6 9ET
☎ *0268 780991*
Porcelain flowers, in baskets etc.
Tuition given at Centre. Porcelain
supplies. C/W
Wed-Sun 10am-5pm.
Directions: see under Craft Centres.

EGG DECORATING
Kim Stuart
Lower Barn Farm Craft Centre,
London Road, Rayleigh SS6 9ET
☎ *0268 780991*
Goose and duck eggs blown and
decorated. C/R&R. Credit cards.
Wed-Sun 10am-5pm.
Directions: see under Craft Centres.

FABRIC FLOWERS
Mr and Mrs Barry Hill
Lower Barn Farm Craft Centre,
London Road, Rayleigh SS6 9ET
☎ *0268 780991*
Fabric flowers sold individually or
as arrangements.
C/R&R/W/E. Credit cards.
Wed-Sun 10am-5pm.
Directions: see under Craft Centres.

FORGEWORK
Ron Craven
Lower Barn Farm Craft Centre,
London Road, Rayleigh SS6 9ET
☎ *0268 780991*
Traditional blacksmith serving the
local community making hearth
furniture and manufacturing
wrought-ironwork. Tools and
implements repaired.
C/R&R/W/E.
Mon-Fri 8am-4.30pm, Sat 8am-1pm
Directions: see under Craft Centres.

Saffron Walden

CABINET MAKER
David Prue
26 Radwinter Road
Saffron Waldon CB11 3JB
☎ *0799 522558*
Craftsman cabinet maker who
specialises in producing exact
replicas of antique English period
furniture, particularly William and

Mary and Queen Anne pieces in English walnut. C/W/E

Mon-Fri 8am-5.30pm, Sat-Sun by appointment.

FURNITURE RESTORATION
Omega Restoration
(Tony Phillips)
High Street, Newport CB11 3!F
☎ 0799 40720

All kinds of furniture restoration, veneer work, French polishing etc. Workshop is adjacent to shop selling mainly Art Deco and Art Nouveau furniture and objects. R&R

Mon-Fri (closed Thurs) 10am-6pm, Sat 10am-5.30pm, lunchtime closing.

Thaxted

WEAPONS & ARMOUR
Raven Armoury
(Simon Fearnham)
Handley's Farm, Dunmow Road Thaxted CM6 2NX
☎ 0371 870486

Hand-forged swords and knives, weapons and armoury. Sheet metalwork, leatherwork, wood-work, lost-wax and forging. Also jewellery. Work sold all over the world, commissions from museums, including Tower of London Royal Armouries.
C/R&R/W/E. Credit cards.

Mon-Fri 9am-6pm, Sat-Sun 10am-6pm.

Directions: from M11 junc 8 follow A120 to Dunmow, take B184 north towards Thaxted; Raven Armoury on left.

FORGEWORK
Glendale Forge
(Derek and Doreen Tucker)
Monk Street, Thaxted CM6 2NR
☎ 0371 830466

Hand-wrought ornamental iron-work; gates, balustrades, grilles, lanterns, fire furniture, medieval armour, well heads, flower pedestals. Copper work. Exhibition of unusual half-sized vehicles and 24in gauge (George Stephenson's *Rocket*) rides. C/R&R/W/E.

Mon-Fri 9am-5pm, lunchtime closing. Sun 10am-12 noon.

Directions: 1.5 miles south of Thaxted, just off B184.

GLOUCESTERSHIRE CRAFT WORKSHOP CENTRES

Brewery Arts, incorporating Cirencester Workshops & Niccol Centre

Brewery Court, Cirencester GL7 1JH
☎ *0285 657181*
Arts centre with 18 craftworkers running independent businesses: blacksmith, basket maker, letter press printer, bookbinder, jeweller, weaver, potter, antique furniture restorer, carpet restorer, leather worker, fashion embroiderer, textile printer, quilter and designer in leather. Shop, gallery, coffee house, education rooms and theatre.
Mon-Sat 10am-5pm
Directions: adjacent to Brewery car park next to rear entrance of W H Smith & Sons. Other car parks: Old Cattle Market and The Beeches.

Fodca Crafts

(The Forest of Dean Craftworkers Association Ltd)
New Road, Parkend, Lydney GL15 4JA
☎ *0594 564222*

Co-operative of 32 local craft-workers selling from the Forest Craft Shop at Parkend. Craftwork includes: wood carving and turning, pottery and ceramics, textiles, paintings, silk, jewellery, pyrography, picture framing, tapestry, cider making.
Open daily, 9.30am-5pm (closed Xmas-New Year).
Directions: Parkend is on B4234 between Lydney and Lydbrook.

Winchcombe Pottery Workshops

Becketts Lane, Greet
Winchcombe GL54 5NU
☎ *0242 602462*
Pottery workshop employing six craftworkers and showroom. Also craft workshops including a furniture maker, furniture and decorative painter, upholsterer and sculptor.
Opening times: see individual entries.
Directions: 1 mile north of Winchcombe, just off B4632.

GLOUCESTERSHIRE CRAFT WORKSHOPS

Bourton-on-the-Water

CERAMICS
Bourton Pottery
(John and Judy Jelfs)
Clapton Row
Bourton-on-the-Water GL54 2DN
☎ *0451 20173*
Functional tableware and oven-proof kitchenware, plus large and small one-off individual pieces. W/E. Credit cards.
Mon-Fri 9am-5.30pm, Sat and some Suns 10am-5pm.

Cheltenham

FORGEWORK
Nigel B King
285 Old Bath Road
Cheltenham GL53 9AJ
☎ *0242 524274*
Architectural metal work; argon welding, mobile welding, gates, railings and balustrades, weather vanes. One-off specials. C/R&R/E
Mon-Fri 8am-5.30pm, lunchtime closing. Sat 8am-12noon. Advisable to telephone first.

CERAMICS
Prestbury Pottery
(Tony and Sue Davies)
31 New Barn Lane
Prestbury GL52 3LB
☎ 0242 528156
Pottery workshop with showroom and garden. Handmade pottery, mainly domestic earthenware and planters. Commemorative wares for sports clubs, weddings etc. C/W/E
Mon-Fri 9am-5pm, Sat 9am-1pm.
Visit by appointment.

WOODTURNING
Woodcom
(Jonathan Foster)
The Workshops, The Airfield
Chedworth GL54 4NX
☎ 0242 89315
Two woodturners supplying joinery shops and furniture manufacturers. C/W/E
Mon-Sat 8am-5pm
Directions: workshops at rear of Foxley Cottages, The Airfield.

PICTURE FRAMING
Brian Jones
The Workshops, The Airfield
Chedworth GL54 4NX
☎ 0242 890275
Picture framing and mount cutting. Large selection of prints and pictures available. Commissions taken for local artists; houses, animal portraits and human portraits.
C/R&R/W/E.
Mon-Sat 10am-7pm, Sun 10am-1pm.
(Closed 1st week in Aug & Sept).
Directions: workshops at rear of Foxley Cottages, The Airfield.

Chipping Campden

ANTIQUE FURNITURE RESTORATION
Campden Antique Restorations
(Michael Garrity)
Unit One, Cutts Yard, High Street
Chipping Campden GL55 6AT
☎ 0386 841415
Furniture restoration; copying and making of furniture. Turning, traditional hand French/wax polishing. C/R&R/E.
Mon-Fri 8.30am-6pm, Sat 9am-1pm.

SILVERSMITHING
Guild of Handicraft
(David Hart, Derek Elliott)
The Guild, Sheep Street
Chipping Campden GL55 6DS
☎ 0386 841100
Family silversmith business spanning more than 80 years, producing work for cathedrals and churches world-wide. Also private commissions for royalty; prestigious trophies made. C/R&R/E
Mon-Fri 9am-5pm, Sat 9am-12.30pm
or by arrangement.
Charge for visiting groups £1 per head.
Directions: off main High Street, 50yd along Sheep Street set back on right-hand side. Workshop on 1st floor.

Churchdown

WILDFOWL CARVING
Michael Lythgoe
71 Chosen Dr, Churchdown GL3 2QS
☎ 0452 712080
Wood sculptures of wildfowl and game birds by winner of British and European championships. Each individual carving handmade and painted in exacting detail. C/W/E
Mon-Fri 9am-5pm, Sat-Sun 9am-12noon.

Cinderford

GARDEN STONEWARE
Forest Stonecraft
(Audrey Phillips and Lesley Drew)
Yew Tree Cottage, Bradley Hill
Soudley GL14 2UQ
☎ *0594 824823*
A full range of ornaments for house and garden including stone tubs, pots and troughs, garden seats, figures, stepping stones etc. Also supply to garden centres. C/R&R/W
Open daily 9am-7pm (closed Jan/Feb).

Cirencester

FURNITURE & RUSHWORK
Paul Spriggs
The Croft, Silver Street
South Cerney GL7 5TR
☎ *0285 860296*
Designer and maker of furniture in temperate hardwoods; domestic, ecclesiastical and commercial. One-off pieces a speciality. Rush seated chairs made to designs inspired by Ernest Gimson, some always on display. Chairs reseated. C
Visit by appointment at any reasonable time.
Directions: off A419 south of Cirencester, turn west at crossroad (Dogotel & Cattery on corner), left at T junc, The Croft is 50yd on the left.

SADDLERY
Sydney Free Ltd
(S W Free, L J Pearman & A E J Lewis)
54 Querns Lane, Cirencester GL7 1RH
☎ *0285 656107*
Handmade saddles and bridles, stock items held and work to commission. General saddlery and repairs. C/R&R/E. Credit cards.
Mon-Fri 7.30am-5.30pm, Sat 7.30am-1pm.

THATCHING
David White
180 North Home Road
Cirencester GL7 1DX
☎ *0285 657127*
House thatching; specialist in combed wheat reed. C/R&R

CERAMICS
Western-Ville Pottery
(Miss Carol Butler, BA Hons Ceramics)
Kemble GL7 6AW
☎ *0285 770651*
Range of domestic ware (glazed stoneware) and unglazed terracotta kitchen/garden ware. C
Open daily 10am-5pm. Advisable to telephone first.

FURNITURE RESTORATION
Hunt & Lomas
(Christian M Hunt and John V Lomas)
Unit 3 Village Farm
Preston Village GL7 5PR
☎ *0285 640111*
Furniture restoration and cabinet making. French polishing. Fitted kitchens. C/R&R/E
Mon-Fri 8.30am-5.30pm, Sat 10am-2pm.

ANTIQUE FURNITURE RESTORATION
Stephen Hill
Unit 5, Cirencester Workshops,
Brewery Court, Cirencester GL7 1JH
☎ *0285 658817*
All types of furniture restoration, re-polishing, cane and rush work, upholstery. Furniture made to commission; chair matching. C/R&R
Mon-Fri 9am-5.30pm, lunchtime closing. Sat 9am-1pm.
Directions: see under Craft Centres.

BASKET MAKING
Mike Smith
*Unit 6, Cirencester Workshops,
Brewery Court, Cirencester GL7 1JH*
☎ *0285 640356*
Basket-making in willow, some being home-grown. Award-winning work widely exhibited. C/E.
Mon-Fri 9am-5pm, Sat 10am-3pm, please telephone first.
Directions: see under Craft Centres.

RUG RESTORATION
Elizabeth Taylor
*Unit 9, Cirencester Workshops,
Brewery Court, Cirencester GL7 1JH*
☎ *0285 641177*
Restoration of antique rugs and carpets. R&R
Mon-Fri 9am-5pm, lunchtime closing.
Directions: see under Craft Centres.

BOOKBINDING
Pamela Richmond
*Unit 9, Cirencester Workshops,
Brewery Court, Cirencester GL7 1JH*
☎ *0285 658946*
Fine bookbinding and design binding; presentation binding. Award winner and author of *Bookbinding — A Manual of Technique* (Crowood Press). C/R&R/E.
Mon-Fri 10am-5.30pm, lunchtime closing. Sat 11am-4pm by appointment.
Directions: see under Craft Centres.

Fairford

NEEDLECRAFT
The Inglestone Collection
*(Jane Greenoff)
The Long Room, Yells Yard,
Cirencester Road, Fairford GL7 4BS*
☎ *0285 712778*
Counted needlework design — kits, books and fabrics. Tuition. Author of three best-selling books. W/E.
Mon-Sat 9.30am-12.30pm, other times by appointment.

Forest of Dean

FURNITURE
Paul Harper
5a Broad Street, Littledean GL14 3NH
☎ *0594 826255*
Craftsman mainly working alone on contemporary pieces, often in collaboration with other craftsmen. Emphasis on design. C/W/E
Visit by appointment only.

Gloucester

FURNITURE AND RESTORATION
Michael White Designs in Wood
Staunton Court, Staunton GL19 3QE
☎ *0452 840647*
13th century house provides setting for the production of fine quality English furniture in 17th century barns. Mainly copies of 18th century furniture. Antique furniture restoration; French polishing and woodturning. C/R&R/E.
Mon-Fri 8am-6.30pm, Sat 8am-5pm, Sun by appointment.
Directions: Staunton Court is set midway between Ledbury and Gloucester on A417, 4 miles from M50 junc 2.

HANDWEAVING & EMBROIDERY
Colin Squire and Janice Williams
Sheldon Cottage, Epney, Saul GL2 7LN
☎ *0452 740639*
Handwoven rugs, furnishings and hangings, mainly to commission. Embroidery — mainly metal thread work usually for ecclesiastical commissions. Many commissions in public places, particularly churches, also private collections. C/W/E
Visitors welcome by appointment only.

THATCHING
P K Battrick
Holly Cottage, Framilode Passage
Saul GL4 7LE
☎ 0452 740257
Thatching work and restoration.

Lechlade

CERAMICS & TOPIARY TREES
Old Bell Pottery
(Keith Broley)
High Street, Lechlade GL7 3AD
☎ 0367 52608
Ceramics and garden pots made
and sold. Also topiary trees in box
and yew. Credit cards.
Open daily 9.30am-7pm.

TRADITIONAL RUSH CHAIRS
Payne & Poole
(Geoff Payne and Robert Poole)
The Calf Pens, Cross Tree
Filkins GL7 3JL
☎ 0367 860522
Two-man 20-year partnership
specialising in traditional chair-
making and rush seating.
C/R&R/W/E.
Mon-Fri 8.30am-5pm, lunchtime
closing. Sat 9am-12.30pm.
Directions: between Burford (6 miles)
and Lechlade (2 miles) off A361, The
Calf Pens is in centre of Filkins village.

TEXTILE DYEING
The Cotswold Natural Dyers
(Judith Fay)
Filkins Farmhouse, Filkins GL7 3JJ
☎ 0367 860253
Cotswold Yarn for needlepoint and
weaving, hand-dyed with natural
materials. Colour-fast. Needlepoint
kits in traditional Florentine designs
for cushions, chair seats etc (individual orders taken). Yarn (and

designs) for church wall-hangings,
or commissions to make them.
Mon-Fri, visit by appointment.

SADDLERY
John Hayes Products
(John Hayes)
No 1 Oxleaze Farm Workshops
Filkins GL7 3RB
☎ 0367 85472
Leather workshop, saddles, bridles
and general saddlery work and
repairs. C/R&R/W/E.
Mon-Fri 9am-5pm, Sat-Sun by
appointment.

WOOLLEN MILL
Cotswold Woollen Weavers
(Richard and Jane Martin)
Filkins GL7 2JJ
☎ 0367 860491
Working woollen mill where
visitors are welcome to watch work
in progress. Makers of a wide range
of woollen cloth, garments, rugs,
ties etc, all for sale in large mill
shop. Exhibition area, coffeee shop,
picnic area. Free admission; large
groups by appointment only.
C/W. Credit cards.
Mon-Sat 10am-6pm, Sun 2-6pm.
(Closed Xmas-New Year).

Lydney

CABINET MAKING
Andrew J Pyke — Craftsman in
   Wood
(Andrew and Susan Pyke)
Grove Farm, Brockweir Lane
Hewelsfield GL15 6UU
☎ 0594 530924
Country workshop using carefully
selected timbers seasoned and
dried, traditionally jointed to make

furniture of lasting quality with a choice of finishes and quality brass fittings. C/R&R
Mon-Sat 8am-8pm.
Directions: from B4228 Chepstow/Coleford road, take turning signed Brockweir at Hewelsfield crossroads. Or turn off A466 Chepstow/Monmouth road over Brockweir bridge and proceed up the main lane.

CERAMICS & SILVERWORK
St Briavels Pottery & Silver
(Gill McCubbin)
St Briavels GL15 6TQ
☎ *0594 530297*
A comprehensive range of domestic stoneware. Modern silver jewellery including rings, earrings etc.
Easter-Oct open daily (closed Tues) 9.30am-6pm (lunchtime closing). Nov-Dec 9.30am-4pm, Jan-Easter Sat-Sun only 9.30am-4pm. Other times by appointment.

Moreton-in-Marsh

WOODCARVING & FURNITURE
Julian Stanley
Unit 5, The Sitch
Longborough GL56 0QJ
☎ *0451 31122*
High quality woodcarving and furniture. Reproductions and one-off designs for mirrors, portrait busts and large-scale architectural pieces. Contemporary furniture, mainly to commission. C/R&R/E.
Visitors welcome by appointment.

Nailsworth

FORGEWORK
Tinker, Tailor...
(Mr and Mrs D Tigwell)
Ladywood Cottage, Whips Lane
Watledge GL6 0BB

(Showroom: Millbrook House, George Street, Nailsworth)
☎ *0453 834704*
Husband and wife team designing and making blacksmithery and metalwork. Original designs and specialists in 'Arts & Craft' metalwork to E W Gimson's designs. C/R&R/W. Credit cards.
Mon-Sat 9.30am-5pm.
Directions: to workshop from showroom, go towards Michinhampton, over cattle grid. Left to Watledge, after half-mile right into Whips Lane, after 50yd right onto track to last cottage.

FURNITURE
Atelier Damien Dewing
(Damien Dewing)
7C Nailsworth Mills Estate
Nailsworth GL6 0AG
☎ *0453 834100*
Design and making furniture using antique building materials from old French barns and 18th century Welsh and English oak. Entire buildings refashioned and individual furniture made. Restoration of timber structures. C/R&R/W/E.
Visitors welcome by appointment.

Newent

JEWELLERY
Newent Silver & Gold
(K I Vowles)
15 Broad Street, Newent GL18 1AQ
☎ *0531 822055*
Working craftsman on view to the public, making mainly gold and silver jewellery with stone setting. C/R&R/E. Credit cards.
Tues-Sat 9.15am-5pm (early closing Wed). Lunchtime closing except Sat.
Directions: workshop is almost opposite Barclays Bank in Newent.

GLASSBLOWING
The Glassbarn
(P C Solven and H L Cowdy)
31 Culver Street, Newent GL18 1DB
☎ 0531 821173
Contemporary glass made at glass-blowing workshop, with show-room. Original, often colourful glass ranging from vases, bowls, paper-weights and wine glasses; one-off signed pieces. Advisable to check schedules for glass-blowing. W/E. Credit cards.
Mon-Fri 10am-5pm (lunchtime closing 12.30-1.30pm). Sat (gallery only) 10am-12.30pm.
Charge for glass-blowing demonstrations: £1 adults, 75p children and OAPs (includes information sheet).

Newham

UPHOLSTERY & SOFT FURNISHING
Cottonwood
(Alan Curtis)
High Street, Newnham GL14 1BW
☎ 0594 516633
Traditional and modern upholstery and furnishing designers. Contract furnishers and manufacturing upholsterers; specialist range of furniture for the elderly. Also curtains, loose covers etc. C/R&R/W/E. Credit cards.
Mon-Fri 9.30am-5pm, lunchtime closing.

Staunton

FURNITURE & KITCHENS
Belbic Designs
(G D Crawford and M Bick)
The Hawthorns
Pillows Green GL19 3NY
☎ 0452 840791
Established partnership in country workshop designing and making

exclusive kitchens and furniture for the home and office in hard and soft woods. Work widely exhibited. Guild members. C/R&R/W/E
Mon-Fri 8.30am-5.30pm, advisable to telephone first. Sat-Sun by appointment.

Stroud

FURNITURE
Ian D F Sim
Unit 10 Piccadilly Mill, Lower Street Stroud GL5 2HT
☎ 0453 753287
Small well-equipped workshop taking commissions of all types. Handmade bespoke furniture, antique furniture restoration, cane and rush seating, upholstery. Small batch production. C/R&R/W
Mon-Thurs 2-4pm, advisable to telephone first.

WOODWORK
Dennis French Woodware
The Craft Shop and Workshop
Brimscombe Hill GL5 2QR
☎ 0453 883054
Produced by the principal British demonstrator at the International Woodturning Seminar in 1989, a comprehensive range of carefully hand-finished domestic woodware made in English hardwoods, including kitchen boards, bowls and platters, table lamps etc. C/W
Showroom: Tues-Fri 9am-5pm, Sat 8am-4pm; Workshop by appointment only. (Woodturning demonstrations at newly opened showrooms: Painswick Woodcrafts, New Street, Painswick. Telephone for details.)
Directions: off A419 between Thrupp and Chalford, turn south to Brims-combe over railway; workshop and showroom near parish church.

BRASS & COPPER FORGEWORK
Michael E Roberts — Designer Metalworker
Anvil Barn, Miserden GL6 7JD
☎ *0285 821244*
Master Blacksmith with 26 years experience trained in Switzerland, specialist in brass, bronze, copper, alloys and steel forgework. Visitors are welcome to visit this working forge in a 17th century Cotswold barn. C/R&R/E.
Mon-Fri visit by appointment only. Visitors welcome Sat-Sun.
Directions: 9 miles from Stroud and Cirencester.

ANTIQUE FURNITURE RESTORATION
Relda Restorations
(John Selby Greene)
8 Piccadilly Mill, Lower Street Stroud GL5 2HT
☎ *0453 755538*
Restoration work on all types of antique woodwork including tea caddies, boxes, etc. Also burr walnut dashboards. French polishing, colouring, and general repair work. C/R&R/E.
Mon-Fri 8.30am-5pm, lunchtime closing. Sat-Sun by appointment.

CERAMICS
Michael and Barbara Hawkins
The Pottery, Rooksmoor Mills Bath Road Stroud GL5 5ND
☎ *0453 873322*
Small pottery workshop with showroom. Handmade, finely painted lamps, bowls, vases etc. Work sold in galleries throughout UK. C/W/E. Credit cards.
Mon-Fri 9am-5pm, Sat-Sun 10am-4pm.

Tetbury

CERAMICS RESTORATION
The China Repairers
(Mrs P J Targett and Mrs A J Chalmers)
Workshop One, Street Farm Workshops Doughton GL8 8TQ
☎ *0666 503551*
Three craftspeople specialising in the restoration of antique china and all other ceramics, glassware, alabaster, marble, jade, tiles, terracotta etc. C/R&R/E.
Mon-Fri 9am-5pm, Sat 10am-12noon
Directions: south west of Tetbury on A433.

FURNITURE
Colin Clark Furniture
The Old Forge, Hampton Street Tetbury GL8 8JN
☎ *0666 504838*
Three employed in furniture making, working to commission, mainly in native oak. Oak panelling, hand-carving and woodturning. C/R&R/E.
Mon-Fri 8am-5pm, lunchtime closing. Sat-Sun by appointment.

LEATHER WORK
MacGregor and Michael
(Neil MacGregor and Valerie Michael)
37 Silver Street, Tetbury GL8 8DL
☎ *0666 502179*
High quality, hand-stitched leather goods including belts, purses, bags, briefcases, attaché cases. Work displayed in leather craft museums. Guild members. C/W/E
Mon-Fri 9am-5.30pm, lunchtime closing. Sat by arrangement.

CERAMICS
Hookshouse Pottery
(Christopher White)
Westonbirt GL8 8TZ

☎ *0666 880297*
Hand-thrown stoneware pottery
including domestic ware, one-off
pieces and garden pots.
C/W/E. Credit cards.
Open daily 10am-6pm.

ANTIQUE FURNITURE
RESTORATION
Andrew Lelliott
6 Tetbury Hill, Avening GL8 8LT
☎ *0453 835783*
Cabinet makers, graduates from
West Dean (BADA diploma) and
Rycotewood Colleges, listed in
Conservation Unit register. C/R&R
*Mon-Fri 9am-5pm, telephone for
appointment.*

CABINET MAKING
William Cook
*Primrose Cottage, 11 Northfield Road
Tetbury GL8 8HP*
☎ *0666 502877*
Specialist in clock-case restoration
and gilding (water and oil), making
clock-cases for old movements and
restoration of gilded mirrors and
frames. C/R&R/E.
Visitors welcome by appointment only.

WOOD SCULPTURE AND
WATERCOLOURS
Woodpecker & Kadecraft
*(Roy Bishop and Kath Eamer)
5 Street Farm Workshops,
Doughton GL8 8TQ*
☎ *0666 503400*
Wood carving including wildlife
sculpture, heraldry, industrial
carving, lettering. Also watercol-
ours and embroidery, cards and
framed prints.
C/R&R/W/E. Credit cards.
Visitors welcome by appointment.

Tewkesbury

SILK PRINTING
Beckford Silk Ltd
*(James and Marthe Gardner)
Ashton Road, Beckford GL20 7AD*
☎ *0386 881507*
All processes involved in the hand
screen-printing of silk can be
observed in the print shop. The
finished articles, all exclusively
designed and made to a high
standard, are available in the Silk
Store next door. Coffee shop.
C/W. Credit cards.
Mon-Sat 9am-5.30pm.

CERAMICS
Conderton Pottery
*(Toff and Georgina Milway)
The Old Forge, Conderton GL20 7PP*
☎ *0386 89387*
Handmade salt-glazed studio stone-
ware pottery; kitchen/tableware
and garden pots all made by Toff
Milway. All processes can be seen
in the pottery. C/E. Credit cards.
*Mon-Sat from 9am-5.30pm, other
times by appointment.*

FORGEWORK
Pimlico Forge
*(Chris Hunting)
Pimlico Cottage
Little Beckford GL20 7AL*
☎ *0386 881836*
Hand-forged ironwork, welding
and repairs. General blacksmithing.
Commissioned to replace ironwork
in Cheltenham removed during the
war. C/R&R
*Mon-Fri 8am-5.30pm, lunchtime
closing. Sat 8am-1pm. Advisable to
telephone first.*
Directions: 1 mile north east of
Teddington Hands roundabout.

Winchcombe

CERAMICS
Winchcombe Pottery Ltd
(Ray and Mike Finch)
Broadway Road, Winchcombe GL54 5NU
☎ *0242 602462*
Established in 1926 on the site of an old country pottery. Six potters under the leadership of Ray Finch, the Master Potter, handmaking decorative and functional wood-fired stoneware. Work in art museums and private collections worldwide. Guided tours for groups by appointment; charge of £10 for up to 20 people.
C/W/E. Credit cards.
Mon-Fri 8am-5pm, Sat 9am-4pm, May-Sept Sun 12noon-4pm.
Directions: see Craft Centres.

JEWELLERY & BOXES
Colin Clark — Goldsmith
Winchcombe Pottery, Broadway Road
Winchcombe GL54 5NU
☎ *0242 603059*
Fine, hand-crafted designer work in gold, silver and finely carved hardwoods. C/R&R/W/E
Mon-Fri 9am-6pm, Sat by appointment.
Directions: see Craft Centres.

FURNITURE
William Hall Furniture
Winchcombe Pottery, Broadway Road
Winchcombe GL54 5NU
☎ *0242 603059*
Small business making individual furniture to commission (private customers, architects, designers, churches etc). Work includes seating for Cheltenham Gallery & Museum; chapel furniture, Moreton Fire College; chairs for Lutyens Design Associates. C/W/E

Mon-Fri 8am-6pm (lunchtime closing 12.30-1.30pm) Sat 8.30am-1pm. Please telephone before visiting.
Directions: see Craft Centres.

DECORATIVE PAINT FINISHES
Decorative Painting and Signs
(Katie Morgan)
Winchcombe Pottery, Broadway Road
Winchcombe GL54 5NU
☎ *0242 603059/513925*
Decorative painting on furniture etc. Fairground signs and art. Marbling and graining. C/R&R/E
Open during normal working hours, but often working on site therefore advisable to telephone first.
Directions: see Craft Centres.

Woodchester

SPECIALISED GLASS ETCHING
John Williams & Partners
Unit E5, Inchbrook Trading Estate,
Bath Road, Woodchester GL5 5EY
☎ *045383 5869*
Two craftsmen undertaking highly specialised etching work recognised as being of the highest quality, using unique processes. Stained glass on commission. C/R&R/E
Mon-Fri 9am-5pm.

Wotton-Under-Edge

CERAMICS
English Country Pottery Ltd
(P Hyde, J Collett and A Parkinson)
Station Road, Wickwar GL12 8NB
☎ *0454 299100*
Pottery making reasonably priced hand-decorated pottery, lampbases, vases; kitchen and bathroom ranges. Showroom. C/W/E.
Mon-Fri 8.30am-4pm.
Directions: Wickwar is 4 miles from M5 junc 14, between Bristol and Gloucester.

HAMPSHIRE CRAFT WORKSHOP CENTRES

Alderholt Mill

(John and Ann Pye)
Sandleheath Road, Alderholt
Fordingbridge SP6 1PU
☎ *0425 653130*

Picturesque watermill housing all kinds of craft including woodwork, glass, ceramics, leatherwork, wooden toys and walking sticks. Three-weekly exhibitions of artwork in gallery. Milling on Sunday afternoons.

Easter-Sept/mid Nov-Xmas Tue-Sun 2-6pm, Sat & Bank Holidays 10am-6pm. Closed Oct-mid Nov/Xmas-Easter.

Viables Craft Centre

(Viables Activities Trust Ltd)
Harrow Way, Basingstoke RG22 4BJ
☎ *0256 811911*

19th century farm converted to provide craft workshops including model locomotives with miniature railway, bronze sculptures, cracker making, engraving, dried and fabric floral art, garden furniture, picture framing, knitwear, oriental pictures in lacquered wood, gold and silver smithing, pottery, woodturning, glass engraving. Wine bar, restaurant and tea rooms.

Tues-Fri 1-4pm, Sat-Sun 2-5pm.
Charges: Free entry and parking
(small entry fee for special functions).

The Old Granary

Bank Street, Bishop's Waltham SO3 1AE
☎ *0489 894595*

A group of independent designer craftsmen working in an old converted granary. Crafts include: picture framing, pottery, knitwear and calligraphy. Art gallery and restaurant.

Tues-Sat 10am-4pm.

HAMPSHIRE CRAFT WORKSHOPS

Alresford

HAND-WOVEN TEXTILES & TAPESTRIES
Sophie Pattinson
No 3, Sunnyside Cottages
Church Street, Ropley SO24 9SA
☎ *0962 772516*

Hand-woven and painted textiles incorporating strips of painted wood, wool and linen, producing colourful wall hangings and tapestries. Commissions for offices, public buildings etc and private sales. Exports to Japan. C/W/E
Mon-Fri 9am-5pm, Sat-Sun by appointment.

GLASS ENGRAVING
Tony Gilliam
29 West Street, Alresford SO24 9AB
☎ *0962 734504*

Glass engraving by diamond point (stippling), drill and sandblasting on goblets, bowls etc. Also windows, doors and screens for churches and architectural situations. Award winner; work in major exhibitions in UK; Fellow of Glass Engravers Guild. C/R&R/E
Tues-Sat 9am-5.30pm, lunchtime closing.
Directions: West Street is on A31 Winchester to Alton road. (Go straight into Alresford, not on by-pass).

Alton

FURNITURE
Hugo Egleston Furniture
New Farm Buildings
Lasham GU34 5RY
☎ *0256 381368*
Furniture and woodwork of quality, designed and made to order in fine timbers such as walnut, oak, yew and cherry. Small workshop relying on handwork and carefully selected solid wood to produce individuality and distinction. C/E. Credit cards.
Mon-Fri 10am-5.30pm (advisable to telephone), Sat-Sun by appointment.

CERAMIC SCULPTURE
Audrey Richardson Ceramics
Neatham Mill Workshops
Lower Neatham Lane
Holybourne GU34 4ET
☎ *0420 542021*
Ceramic work; one-off pieces and sculpture, including garden sculpture. C/W/E
Mon, Wed, Fri 9.30am-6pm. Some Sat & Suns.
Directions: From Holybourne PO take 1st right (Lower Neatham Lane) to end. Park and walk under A31 bridge; workshop on right over stream.

FORGEWORK
Robert Smith
The Forge, Gosport Road
Farringdon GU34 3DL
☎ *042 058 7233*
Traditional forge making ironwork to order. Gates, fire tools, housenames and door furniture. C/R&R/W/E
Mon-Fri 8am-5.30pm, advisable to telephone first.
Directions: on A32 Farringdon crossroads, south of Alton.

CERAMICS
Robert Goldsmith Pottery
Lower Neatham Mill
Holybourne GU34 4ET
☎ *0420 80915/87597*
Hand-thrown domestic stoneware decorated with pigments of cobalt, iron and copper to give distinctive lustrous blues, browns, reds and gold. Work sold through Crafts Council listed galleries and shops throughout UK and abroad. C/W/E. Credit cards.
Mon-Sat 10am-6pm, Sat-Sun by appointment.

PIPE MAKING
David Cooper
Candovers Barn
Hartley Mauditt GU34 3BP
☎ *0420 50293*
Smoke and bubble pipe maker also demonstrating at craft fairs and at Amberley Chalk Pits Museum near Arundel, Sussex. W/E.
Closed Nov-Mar. Ring for appointment. Charges: Adults £3.70, OAPs £2.70, children £1.50.

CABINET MAKING
Guy Bagshaw
Plain Farm Old Dairy
East Tisted GU34 3RT
☎ *0420 58362*
Family firm of cabinet makers and restorers of 17th/18th century furniture. Carving, veneering, polishing, leatherwork, marquetry and gilding. Weekend restoration tuition. C/R&R/E
Mon-Fri 8am-6pm, Sat-Sun by arrangement.
Directions: from Alton towards Fareham on A32, cross A31, go past Farringdon & East Tisted. Take next turn right, then first left.

CERAMICS
Froyle Pottery & Tiles
(Rupert Spira, BA Hons)
Lower Froyle GU34 4LL
☎ *0420 23693*
Tiles made by traditional pressed clay method, decorated with glazes prepared from minerals and metal oxides (no lead). Also all aspects of functional pottery; work in many notable private collections. Awards and features in publications. C/W/E
Mon-Fri 9am-5pm (lunchtime closing). Sat-Sun by appointment.
Directions: from A31 Alton-Farnham, take turning to Upper and Lower Froyle near Bentley. Go past The Anchor PH and Prince of Wales PH; pottery on left.

Andover

SADDLERY
Kestan Horses
(Mrs Kay Hastilow)
The Old Dairy, Chute Forest SP11 0EA
☎ *0264 70367*
Specialist saddlery workshop employing eight. Handmade bridlework for professional riders. Specialist polo section and eventing. Expert fitting service by Master Saddler with 25 years' experience. C/R&R/E. Credit cards.
Mon-Fri 9am-6pm, Sat 9am-2pm.
Very rural setting on Hants/Wilts border — telephone for directions.

Basingstoke

FURNITURE MAKING
Wood Works
(Mark Buckner and David Titcombe)
Newnham Lane
Old Basing RG24 0AT
☎ *0256 812962*
Graduates from Rycotewood College; specialist woodworking to customers' individual requirements. Entry in *The Yellow Book* (best of student art and design). C/E
Mon-Fri 10am-4pm, Sat-Sun by appointment.
Directions: From Old Basing take Newnham Lane towards Newnham village for approx quarter mile. Wildwood Farm is 1st lane on right after last house; workshop is at end of lane.

WOODTURNING
Centric Studio
(Brian Hannam)
Unit 9, Viables Craft Centre
Harrow Way
Basingstoke RG22 4BJ
☎ *0256 811911*
High quality decorative woodturning. Also quilting frames and other specialist work to order. House signs. Winner of Hampshire Woodturners' Association competition. C/R&R/E.
Mon-Fri 9am-5pm, Sat-Sun 2-5pm. (Closed 2 weeks from Xmas).

CERAMICS
Cairncraft Pottery
(David Cairn)
Viables Craft Centre, Harrow Way
Basingstoke RG22 4BJ
☎ *0256 54663*
Established pottery; house signs, presentation plates, fretwork lamps, domestic and decorative pottery. C
Tues-Fri 9am-5pm, Sat 10am-5pm, lunchtime closing. Sun 2-5pm.

ENGRAVING
Jeanette's Hand & Machine Engravers
(Jeanette Moore and Paul Shepherd)
Unit 16, Viables Craft Centre,
Harrow Way, Basingstoke RG22 4BJ

☎ 0256 474009
Engraving; trophies, gifts and
awards. C/R&R/E.
Mon-Fri 9am-5pm, lunchtime closing.
Sat-Sun 2-5pm.

KNITWEAR
Knitique
(Margaret Pallant)
Unit 7, Viables Craft Centre
Harrow Way, Basingstoke RG22 4BJ
☎ 0256 29229
Workshop where machine knitting,
spinning and weaving is demon-
strated. Sale of knitwear, yarns,
spinning and weaving supplies. C/E
Tues-Thur 9.30am-5pm, Fri 12noon-
5pm, Sat-Sun 2-5pm.

GLASS ENGRAVING
Mike Burrows
Unit 14/15, Viables Craft Centre,
Harrow Way, Basingstoke RG22 4BJ
☎ 0256 50983
Established hand-engraved glass-
ware workshop. A wide range of
glass for individual engraving.C
Tues-Fri 1-5.30pm, Sat 11am-5pm, Sun
(most) 2-5pm. Closed Xmas-New Year.

GOLD & SILVER JEWELLERY
Petra's
Unit 8, Viables Craft Centre
Harrow Way, Basingstoke RG22 4BJ
☎ 0256 55945
Award winning gold and silver-
smith undertaking commission
work, repairs and alterations of
jewellery. C/R&R/E.
Tues-Thur 11am-4pm, Sat-Sun 2-5pm.

CRACKER MAKING
Snap Happy
(Ruth Wiseman)
Unit 5, Viables Craft Centre
Harrow Way, Basingstoke RG22 4BJ

☎ 0256 21012
Custom made crackers for all
occasions. Specialities include
'keepsake crackers' which can be re-
used. Over 300 suitable cracker gifts
or small presents. Tuition given and
materials available.
C/R&R/E. Credit cards.
Tues-Fri 1-4pm, Sat-Sun 2-5pm.
Closed January.

LACQUERED WOOD
Pictures in Wood
(David Glover)
Unit 25, Viables Craft Centre,
Harrow Way, Basingstoke RG22 4BJ
☎ 0256 24144
Pictures in lacquered wood, to
commission. C/R&R/E. Credit cards.
Tues-Fri 9am-5pm, Sat-Sun 2-5pm.

Beaulieu

CERAMICS
Kristen Pottery
(Mr and Mrs K Pratt)
High Street, Beaulieu SO42 7YA
☎ 0590 612064
Hand-thrown and handmade
pottery including candles, dolls,
fossils, perfume burners, flower-
embossed vases etc. Individual
house signs/numbers, mirrors,
garden pots. C/W. Credit cards.
Open daily 10am-5.30pm.

RUSH & CANE WORK
Cane & Woodcraft Centre
(John and Christine Hayward)
57 High Street, Beaulieu SO42 7YA
☎ 0590 612211
Cane, rush and seagrass furniture
restoration. Also manufacture of
stools and childrens' chairs. Guild
member; work commissioned by
stately homes, museums and

furniture trade. C/R&R. Credit cards. *Wed-Sat 10am-5pm, lunchtime closing. Sun 12noon-5pm. Closed Xmas-mid Jan.*

Bordon

CONTEMPORARY FORGEWORK
Forged Metalwork
(Ani Duckworth BA Hons and Paul Gulati HND BA Hons)
Garage No 5, 1-3 Petersfield Road
Whitehill GU35 9AW
☎ *0420 476017*
Hand-forged steel/aluminium interior and architectural metalwork. Commissions for any scale of work. Gates, grilles, railings, sculptural pieces etc. Also repair and reproduction work. Exhibitions, including Sotherbys' Iron-work Auction. C/R&R/W/E . Credit cards. *Mon-Fri 9.30am-6pm.*

Emsworth

WOODEN SCREENS
Screens Gallery
(Bev Houlding, B Ed and Mark Houlding, MA, RCA)
The Malthouse, Bridgefoot Path
Emsworth PO10 7EB
☎ *0243 377334*
Room screens made to innovative designs ranging from carved and painted wood to handworked steel, richly decorated for use in the home, office and leisure environments. Work also sold abroad and collected by museums including Portsmouth Museum. C/W/E
Visitors welcome by appointment.
Directions: gallery is next to Western Millpond.

CERAMICS
Richard Boswell Ceramics
The Malthouse, Bridgefoot Path
Emsworth PO10 7EB
☎ *0329 284701*
Specialist hand-thrown earthenware decorated with inlays of coloured clays. The crisp and precise decoration is unique to each individually signed piece. Work widely exhibited in UK and held in many private collections abroad. Professional member of the Craft Potters Association. C/W/E.
Advisable to telephone before visiting.
Directions: situated on Emsworth Quay Mill Pond, next to Emsworth Slipper Sailing Club.

Fordingbridge

PORCELAIN FACTORY
Branksome China Works
(E Baggaley Ltd)
Shaftesbury Street
Fordingbridge SP6 1JF
☎ *0425 652010*
Business established nearly 50 years ago, with nine skilled staff making quality tableware unique in recipe and design. Fine china, tea and dinner sets and animal studies sold to private customers worldwide. Shop and free factory tours on weekdays. E
Mon-Thur 9am-5pm, Fri-Sat 10am-4.30pm.

FURNITURE
The Walnut Tree Workshop
(Mike Williamson)
The Old Dairy, Station Approach
Breamore SP6 2AB
☎ *0725 22165*
Workshop housed in a Victorian dairy where visitors are welcome to

watch work carried out. Design and handmaking of furniture in many different hardwoods. Also short batch production runs and fitted furniture. C/R&R/W/E

Mon-Fri 8.30am-5.30pm, Sat-Sun by arrangement.

Directions: from Salisbury on A338 south to Ringwood, past Bat and Ball PH on left; take left branch to Wood-green, take slip-road to workshop.

ANTIQUE FURNITURE RESTORATION & UPHOLSTERY
Seagers Restorations
(Bernard A Cheater)
Seagers Farm, Stuckton SP6 2HG
☎ *0425 652245*
Specialists in repair and restoration work of antique furniture and allied upholstery. R&R
Mon-Fri 9am-5.30pm, lunchtime closing.
Directions: 1 mile east of Fordingbridge.

THATCHING
M J Jones
Appleslade, Sandleheath SP6 1PP
☎ *0425 652475*
Member of regional and national Master Thatchers' Societies. Renovation of thatch, plain and ornamental ridging using wheat reed, water reed and long straw.

Hartley Wintney

CERAMICS
Winchfield Pottery
(Susan Ferraby)
The Chase, Winchfield RG27 8BX
☎ *0252 842476*
Hand-thrown domestic stoneware with some semi-porcelain decorated with Chinese brushwork. Workshop and showroom in Victorian stable. C/E.

Advisable to telephone first for appointment.

Directions: between Odiham (A287) and Hartley Wintney (A30). Follow signs for Winchfield Station on B3016; pottery is on corner of turning to station.

Lymington

FORGEWORK
Milton Forge/Milton Metalcrafts
(Gordon and Sandra Brownen)
Workshop: Little Gordleton Farm Silver Street, South Sway
☎ *0425 610610*
General blacksmithing work including handmade gates, balustrades, fire baskets etc. Also some repair work. Showroom at 92 Old Milton Road, New Milton. C/R&R
Workshop: Mon-Sat 9am-6pm, advisable to telephone on previous day. Showroom open Tues, Fri & Sat 9.30am-5pm.

Lyndhurst

CERAMICS
Angels Farm Pottery
(Joanna Osman)
Pinkney Lane, Lyndhurst SO43 7FE
☎ *0703 282185*
A range of simply decorated reduction-fired pots for table and kitchen use. Also some flower pots and vases. C/W
Mon-Fri 9.30am-4.30pm, Sat 10am-4pm.
Directions: on Lyndhurst one-way system

Petersfield

DOLLS' HOUSES & MINIATURES
D R Designs
(Ronald Nader)
12 The Close, Langrish GU32 1RH
☎ *0730 61040*
One-twelfth scale model period

homes; collectors' items in exquisite woods eg vintage aircraft, cars, butter churns, toys etc. C/R&R/W/E
Mon-Fri 9am-6pm, lunchtime closing. Sat-Sun 9am-6pm by appointment.
Directions: from Winchester/ Petersfield A272 road at Langrish turn south to East Meon/HMS Mercury. Turn right opposite white thatched cottage, into The Close (unmarked).

BOOKBINDING & CALLIGRAPHY
Rose Voelcker
Wolverton, West Meon GU32 1LQ
☎ 0730 829267
Cloth, paper, leather and vellum bindings, portfolios and boxes, slip cases, scroll cases, restorationof books and manuscripts. Calligraphy: certificates, greeting cards, invitations, labels, logos, menus, posters, stationery etc. C/R&R/E.
Visitors welcome by appointment.
Directions: 1st house on right coming into West Meon on A32 from Alton to Fareham, opposite bus lay-by on left.

CABINET MAKING
Edward Barnsley Educational Workshop
Cockshott Lane, Froxfield GU32 1BB
☎ 073 084 233/329
Fine cabinet making and designing to train young people in traditional apprenticeships. Pupils work along side craftsmen, making furniture to Edward Barnsley designs and adaptions following the 'Cotswold School' tradition. C/E
Mon-Fri 8am-5pm, Sat-Sun 9am-6pm, visit by appointment only.

FORGEWORK
F M Engineering/Forge Crafts
(D G Mustchin and A L Fry)
The Forge, High Street
East Meon GU32 1QD
☎ 0730 87527
General ironwork, blacksmithing and sheet fabrication. Gates, railings, balconies, balustrades, fire baskets and guards, steel and copper canopies, lanterns, wall lights and garden furniture. C/R&R
Mon-Fri 7.30am-5pm, Sat 9am-5pm.

FORGEWORK
The Forge
(Steve Pibworth)
Swan Street, Petersfield GU32 3AJ
☎ 0730 65198
General blacksmithing, specialising in church ironwork of all types. Gates, railings, weather vanes, fire hoods in all metals, fire backs and baskets. Restoration work, including ancient water pumps. C/R&R/E.
Mon-Fri 8am-5pm (lunchtime closing 1-2.30pm). Sat 8am-12noon.
Directions: in town centre adjacent to New Petersfield Hospital.

Portsmouth

CERAMICS
Denmead Pottery Ltd
(M J Ryan)
Forest Road, Denmead PO7 6TT
☎ 0705 261942
Oven-to-table ware, lamps and noveltyearthenware items . Guided tours arranged. Credit cards.
Mon-Fri 9am-5.30pm, Sat-Sun 9am-5pm. Entry charge: 40p.

Ringwood

CERAMICS
Burley Pottery
(M Crowther)
3 Ringwood Road, Burley BH24 4AD
☎ 04253 3205
Selection of hand-thrown stoneware pottery for oven and table, ceramic

pictures, cottages, clowns and figures (especially to commission). Also macramé work. C/W
Open daily 10am-5.30pm.

SADDLERY
Glenn Hasker
(Glenn, Martin and Norman Hasker)
31 Southampton Road
Ringwood BH24 1HB
☎ *0425 476092*
Highly qualified saddler (past President, Society of Master Saddlers) with three other craftsmen. Saddles to order. C/R&R/E. Credit cards.
Mon-Sat 8.30am-5.30pm.

Romsey

HANDMADE BRICKS
Michelmersh Brick Co Ltd
(D A and Q Hill)
Hillview Road
Michelmersh SO51 0NN
☎ *0794 68506*
Handmade, kiln-fired bricks, briquettes, paviors (paving bricks), sawn-gauged arches and specially shaped bricks in a range of colours. Fireplace kits and fittings. C/R&R/W/E. Credit cards.
Mon-Fri 9am-5pm, Sat 8-11.30am.
Bank holidays by appointment.
Directions: turn right off A3057 4 miles north of Romsey.

GARDEN FURNITURE
Heritage Woodcraft
(Anne and David Smith, June and David Finch)
Unit 5, Shelley Farm, Ower SO51 6AS
☎ *0703 814145*
Wooden garden furniture, traditional wooden wheelbarrows and wheel-making in hard and softwoods using traditional methods. C/E. Credit cards.

Mon-Fri 8am-5pm, visit by appointment only.

Soberton

CERAMICS
Elizabeth Gale Ceramics
Taplands Farm Cottage
Taplands Corner, High Street
Webbs Green SO3 1PY
☎ *0705 632686*
Stoneware reduction domestic ware and individual pieces. Professional Member of the Craft Potters Association. C/W/E. Credit cards.

Southampton

GOLD & SILVERSMITHING
Chris Haslett
Arundel House, The Avenue
Bishop's Waltham SO3 1BN
☎ *0489 895844*
Gold and silver work specialising in individual pieces of jewellery, spoons and miniatures. Jewellery repair work undertaken. Regular exhibitor in the south of England. C/R&R. Credit cards.
Visitors welcome by appointment.

CERAMICS
W Teal
The Old Granary, Bank Street
Bishop's Waltham SO3 1AE
☎ *0489 894595*
Pottery workshop making presentation plaques, bread crocks, parsley pot men, house plaques and Roman entrance plaques. Regular exhibitions; work on sale in Salisbury and Winchester Cathedrals. C/W
Tues-Sat 10am-4pm, lunchtime closing.

KNITWARE
Northern Light Knitwear
(Liz Harvey)
The Old Granary, Bank Street
Bishop's Waltham SO3 1AE
☎ *0489 894595*
Wool, mohair and cotton knitwear
based on traditional Norwegian
patterns. C/W/E. Credit cards.
Tues-Sat 10am-4pm.

PICTURE FRAMING
Mussellwhite Framing
(Robert John Mussellwhite)
The Old Granary, Bank Street
Bishop's Waltham SO3 1AE
☎ *0489 895973*
Creative picture framing; mount-
cutting, wash-lining, dry mounting,
tapestry stretching. Free collection/
delivery service. C/W/E. Credit cards.
Tues-Sat 10am-4pm.

CALLIGRAPHY
Barbara Bundy
The Old Granary, Bank Street
Bishop's Waltham SO3 1AE
☎ *0489 894595*
Calligraphy, illumination, heraldic
art, hand-drawn lettering. Use of
traditional materials in a contemporary way, 'painting with words' for
exhibition work. Two books
published: *Colour Calligraphy* and
Illuminated Lettering and Heraldry. C/E.
Tues-Sat 10am-4pm.

Southsea

TEXTILE ARTIST
Textiles and Education
(Ewa Kuniczak)
Wilton House, 13 Norfolk Street
Southsea PO5 4DR
☎ *0705 731634*

Handmade felt, paper, embroidery,
printing and dyeing textiles. Work
undertaken for exhibitions and
commissions. Tuition: Saturday
workshops periodically throughout
year (write for details). C
Sat 10am-4.30pm or by appointment.

Waterlooville

THATCHING
Matthew Holt
14 Hillside Avenue
Widley PO7 5BB
☎ *0705 382085*
Thatching contractor employing
three full-time thatchers. Member of
Master Thatcher associations.

CERAMICS
Clanfield Pottery
(Roger and Sarah Mulley)
131 Chalton Lane, Clanfield PO8 0RG
☎ *0705 595144*
Four potters making ceramic name
plaques and terracotta name pots.
Also a large variety of garden pots
(terracotta and glazed). Ceramic
lamp bases and jardinieres. C/W/E
Shop: Wed-Fri 10.30am-6pm, Sat-Sun
10.30am-4pm. Workshop open Sat-
Sun only.

CABINET MAKING & ANTIQUE
FURNITURE RESTORATION
Horndean Antiques
(F W Scowen)
69 London Road, Horndean PO8 0BW
☎ *0705 592989*
Traditional cabinet making and
antique furniture restoration. Also
retail furniture. C/R&R
Mon-Fri 9am-5.30pm, lunchtime
closing. Sat 9am-1pm.

Whitchurch

SILK MILL
Whitchurch Silk Mill
28 Winchester Street
Whitchurch RG28 7AL
☎ *0256 892065 (mill)*
893882(shop & visitors)
Working silk mill on the River Test.
Silk woven on historic looms, the
winding machines powered by
water wheel. Silk sold retail and
woven to order for furnishings,
costume, legal and academic
gowns. Also costume exhibition,
gardens, tea rooms.
C/W/E. Credit cards.
Tues-Sun 10.30am-5pm. Closed Xmas-
New Year.
Charges: £2 Adults, £1.50 OAPs/
students, 50p children.
Directions: Follow brown ETB signs
from A34 Winchester-Newbury road.

Winchester

SADDLERY
Calcutt & Sons Ltd
(I K Compton)
Bullington Lane
Sutton Scotney SO21 3RA
☎ *0962 760210*
Saddlers and riding outfitters
providing everything for horse and
rider. Manufacturers of leatherwork
and saddlery employing 13 workers
C/R&R/W/E. Credit cards.
Mon-Thur 9am-6pm, Fri 9am-7pm,
Sat 9am-5pm.

FORGEWORK
Wheely Down Forge
(Charles Normandale)
Warnford SO3 1LG
☎ *0730 829300*
Fine contemporary ironwork for
architectural or domestic settings.
Also traditional forgework and
restoration work undertaken. Work
mostly to commission, is widely
exhibited and has won national
awards. C/R&R/W/E
Mon-Fri 8.30am-5pm, Sat 9am-12noon.

HEREFORD & WORCESTER CRAFT WORKSHOP CENTRES

Burcot Forge
338 Alcester Road
Burcot, Nr Bromsgrove B60 1BH
Craft workshops (see separate entries) include: cabinet making, forgework, stained glass, picture framing, signwriting and calligraphy.

Jinney Ring Craft Centre
(Richard and Jenny Greatwood)
Hanbury, Nr Bromsgrove B60 4BU
☎ *0527 821272*
Craft workshops in old farm buildings include: crystal glass, millinery, cast house signs, artist, sculpture, photography, pottery, jewellery, stained glass, leatherwork, fashion design. Gallery with exhibitions throughout the year. Gift shop, licenced restaurant.

Wed-Sat & Bank Holidays 10.30am-5pm, Sun 2-5.30pm.
Admission charges for special events only.
Directions: on B4091 Hanbury-Droitwich road.

Wobage Farm Workshops
Upton Bishop
Ross-on-Wye HR9 7QP
☎ *0989 85495*
Degree qualified craftspeople with own workshops including furniture/woodworking, wood carving, jewellery and several potters. Work displayed in large showroom; exhibitions held.
Sat-Sun 10am-5pm, other times by appointment with individual craft worker.

HEREFORD & WORCESTER CRAFT WORKSHOPS

Broadway

THATCHING
Master Thatchers (Vale of Evesham)
(Alistair Duncan Holloway)
8 Ley Orchard, Willersey WR12 7PW
☎ *0386 858676*
Thatching contractor, member of National Society of Master Thatchers and Guild of Master Craftsmen.

Bromesberrow Heath

FURNITURE
Terence Diss Furniture
Lintridge Farm
Bromesberrow Heath HR8 1PB
☎ *0531 650598*

Craftsman with over 30 years' experience producing quality domestic furniture made in the traditional way. C/R&R/E.
Mon-Fri 9am-5pm, lunchtime closing. Closed during July.

Bromsgrove

CERAMICS
Daub & Wattle (Ceramics) Ltd
(R P Cook)
Windsor Street, Bromsgrove B61 8HG
☎ *0527 79979*
Distinctive handmade ceramic gifts; clocks, mirrors, lamps, vases and bowls. C. Credit cards.
Mon-Sat 9.30am-5.30pm (Closed Bank Holidays).

CERAMICS
Bridget Drakeford Porcelain
Studio 1, Jinney Ring Centre
Hanbury B60 4BU
☎ 0527 821676
Hand-thrown porcelain decorated
with rich lustre glazes. Work
inspired by Chinese and Egyptian
ceramics. W/E. Credit cards.
Wed-Sat 10.30am-5pm, Sun 2-5.30pm
Directions: see under Craft Centres.

CALLIGRAPHY & SIGNWRITING
A T Daniel FRSA, FCID
Unit 6, Burcot Forge
338 Alcester Road, Burcot B60 1BH
☎ 021 445 1375
Specialist lettering for architects and
commerce, pictorial sign work,
heraldic panels throughout the
Midlands. Lecturer and examiner
for C&G. C/W
Mon-Fri 9.15am-5.30pm, lunchtime
closing.

PICTURE FRAMING
Michael S Plank
Unit 6, Burcot Forge
338 Alcester Road, Burcot B60 1BH
☎ 0527 70140
Bespoke picture framing and
mounting, including embroidery/
tapestry. Small gallery with framed
prints/originals. Prints and posters
to order. Guild member, work for
local authorities. C/R&R/E.
Mon-Fri 9.15am-5.30pm, Sat 9.15am-
1pm.

STAINED GLASS
Paul Phillips
Unit 5, Burcot Forge
338 Alcester Road, Burcot B60 1BH
☎ 0527 579445
Design and manufacture of stained
glass windows. Also restoration

and repair. Glass painting and
firing. Commissioned work; three-
dimensional sculpture at Evesham
Public Library. C/R&R/E.
Mon-Fri 10am-6pm, Sat 10am-4pm.

CABINET MAKING
Andrew Biggs
Unit 1, Burcot Forge
338 Alcester Road, Burcot B60 1BH
☎ 0527 74873
Furniture making and antique
furniture restoration. C/R&R
Mon-Fri (Tues half-day closing) 9am-
6pm, lunchtime closing.

FORGEWORK
Ken White
Unit 2, Burcot Forge
338 Alcester Road, Burcot B60 1BH
☎ 021 445 1211 *(evenings)*
Wrought ironwork, copper work.
Commissioned work for public
buildings, churches etc. Lecturer in
UK and USA and national show
judge. C/R&R
Mon-Fri 8.30am-5.30pm, lunchtime
closing. Sat 8.30am-1pm.

HEDGELAYING
Harry Snutkever
Wythwood Farm, Wythall B47 6LG
☎ 0564 823350
Member of Hedgelayers Federation
with 40 years' experience; prizewinner in national championships.

Clifford

TRADITIONAL CHAIRS
Chris Armstrong Country Furniture
(Chris and Claire Armstrong)
Paddock House, Clifford HR3 5HB
☎ 04973 561
Traditional handmade chairs,
tables, children's cots, stools. Rush

seated ladderback chairs. Furniture polished with natural beeswax or special resistant finish. Happy to discuss special requests and variations. Export.
Mon-Fri 8.30am-6.30pm. Sat-Sun 9am-5pm, advisable to telephone first.
Directions: workshop on B4352 Hay-Bredwardine road quarter-mile east of Castlefield Inn.

Droitwich

FURNITURE
Brian Maiden Fine Furniture
Valley Farm, Elmley Lovett WR9 7HW
☎ *0299 23341*
Small rural workshop employing two craftsmen, design-led, specialising in one-off high quality contemporary work. Free-standing and built-in furniture. Quality hardwoods, solid and veneer work. Keen to experiment. C/E.
Mon-Fri 8.30am-5.30pm, Sat- some Suns 9.30am-5pm.
Directions: Worcester 11 miles, Kidderminster 6 miles; at roundabout on A449 turn down Crown Lane towards railway line, turn right into Gated Road; Valley Farm ahead.

Evesham

WOODWORK
Lenchcraft
(Graham Whitehouse)
Robin Hill
Atch Lench WR11 5SP
☎ *0386 870734*
Workshop producing various turned items including plant stands, small tables, also Victorian conservatories and fitted kitchens. C/R&R.
Mon-Fri 9am-5pm, telephone to confirm.

BASKETWORK
The Basket Maker
(Russell A Rogers)
Main Street, South Littleton WR11 5TJ
☎ *0386 830504*
Traditional English willow and cane baskets of a high quality are made on premises. Showroom . C/R&R.
Mon-Sat 9am-6pm, other times by appointment.

MOSAICS & HERALDRY WORK
Dereford of Badsey Hall
(George Dereford)
Badsey WR11 5EW
☎ *0386 831070*
Murals, reliefs, sculptures in various materials. Coats of arms in various materials. Mosaic/ceramic experts. C/R&R/W/E.
Visitors welcome by appointment only.

Hay-on-Wye

WOODWORK
Design Link
(Duncan Linklater)
Still Point, Cusop HR3 5RQ
☎ *0497 820058*
Woodturning, carving, cabinet making. Decorative architectural joinery; panelling screens, balustrades, timber ceilings etc. Also calligraphy on wood; inscribed dishes, chargers and platters, astrological natal bowls. C.
Flexible hours; please telephone first.

Hereford

CERAMICS
Hill Cottage Pottery
(Wendy Nolan)
Wellington HR4 8BE
☎ *0432 71236*
A wide selection of ceramics from

garden ware, jardiniers etc, to teapots and tableware. House names and commemorative plates to commission. Also portraits in terracotta. C/R&R/W
Open daily 10am-8pm.
Directions: 6 miles north of Hereford on A49.

GLASSWORK
H J and D Hobbs & Co
Karinya, Tump Lane
Much Birch HR2 8HP
☎ *0981 540516*
Manufacturers of leaded light window panels, lights and screens. Glass can be stained, painted, etched or blasted using traditional or modern designs. C/R&R/W/E
Mon-Fri 8am-5pm, lunchtime closing. Sat 8.30-12noon.

PEWTER MINIATURES
Holly Tree Miniatures
(Colin and Jean Oram)
Holly Tree Cottage
Burley Gate HR1 3QS
☎ *0432 820493*
Exquisite hand-painted one-twelfth scale miniatures; tea/coffee sets, decorative plates, tureens etc representing the finest porcelain in miniature. Also thimbles, badges and brooches. Featured in many magazines and ABC television, USA. Collected world-wide. W/E.
Mon-Fri 9.30am-5.30pm, Sat 8.30am-12noon.

CERAMICS
Backwater Pottery
(Mary S Boardman)
Capuchin Yard, Church Street
Hereford
☎ *0432 274269*

Small studio with potter producing functional pots; subtle stoneware, colourful earthenware, pristine porcelains and Raku. C/W
Mon-Fri Easter-Xmas 10am-5pm, Xmas-Easter 11am-4pm. Sat 10am-5pm.
Directions: Capuchin Yard is off Church Street leading to the Cathedral.

ETCHING
Frances St Clair Miller
Pope Place Studio
Preston-on-Wye HR2 9JX
☎ *0981 7240*
Limited edition etchings, particularly pastoral scenes. Frances St Clair Miller studied at the Slade School of Fine art and exhibits world-wide. C/W/E
Mon-Fri 9am-5pm, Sat-Sun by arrangement.
Directions: 6 miles from Hereford on Brecon road A438, turn left signed Preston-on-Wye, right after 1 mile. Turn left in village to Blakemere; studio half-mile on left.

Kington

CABINET MAKING
English Oak Furniture
(Terence William Clegg)
8 Headbrook, Kington HR5 3DZ
☎ *0544 230208*
Experienced craftsman and cabinet making apprentices producing fine quality country oak funiture and replica period walnut veneered furniture using traditional methods Restoration, polishing, carving, wood turning. Tuition. C/R&R/E
Mon-Fri 8am-5pm, Sat 9am-5pm.

CARRIAGE BUILDING
Montfort Carriage Builders
(Michael Saunders)
Hergest HR5 3EL

☎ 0544 230629
Small family business building horse-drawn carriages. Award won for horse-drawn carriage for use by wheelchair disabled. C/R&R/W/E
Mon-Fri 9am-6pm, Sat 9am-5pm.

Ledbury

WOODTURNING
David Nye — Quality Woodturner
2 Kimbrose Cottages, Falcon Lane
Ledbury HR8 2JN
☎ 0531 3444
Clocks and barometers, lamps, bowls and kitchenware, cheese boards, Lazy Susans, stools, tables; made from hardwoods with a hard-wearing polished finish. C/R&R
Mon-Fri 9am-5pm, lunchtime closing.

CERAMICS
Homend Pottery
(Mark and Caroline Owen-Thomas)
205 The Homend, Ledbury HR8 1BS
☎ 0531 4571
Hand-thrown decorated slipware and terracotta pots for the garden, conservatory and house. Guild members. C. Credit cards.
Tues-Sat 10.30am-5pm (Wed early closing).

CERAMICS
Hollybush Pottery
(Jerry Fryman)
The Cottage, Hollybush HR8 1ET
☎ 0531 2316
Individual workshop specialising in a range of ceramics that exploit the effects of wood-fire and subtle glazes. Experienced tutor producing functional stoneware. W/E
Wed/Fri/Sat/Sun 9am-6pm, advisable to telephone first.
Directions: on A438 at top of Malvern Hills.

Leominster

CIDER MAKING
Dunkertons Cider Co
(Susie and Ivor Dunkerton)
Luntley, Nr Pembridge HR6 9ED
☎ 05447 653
Traditional ciders and perry made on the premises from un-sprayed, locally-grown cider apples and perry pears, some rare varieties. Draught cider available. W/E
Mon-Sat 10am-6pm.

FORGEWORK
Sam Thompson
The Forge, Downes' Yard
70 South Street, Leominster
☎ 0568 614140
General blacksmithing, welding etc. Metalwork from lawnmower handles to large gates, hinges, domestic and architectural fittings etc. C/R&R/W/E
Mon-Fri 9.30am-5.30pm. Some Sats.

Malvern

MUSICAL INSTRUMENT MAKING
Lionel K Hepplewhite
(Lionel and Michael Hepplewhite)
Soundpost, 8 North End Lane
Malvern WR14 2ES
☎ 06845 62203
Violins, violas and cellos handmade in the traditional manner to professional standards. Only the finest woods used. C/R&R/E
Visit by appointment only.

MUSICAL INSTRUMENT MAKING
Hibernian Violins
(Padraig ó Dubhlaoidn)
24 Players Avenue
Malvern WR14 1DU

☎ 0684 562947
Making and restoration of stringed
instruments and bows. C/R&R/E
Mon-Fri 9am-5pm, lunchtime closing.
Directions: approaching Malvern from
the south on A449, turn right into
Lower Howsell Road, then 4th turn
left, workshop at end on right.

HANDLOOM WEAVING
Margaretha Bruce-Morgan
Guarlford Court
Guarlford WR13 6NX
☎ 0684 564211
Small hand-weaving studio
specialising in coverlets, runners
and rugs. C/R&R/E
Visitors welcome by appointment only.
Directions: 2 miles out of Malvern on
B4211, opposite Guarlford church.

CERAMICS
Malvern Arts Workshop
(J McGuffie)
90 Worcester Rd, Malvern WR14 1NY
☎ 0684 568993
Very fine Raku ceramics by Peter
Sparrey. Work widely exhibited.
Member of Craftsman Potters
Association. C/E
Wed-Fri 10am-5pm, Sat-Sun 10.30am-5pm. Closed 1st fortnight in January.
Directions: on A449 Worcester Road,
opposite church.

Pershore

WOODWORK
Valtone Woodcraft
(Geoff and Kathy Valentine)
Unit 12, Lyttleton Road
Racecourse Road Estate
Pershore WR10 2DL
☎ 0386 554759
Business employing nine carpenters
and joiners; custom made and

designed conservatories, windows,
doors, porches and patios. Awards
at Hampton Court International
Flower Show. W/E. Credit cards.
Mon-Fri 8am-5pm, lunchtime closing.
Sat 9am-1pm.
Directions: entering Pershore from
Worcester on A44, at traffic lights turn
left into Station Road. Past high school
on right, turn right into Racecourse
Road Estate, take 1st right into
Cobham Road. Workshop on left.

Ross-on-Wye

CABINET MAKING
Morel & Partners
*(K Mooney, P Russell, T Underwood
and H Eldridge)*
Old Gloucester Road
Ross-on-Wye HR9 5JG
☎ 0989 67750
Cabinet makers specialising in
custom-built furniture for kitchens
and bedrooms. Also tables, chairs,
cabinets and fire-surrounds. C
Mon-Fri 8am-6pm, Sat 9am-1pm.

FURNITURE MAKING &
WOODTURNING
**J Arthur Wells — Craftsman in
Wood**
Gatsford HR9 7QL
☎ 0989 62595
Furniture maker and woodturner
producing purpose-made wood-
work of all descriptions. A selection
of turned items in stock. C/R&R
*Open daily 9am-9pm, advisable to
telephone first.*
Directions: 400yd from end of M50 on
A449.

CERAMICS
Jack & Joan Doherty — Porcelain
Hook's Cottage
Lea Bailey HR9 5TY

☎ *0989 750644*
Studio workshop making porcelain functional tableware and Raku sculptural pieces. Work exhibited widely in England and overseas. Showroom. C/W/E
Mon-Sat 9am-5.30pm. Please telephone before visiting.
Directions: from centre of Lea on A40 follow signs for Micheldean. Turn right signposted Drybrook (after new housing); workshop on right after few hundred yards. Park in the road.

STONE MASONRY, FIREPLACES & GILDING
A W Ursell Ltd
(David Ursell)
Waterloo Works, Cantilupe Road
Ross-on-Wye HR9 7AN
☎ *0989 62530*
Hand-cut lettering, stone carving, monument and fireplace making. Work in stone, marble, slate, cast bronze etc. Also painted or panelled finishes, gold leaf work. C/R&R/E.
Mon-Fri 8am-5pm, lunchtime closing. Sat 9am-12noon.

FURNITURE
Dirk Jan Driessen — Fine Furniture
The Rowans, Upper Grove Common
Sellack HR9 6LY
☎ *0989 87404*
Cabinet making, contemporary design and reproduction; one-off commissions and small batches to the highest standards. E
Mon-Fri 8am-6pm and Sat-Sun; visit by appointment.
Directions: from Ross-on-Wye to Hereford on A49, cross M50/A40 at Wilton roundabout, take 2nd turn right signposted Sellack. Follow road for 2.5 miles to stone cross on left of road, turn left; last (white) house on right.

CERAMICS
Wobage Farm Pottery
(Michael and Sheila Casson)
Wobage Farm
Upton Bishop HR9 7QP
☎ *0989 85233*
Functional and decorative pottery, wheel-made, wood and gas fired, mostly salt glazed. Crafts Council and guild members. W/E
Sat-Sun 10am-5pm, other times and weekdays by appointment. Groups (fee charged) by appointment only.
Directions: see under Craft Centres.

WOODCARVER
Lynn Hodgson
The Workshop, Wobage Farm
Upton Bishop HR9 7QP
☎ *0989 85495*
Working mainly to commission and exhibition. C/W/E
Sat-Sun 10am-5pm.
Directions: see under Craft Centres.

JEWELLERY
Clair Hodgson — Jeweller/ Designer
The Workshop, Wobage Farm
Upton Bishop HR9 7QP
☎ *0989 85495*
C/W/E
Sat-Sun 10am-5pm.
Directions: see under Craft Centres.

FURNITURE MAKING
Ben Casson
The Workshop, Wobage Farm
Upton Bishop HR9 7QP
☎ *0989 85495*
Designer and maker of original furniture in native timbers. C/W/E
Sat-Sun 10am-5pm, other times by appointment.
Directions: see under Craft Centres.

CERAMICS
Rachel Kyle
The Workshop, Wobage Farm
Upton Bishop HR9 7QP
☎ *0989 85495*
Thrown domestic ware pottery.
C/W/E
Sat-Sun 10am-5pm, other times by appointment.
Directions: see under Craft Centres.

CERAMICS
Patia Davis
The Workshop, Wobage Farm
Upton Bishop HR9 7QP
☎ *0989 85495*
One-off items in porcelain for domestic use. C/W/E
Sat-Sun 10am-5pm, other times by appointment.
Directions: see under Craft Centres.

Stourbridge

FORGEWORK
Forging Ahead
(Paul Margetts)
Field House Farm, Dark Lane
Belbroughton DY9 9SS
☎ *0562 730003*
Contemporary hot-forged artistic blacksmithing; gates, fire baskets, weather vanes, lighting, sculpture etc. Consistent prize-winner, Royal Show. C/R&R/W/E
Mon-Fri 8am-5.30pm, Sat 8am-1pm.
Directions: from centre of Belbroughton follow signs for

Stourbridge for quarter-mile, cross brook and turn right up Dark Lane. Field House Farm is 300yd on right.

Upton-upon-Severn

WEAVING
Midsummer Weavers
(Sigrid Gonnsen)
London Lane
Upton-upon-Severn WR8 0HH
☎ *0684 593503*
Weaving workshop and showroom; fabrics and accessories in natural fibres. Stock spinning equipment, fleeces and other locally made craft products also for sale.
W/E. Credit cards.
Tues-Sat 9.30am-5.30pm, lunchtime closing.

Worcester

WOODCARVING
Custom Woodcarving
(Stan Greer)
Langland, Church Lane
Whittington WR5 2RQ
☎ *0905 360404*
Carousel-style rocking horses with carved saddles, manes and tails; traditional rocking chairs for children and adults with carved name/initials to commemorate birthday or retirement etc. Inscription plaques, signs in wood, incised/raised lettering. C/R&R/E.
Mon-Fri 9am-5pm.

HERTFORDSHIRE CRAFT WORKSHOPS

Baldock

CERAMICS
Seven Springs Gallery & Pottery
9-11 Mill Street, Ashwell SG7 5LY
☎ 0462 742564
A variety of ceramic work, particularly figure groups, animals and buildings. C
Mon, Tues, Fri 10am-5pm, Sat-Sun 10am-6pm, lunchtime closing. Open daily during December.

CERAMICS
Susan Morris
50 High Street, Ashwell SG7 5NR
☎ 0462 742547
Thrown and hand-built oxidised stoneware pots for domestic use, also individual pieces. C
Telephone for opening hours.

Berkhamsted

FABRIC WEAVING
Pheasant House
(Elissa F Dereford)
Little Heath Lane, Berkhamsted HP4 2RT
☎ 0442 871485
Distinctive fabrics woven in natural fibres including mohair, cashmere, silk, wools and cotton. Available by the metre, or as exclusive made-to-measure garment. Unusual ideas a speciality (eg silk-lined dressing gowns, carriage-driving outfits). C/W/E. Credit cards.
Visitors welcome by appointment.

Hitchin

CERAMICS
Lannock Pottery
(Andrew and Cressida Watts)
Hitchin Road, Weston SG4 7EE
☎ 0462 79356
Thrown and pressed high fired reduction stoneware. Most products associated with gardens and plants. Also small range of oven-to-table ware and table lamps.C/W/E. Credit cards.
Mon-Fri 9am-5.30pm, Sat-Sun by appointment.

CERAMICS
Louise Shotter — The Tea Potter
108 London Road, St Ippollitts SG4 7RD
☎ 0462 455789
Potter making individually designed and decorated tableware, teapots and tiles. Pieces decorated with animals, figures and flowers. Also white earthenware. C/W
Open during normal working hours and at weekends; advisable to telephone first.
Directions: from Hitchin take Codicote route onto main road (London Road B656) for about 1 mile. On right just before crossroads and PO.

ANTIQUE FURNITURE RESTORATION
R & V Restorers
(Robert and Veronica Reid)
Mill Barn, High Street
Whitwell SG4 8AG
☎ 0438 871277
Furniture restoration work and furniture copied by craftsman with 25 years' experience. Also furniture designed and made to customers' specification. Gilding work, mouldings etc, upholstery, rush and cane work. C/R&R
Mon-Fri 9am-7pm, Sat-Sun 9am-1pm by appointment.

Much Hadham

FORGEWORK
Neil Stuart Design Blacksmith
The Forge, High Street
Much Hadham SG10 6BQ
☎ 027984 3856
Independent forge, part of Forge
Museum set up by Hertfordshire
Building Preservation Trust. Blacksmithing (worked by bellows),
mobile welding, cutting and
grinding. C/R&R/E. Credit cards.
*Tues-Sat 8.30am-5.30pm. Museum
charges: 80p adults, 40p OAPs and
children. No charge for workshop and
bellows room. Groups by appointment.*

Royston

WOODTURNING & FURNITURE
The Maltings
(M D and B Gratch)
99 North End, Meldreth SG8 6NY
☎ 0763 261615
Two-person partnership working
in picturesque village. Hand-turned
wooden bowls, platters, fruit bowls
etc. Exclusive bedroom/dining
room and occasional furniture
made to order in English hardwoods. C/R&R/W/E. Credit cards.
*Mon-Fri 8.30am-5.30pm, lunchtime
closing. Sat-Sun by appointment.*

STAINED GLASS
Foxhall Studio
(Gill Rennison)
Kelshall SG8 9SE
☎ 076387 209
Stained glass workshop and gallery
in annex to 17th century cottage.
Windows, doors, fanlights, lampshades and gifts made using
predominantly the copper foil
technique. Also church window

panels and restoration. C/R&R
*Open daily 9am-5pm; advisable to
telephone first.*
Directions: Kelshall is approx 2 miles
east of A505 road to Royston.

St Albans

ANTIQUE FURNITURE
RESTORATION
Collins Antiques
(Sam and Michael Collins)
Corner House
Wheathampstead AL4 8AP
☎ 0582 833111
Cabinet maker restoring 17th, 18th
and 19th century furniture. 18th/
19th century chairs copied to match
sets. C/R&R/W/E. Credit cards.
*Mon-Fri 9am-5.30pm, Sat 9.30am-
5.30pm, lunchtime closing. Advisable
to telephone for appointment.*

Welwyn Garden City

PAINTING/ILLUSTRATING &
WALL DECORATING
Bridge House Designs
(Lorette Roberts)
Bridge House, Lemsford Village AL8 7TN
☎ 0707 323741
Artist and illustrator specialising in
botanical paintings, collage and
abstract work. Pictures, mats, mugs,
cards etc. for individuals and
companies, work sold world-wide.
Tuition courses in watercolour
painting, wall decorating and
marbling. C/W/E. Credit cards.
*Open daily 10am-4pm, please
telephone first.*
Directions: from A1(M) junc 4 to
Welwyn Garden City A6129, under
motorway, then to Wheathampstead
B653. Half mile to church, turn right
into Lemsford. Quarter mile down hill
next to The Sun PH.

HUMBERSIDE CRAFT WORKSHOP CENTRES

Abbeygate Antique and Craft Centre
(Mrs Jeanne Chatburn)
14 Abbeygate, Grimsby DN31 1JY
☎ *0472 361129*
Twenty craftspeople manning their
own sections of the craft centre
above the shopping mews of
Abbeygate. Crafts include lace-
making, spinning and weaving,
quilling, woodturning, painting on
slate/agate etc. Commissions
undertaken. Demonstrations of
crafts by arrangement. Tea rooms.
*Mon-Fri 9.30am-4.30pm, Sat 9.15am-
5pm.*

Park Rose Pottery
*Carnaby Court Lane, Carnaby
Bridlington YO15 3QF*

☎ *0262 602823*
Pottery walkabout; watch moulding, glazing and firing of ceramics
in large pottery. Factory shop
selling products. Café. Leisure park.
Open daily 10am-5pm.

The Craft People
(E A Ferguson)
Front Street, Ulceby DN39 6SY
☎ *0469 588774*
Craft centre and gallery in converted farm cottage. Crafts include
leatherwork, sign writing, wooden
toys, polystone and glass. Craft
exhibitions organised. Tea shop and
gardens, parties catered for,
barbecue facilities.
Open daily 10am-8pm.

HUMBERSIDE CRAFT WORKSHOPS

Barton-Upon-Humber

PICTURE FRAMING
Keith Elliott — Bespoke Framer
*The Old Meeting Hall, Maltby Lane
Barton-upon-Humber DN18 5PY*
☎ *0652 660380*
Picture framing and mounting of all
types of artwork; mount cutting and
decorating; swept, oval and circular
frames. Also artists' and framing
supplies and tools. Picture framing
tuition and illustrated lectures by
guild member. C/R&R/W
*Mon-Fri 9am-2pm, Sat-Sun by appoint-
ment. Closed Xmas-New Year.*
Directions: from Brigg-Hull A15, take
A1077 to Barton-on-Humber, over
roundabout, over crossroads. Workshop at end of Maltby Lane on left.

Beverley

TOY MAKING/JEWELLERY
Tocki
(Mr and Mrs Harold W Tock)
Units 5&6
*Riverview Small Business Centre
Riverview Road, Beverley HU17 0LD*
☎ *0482 865630*
Cascading glitter tubes for all ages
produced in large, small and
keyring sizes. Also animal and
flower jewellery and Zodiac range.
Members of National Toy Library
for people with special needs. W/E
Mon-Fri 9am-5pm.

Bridlington

POTTERY AND LEISURE PARK
Park Rose Ltd
(D Hindle and N Rawson)
Carnaby Covert Lane
Carnaby YO15 3QF
☎ *0262 602823*
Manufacturers of ceramic giftware
and lighting; suppliers to high
street retailers. Families welcome,
see under Craft Centres.
W/E. Credit cards.
Open daily 10am-5pm, (closed Xmas
week).

KNITWEAR
Flamborough Marine Ltd
(Mrs Lesley Berry)
The Manor House
Flamborough YO15 1PD
☎ *0262 850943*
Hand-knitted traditional
fishermens' sweaters, specialising in
'ganseys' knitted in one piece on
five steel needles in weatherproof
wool. Choice of local north east
fishing village patterns is available.
'Gansey' knitting kits, other
traditional sweaters and machine
knitted sweaters. All knitwear sold
at The Manor House and by mail
order. C/R&R/E. Credit cards.
Open daily 9.30am-6pm.
Directions: On B1255 from Bridlington
just past St Oswald's Church, on the
corner of Tower Street and Lighthouse
Road. House set back from road.

Brigg

DRIED FLOWERS AND CRAFTWORK
Floribundaful
(Mrs Ann Cadwallader & Mrs C Ladlow)
Grange Farm
Saxby-all-Saints DN20 0P2

☎ *0652 61523*
Small family business dealing
mainly in floral art. Pot pourri,
knitwear, jewellery and ceramics
also available. C/W
Open daily except Wed, 10am-4pm.

FURNITURE
Tony Preston Furniture
rear of 15 Bridge Street
Brigg DN20 8LP
☎ *0652 658515*
Furniture makers and restorers
working to commission; three
craftsmen employed. C/R&R/E
Mon-Fri 9am-5pm, Sat-Sun 9.30am-1pm.

GOLD & SILVERSMITHING
A R Buckingham
Elsham Hall Country Park
Brigg DN20 0RA
☎ *0652 688184*
Handmade silverware and jewellery in silver, gold and plantinum.
One-off designs preferred. Many
important commissions. Award
winning craftsman, Freeman of
Goldsmiths Co and City of London.
C/R&R/W/E
Varied opening hours, usually
Sundays. Advisable to make appoint-
ment, gaining free entry to visit
workshop at Elsham Country Park
(open daily Easter-Mid Sept 11am-
5pm, Sun 11am-4pm only during
winter months — charge £3 adults, £2
children).
Directions: Elsham Hall Country Park
is on B1084 near junction with B1206
Brigg-Humber Bridge road.

SCULPTURE
Ernie Kay Sculptor
Gallery Erotica Bonby
Bonby DN20 0PY
☎ *0652 61709*

Sculpture, figurative and portraiture. Poetic expression sculpted in wood, moulded and cast in alabaster, slate, onyx etc. Drawing and painting. Demonstrations to schools etc. Work exhibited, including Royal Miniature Society. C/W/E
Tues-Fri 10am-8pm, Sat-Sun 12noon-6pm.
Directions: Bonby lies 6 miles south of Humber Bridge on B1204 between South Ferriby and Elsham Hall.

FURNITURE
John Richard Collinson
173 Scawby Road
Scawby Brook
Brigg DN20 9JX
☎ *0652 653436*
Traditional domestic and ecclesiastical furniture handmade in English oak with original adzed finish. Also various small items (ashtrays, cheese boards etc). C/E
Mon-Fri 9am-6pm, Sat-Sun 10am-5pm.
Directions: 1 mile west of Brigg.

Gilberdyke

CERAMICS
Jerry Harper Pottery
South Farm Craft Gallery,
Staddlethorpe Lane
Blacktoft DN14 7XT
☎ *0430 441082 (restaurant 441889)*
Award-winning gallery studio and licensed restaurant in renovated farm buildings. Hand made standard items of pottery and special pots made to order. Gallery stocks crafts, gifts etc. Special demonstrations arranged. C/W/E. Credit cards.
Wed-Sat 10am-5pm, Sun & Bank Holidays 11am-5pm.

Goxhill

FORGEWORK
D G Wrought Iron
(D P and H Gutherless)
Alpha Works, College Road
Goxhill DN19 7HY
☎ *0469 30069/31296*
Specialising in individually designed and manufactured pieces of ornamental ironwork for architects, builders and the general public. Range of work includes railings, security grilles, brackets and garden furniture. C/R&R/W/E
Mon-Fri 9am-5pm, Sat-Sun 10am-4pm.
Directions: from Humber bridge follow signs for New Holland. At crossroads (B1206) on outskirts of Barrow go straight over; forge on left 100yd before next crossroads.

Great Driffield

FORGEWORK
David Athey
The Smithy, West End Works
Garton-on-the-Wolds YO25 0EU
☎ *0377 241723*
Experienced Master Blacksmith using traditional methods of metal casting and fashioning hot iron on the anvil. Decorative and architectural work, sculpture and restoration work. Private and business clients throughout Britain. C/R&R/W/E.
Mon-Fri 8am-5.30pm, Sat 8am-12noon.

FURNITURE MAKING & TIMBER RESTORATION
English Timbers
(Patrick Smith)
1A Main Street, Kirkburn YO25 9DU
☎ *0377 89301*

Specialist timber merchants employing five craftsmen. Native hardwoods dried for use in furniture and flooring. Makers of handmade furniture, panelling, specialist timber floors, specialist mouldings, restoration oak timbers. C/W/E
Mon-Fri 8.30am-5.30pm, lunchtime closing 12.30-1.30pm.

WOODCARVING & CABINETMAKING
Patrick Tite
Galloway Lane
Driffield YO25 7LW
☎ *0377 43689*
Individually designed hardwood furniture and woodcarving for private clients and business etc. Priority given to design and traditional craftsmanship by cabinetmaker with 24 years' experience. C/E
Mon-Fri 9am-5pm, Sat-Sun by appointment.

CLOCKMAKING
Kilham Clocks
(Mr and Mrs John Butterfield)
The Old Ropery
East Street
Kilham YO25 0SG
☎ *0262 82233*
Clock making and restoration. Cases for clocks made to order, antique clocks repaired/restored. Movements and dials made. C/R&R/W/E
Mon-Sat 10am-5pm.

SADDLERY
T Pickering
91 Main Street
Beeford YO25 8AY
☎ *0262 488408*

Three saddlers making leather goods, saddlery and harnesses. C/R&R/E.
Mon-Fri 9am-5pm (lunchtime closing 12noon-1pm). Sat-Sun by appointment.

ANTIQUE FURNITURE RESTORATION
Smith & Smith Designs
(D A Smith, C R Smith and M T Addinall)
58a Middle Street North
Driffield YO25 7SU
☎ *0377 46321*
Antique restoration and furniture making, specialising in antique pine, oak, ash, elm etc. Bespoke items commissioned. Six craftsmen and designer. Restored items available.
C/R&R/W/E. Credit cards.
Mon-Fri 9.30am-5.30pm, Sat 9.45am-5pm, Sun by appointment. (Closed Xmas-New Year).

GLASS BLOWING
Bill Tuffnell
Unit 5, Front Street Court
Middleton-on-the-Wolds YO25 9UA
☎ *0377 217651*
Three craftspeople involved with glass blowing and manufacturing of glass novelties. Members of Guild of Master Craftsmen. C/R&R/W/E.
Mon-Fri 8am-5pm, Sat-Sun by arrangement.
Directions: Front Street Court is on A163 Goole-Driffield road in Middleton-on-the-Wolds.

FURNITURE
Oak Rabbit Crafts
(P G Heap)
Wetwang YO25 9XJ
☎ *0377 86257*
Small family business making

116

dining and bedroom suites, coffee tables, bookcases etc and small articles, in English oak. C/W/E
Mon-Fri 8am-5pm, Sat-Sun 10.30am-5pm.

PICTURE FRAMING
Strides Gallery
(Karen Cross and Alison Botten)
Strides Gallery, 12 Exchange Street
Driffield YO25 7LJ
☎ *0377 241512*
Comprehensive picture framing service. Also gallery exhibiting contemporary craftwork including glass, ceramics, wood, jewellery, limited edition prints and etchings and originals. C. Credit cards.
Mon-Sat (closed Wed) 9.30am-5pm. (Closed Xmas-New Year).
Directions: from the Market Place in Driffield, turn into Exchange Street; Gallery on right.

Hull

FORGEWORK
Preston Forge (Parkers of Preston)
(John Parker)
13 School Road, Preston HU12 8US
☎ *0482 899908*
Experienced forgeworker hand-making all types of wrought iron work and general blacksmithing. Gates, fencing, security grilles and railings, fabrication work, dog grates, net and lobster-pot anchors, pipe clips, brackets etc. C/R&R
Open daily 7.30am-5.30pm, lunchtime closing.

CERAMICS
Kirkholme Crafts
(Michael Smith and Michael Reader)
Westfield Farm Industrial Estate
Sigglesthorne HU11 5QL

☎ *0964 543686*
Porcelain and china giftware; British artists' designs on collector plates, mugs, vases, glassware and other porcelain and china products. Factory shop. C/W/E. Credit cards.
Mon-Fri 9am-5pm, Sun 9.30am-4.30pm, (lunchtime closing 12noon-1.30pm).

Humberstone

PAINTED TILES
Wilton Studios
(Ashley Riggall and David Ward)
Unit 42, Enterprise House, Wilton Road Industrial Estate
Humberstone DN36 4AS
☎ *0472 210820*
Hand-painted tiles and sanitary ware. Work, ranging from individual tiles to large murals and panels, can be seen at studios and showroom; design and sample service available free. C/R&R/W/E
Mon-Fri 9am-5.30pm, Sat-Sun 10am-4pm. (Closed Xmas-New Year).

Market Weighton

ROCKING HORSES
The Rocking Horse Shop
(Anthony and Pat Dew)
Old Road
Holme-upon-Spalding Moor YO4 4AB
☎ *0430 860563*
Makers of rocking horses of all types; workshop employing five craftsmen. Also a wide range of plans for making rocking horses and other projects for amateur woodworkers; rocking horse renovation kits, parts and tools. C/R&R/W/E. Credit cards.
Mon-Fri 9am-5pm, Sat 9am-4pm.

FURNITURE
John M Sugden
Arras Farm
Market Weighton YO4 3RN
☎ *0430 871068*
Craftsman making high quality, solid hardwood furniture mainly to customers' own specification. Some veneered work and marquetry. Both homegrown and imported hardwoods used including maple and black American walnut. C/R&R/E
Mon-Fri 8.30am-5.30pm, Sat-Sun by appointment.
Directions: Market Weighton is on A1079 York-Beverley road.

Scunthorpe

WOODTURNING
Caranda Crafts
(D J Cox)
Cross Lane, Alkborough DH15 9JL
☎ *0724 720614*
Range of wood-turned items including lamps, stair-spindles, clocks, barometers, boxes. Other locally made crafts on sale. C/W. Credit cards.
Mon-Fri 10am-5pm, Sun 2-5pm. Visit by appointment.

FURNITURE MAKING & RESTORATION
Cobweb Furniture
(John Hardy)
Old Chapel Workshops, West Street
West Butterwick DN17 3JZ
☎ *0724 783888*

Woodworking business employing three craftsmen. Work includes French polishing, veneering and restoration work, carving, woodturning and furniture making. C/R&R/W
Mon-Fri 8.30am-5pm, Sat 8.30am-12noon.

FURNITURE
Oakleaf Cabinetmakers
(Kevin Robinson and Stephen Moore)
20 Fieldside
Crowle DN17 4HL
☎ *0724 710937*
Specialists in handmade reproduction furniture from oak and mahogony. Polishing and restoration work. Furniture in solid wood made to commission. Finalists in TV awards. C/R&R/E.
Mon-Fri 9am-5pm, Sat 9am-4.30pm. Closed Xmas-New Year.

Welton

SCULPTURE & JEWELLERY
Jacqueline Stieger
Welton Garth HU15 1NB
☎ *0482 668323*
Jacqueline Stieger has an international reputation as a sculptor and jeweller and has exhibited widely. She works in various materials, particularly bronze, and casts all pieces by the lost-wax method in her own foundry. C. Credit cards.
Visitors welcome by appointment.

ISLE OF WIGHT CRAFT WORKSHOP CENTRES

Brickfields Horsecountry
(Philip Legge)
Newnham Road
Binstead PO33 3TH
☎ *0983 615116/66801*
Shire horse and miniature horse
centre; heritage museum and
carriage collection, working
blacksmith's forge and farriery,
saddlery shop, cider making.
Seasonal farming activities using
modern and vintage tractors.
Twice-daily equine parades,
showjumping Wed evenings.
Miniature rare breeds, children's
farm and play area. Bar and
restaurant.
Open daily 10am-5pm. Entrance
charge adults £3, children £2.
Directions: signed from Binstead
village at top of Quarr Hill. Turn
down road at Brickfields sign (by the
Isle of Wight poultry farm).

Golden Hill Fort
Freshwater PO40 9TF
Several craftspeople working in
studios including: lapidary, fine art,
ceramics, upholstery, fabric
painting and needlecraft.
Directions: from Yarmouth take
A3054 to Freshwater.

Haseley Manor Craft Centre
(R J Young)
Arreton, Newport PO30 3AN
☎ *0983 865420*
Daily demonstrations in the island's
largest pottery studio (children can
make a pottery mouse). Weekday
demonstrations in the sweet-making
factory. Numerous craftspeople
making crafts and gifts. Groups and
children's parties welcome; pets
corner, tea rooms, picnic area.
Easter-end Oct open daily 10am-6pm
Entrance charge.

ISLE OF WIGHT CRAFT WORKSHOPS

Alum Bay

GLASS BLOWING
Alum Bay Glass Ltd
(Michael Rayner)
The Needles Pleasure Park
Alum Bay PO39 0JD
☎ *0983 753473*
Large studio workshop with three
glass-blowers making crystal
glassware incorporating coloured
decoration. Factory shop. Glass-
blowing demonstrations throughout the day between 10am-4pm
(please make appointment for
groups visiting). W/E. Credit cards.

Open daily 9.30am-5pm. Charge for
glass-blowing demonstrations: 40p
adults, 20p children, 30p OAP.

Arreton

BRASS RUBBING
Island Brass Rubbing Centre
(Mrs Isabel Flux)
The Coach House, St George's Church
Arreton PO36 8BA
☎ *0983 402066*
Brass rubbing of facsimiles of the
best known brasses in the country
using traditional materials. Visitors
can either make their own or

purchase prepared rubbings. *Mon-Sat 10am-6pm (closed Oct-Easter).*

Freshwater

LAPIDARY & FOSSIL WORK
Wight Geogems
(Penny Newbery)
Unit 114, Golden Hill Fort Craft Centre
Freshwater PO40 9TF
☎ *0983 754414/294802 (workshop).*
Workshop employing small number of craftspeople cutting and polishing semi-precious stones, fossils etc, especially local material. Established in 1983 by one of Britain's few lapidaries.
C/R&R/W/E. Credit cards.
Easter-Oct open daily 11am-5pm (lunchtime closing), Jan-Feb Sun only.
Entry charge to Golden Hill Fort £1.20 adults, 60p children/OAPs
Directions: see under Craft Centres.

ARTIST
Tram Stop
(Stanley Hider)
Unit 120, Golden Hill Fort Craft Centre
Freshwater PO40 9TF
☎ *0983 62016*
Artist painting on site, creating scenes of bygone transport and street scenes. Also prints and T-shirts. C/W/E
Sat-Sun 10am-5pm during summer months.
Entry charge to Golden Hill Fort £1.20 adults, 60p children/OAPs
Directions: see under Craft Centres.

CERAMICS
Island Ceramics
(G Gordon)
Golden Hill Fort Craft Centre
Freshwater PO40 9TF
☎ *0983 291271/298557*
Hand-thrown and decorated ceramics made on the premises. Also slip case ceramics and ceramic jewellery. C/W/E
Open daily during summer months 11am-5.30pm, Sat-Sun 1.30-5.30pm.
During winter months open Sun pm only.
Entry charge to Golden Hill Fort £1.20 adults, 60p children/OAPs
Directions: see under Craft Centres.

UPHOLSTERY
David Lutas Upholstery
(David and Debbie Lutas)
Golden Hill Fort Craft Centre
Freshwater PO40 9TF
☎ *0983 756100*
All upholstery including three-piece-suites, boat cushions, caravan/bar/car seats — wide range of materials used including leather, carpeting etc. C/R&R
Mon-Fri 8am-5pm, Sat 8am-12noon.
Entry charge to Golden Hill Fort £1.20 adults, 60p children/OAPs
Directions: see under Craft Centres.

FABRIC PAINTING & NEEDLECRAFT
Hobbytex Colourcraft
(J G Cheyne)
Golden Hill Fort Craft Centre
Freshwater PO40 9TF
☎ *0983 756085*
All embroidery; lace making. Fabric painting demonstrated. Tapestry goods stocked. C/R&R
Mon-Fri 10am-5pm, Sat-Sun 11am-4.30pm.
Entry charge to Golden Hill Fort £1.20 adults, 60p children/OAPs
Directions: see under Craft Centres.

Newchurch

POT POURRI MAKING
Isle of Wight Pot Pourri
(Osel Enterprises Ltd)
Sunnycrest Nursery, Wackland Lane
Newchurch PO36 0NB
☎ *0983 862682*
Pot Pourri mainly from own-grown
flowers, made up freshly to order in
a range of eight varieties, with
refresher oils. Sold in gift packs and
drums for wholesale. W/E
*Open Mon-Fri 10am-3pm, (closed
Xmas fortnight).*

Ryde

JEWELLERY
G & M Jewellery
(Guy and Menna Morey)
123 High Street, Ryde PO33 2SU
☎ *0983 611232*

Jewellery designed and made to
order; many gem stones in stock or
customers' own materials used. All
kinds of repair work including re-
stringing, replacing stones and
restoring antique pieces. Old
jewellery remodelled.
C/R&R/E. Credit cards.
Mon-Sat 9.30am-5pm.

Yarmouth

CERAMICS
Chessell Pottery (IOW) Ltd
(John and Sheila Francis)
Chessell, Shalcombe PO41 0UE
☎ *0983 78248*
Decorative porcelain; figurines,
fountains, water gardens, vases etc.
W/E. Credit cards.
*Mon-Fri 9am-5.30pm, Sat-Sun 10am-
5pm (closed Sun Nov-Mar)*
Entry charge: 20p adults, 5p children.

KENT CRAFT WORKSHOP CENTRES

Museum of Kent Life
(Kent County Council)
Lock Lane, Sandling
Maidstone ME14 3AU
☎ 0622 763936
Open air museum on 27-acre site,
telling the story of the Kent
countryside. Craft demonstrations
including chair-making, pottery,
weaving and spinning and blacksmithing featured on most Sundays
and on special event days. Also a
working hop garden, herb garden,
orchard, market garden, livestock
and bees. Shop selling gifts and own
produce. Groups and childrens'
parties welcome; pets corner,
licensed tea rooms, picnic area.
Summer months: open daily 11am-6pm,
closed November-March.
Entry charge: £2 adults, £1 children &
OAPs (except on special event days)
Directions: from M20 Junc 6 to
Maidstone, follow ETB signs from
roundabout. From Maidstone follow
A229 Chatham Road; signposted from
Running Horse roundabout.

The Historic Dockyard
Chatham ME4 4TE
☎ 0634 817721
Museum telling the story of
Britain's greatest fighting ships and
the lives of the dockyard craftsmen.

Rope making and testing, rope and
wire splicing, knots and canvas
work, flag making. Craft Workshops include woodcraft (tuition
given), oil painting, stained glass
(tuition given), ceramics, chandelier
restoration and gilding.
Apr-Oct Wed-Sun & Bank Holidays
10am-6pm. Nov-March Wed, Sat &
Sun 10am-4.30pm. Last admission 1 hr
before closing.
Entry charges to Museums etc. £5.20
adults, £4.50 OAPs, £2.60 children (5-
16 yrs). Family ticket £12. Free entry if
visiting workshops only.
Directions: from M20 Junc 6 follow
signs for Chatham A229.

The Master Makers
(Mary and Nigel Chapman)
Howfield Lane
Chartham, Canterbury CT4 7HQ
☎ 0227 730183
Showrooms and workshops in a
contemporary building where
craftsmen can be seen at work.
Crafts include: ceramics, jewellery,
especially using refractory metals,
children's clothing, stoneware;
pottery to pictures, beads to bellows.
Mon-Sat 9.30am-6pm
Directions: 2 miles from Canterbury,
from A28 Maidstone road turn right
for Chartham Hatch, workshops on left.

KENT CRAFT WORKSHOPS

Ashford

CERAMICS
Robus Pottery & Tiles
(Rosemary and Clive Robus)
Evington Park, Hastingleigh TN25 5JH

☎ 0233 750330
Well established traditional
earthenware pottery employing six
people, producing large garden
pots, floor and wall tiles, architectural ceramics etc. C/R&R/W/E

Mon-Sat 9am-5pm
Directions: from Hythe/Canterbury
road B2068 (Stone Street) turn west
signed Hastingleigh/Wye and follow
road to pottery at crossroads.

CERAMICS
Sellindge Pottery & Crafts
   (Intime Designs)
(T C and P R Huckstepp)
The Pottery, Barrow Hill
Sellindge TN25 6JP
☎ *0303 812204*
Stoneware pottery studio with three
workers and shop, also selling craft
work. Pottery includes clay clocks,
tableware, oil burners, pot pourri
pots etc. Also named store jars,
mugs, jugs, vases, plant pots and
salt pigs. C/W/E. Credit cards.
Mon-Sat 10am-5pm
Directions: on A20 from Ashford to
Folkestone, under M20 bridge, then
pub on right, then under railway
bridge pottery on right before layby.

FORGEWORK
Mr and Mrs R A R Moseley
The Forge, Appledore TN26 2BX
☎ *0233 83358*
Workshop employing four produc-
ing hand-forged quality decorative
ironwork. Many awards won.
Fellow of Worshipful Company of
Blacksmiths. C/R&R
Mon-Thurs 8am-5pm, Fri 8am-4.30pm,
lunchtime closing. Sat 8am-12.30pm.

ROCKING HORSE MAKING
Stevenson Brothers
(M A & A P Stevenson)
The Workshop, Ashford Road
Bethersden TN26 3AP
☎ *0233 820363*
Hand-carved rocking horses with
bow rockers or safety stands. Also

carousel horses, horse sculptures
and fairground animals. Antique
horses restored.
C/R&R/W/E. Credit cards.
Mon-Fri 9am-6pm, Sat 10am-1pm.

Canterbury

STRAWCRAFT
Littlebourne Crafts
(Mrs Jackie Payne)
6-8 High Street, Littlebourne CT3 1UN
☎ *0227 721716*
Traditional corn dollies of all kinds
handmade on the premises. Also
canal art. C/W/E
Visitors welcome by appointment.

FORGEWORK
Nailbourne Forge
(Kenneth Pinnock and Martin
Reeves)
Court Hill
Littlebourne CT3 1TX
☎ *0227 728336*
Established forge employing five,
designing and making ironwork.
General blacksmithing, architec-
tural ironwork, pattern-making for
casting etc. C/R&R/E.
Mon-Fri 8am-6pm, Sat-Sun by
appointment.
Directions: from Canterbury to
Sandwich on A257 turn left into
Jubilee Road, at T junc turn left (Court
Hill). Forge will be found on right.

CERAMICS
Hode Pottery
(Nigel and Mary Chapman)
The Master Makers, Howfield Lane
Chartham CT4 7HQ
☎ *0227 730183*
Pottery employing six; specialists in
large handmade stoneware garden
pots supplied wholesale and retail.

C/R&R/W/E. Credit cards.
Mon-Sat 9.30am-6pm
Directions: see under Craft Centres.

JEWELLERY & METALWORK
Pauline Gainsbury
Unit 5&6, The Master Makers,
Howfield Lane, Chartham CT4 7HQ
☎ 0227 731376
Designer jeweller using precious
metals, specialising in refractory
metals, mainly niobium. Work,
mainly with floral theme, includes:
jewellery, boxes, spoons, napkin
bands. Fellow of Society of Designer
Craftsmen. C/R&R/W/E. Credit cards.
Tues-Fri 10am-5pm, some Sats 10am-
3pm, lunchtime closing.
Directions: see under Craft Centres.

TEDDY BEAR MANUFACTURE
Canterbury Bears Ltd
(Maude and John Blackburn)
The Old Coach House, Court Hill
Littlebourne CT3 1XU
☎ 0227 728238/720802
Established workshops making a
very extensive range of teddy bears
between 10cm-90cm in height, fully
jointed in the traditional way with
wooden joints. Antique replicas
using traditional materials for
collectors, new designs and one-
offs. C/R&R/W/E. Credit cards.
Mon-Fri 9.30am-4.30pm, lunchtime
closing. Visitors welcome by appoint-
ment only.

FORGEWORK
Len Hutton
(Len and Angela Hutton)
The Forge, Bishopsbourne CT4 5HT
☎ 0227 830784
Wrought ironwork ranging from
small ornaments to gates and
balustrades. Emphasis on artistic

design, incorporating animals and
birds, autumn leaves, acorns etc.
Commissioned work includes an
ornate staircase for The Ritz Casino.
Guild awards. C/R&R/W/E.
Mon-Fri 8am-5.30pm, Sat 9am-12noon.
Directions: from A2 Canterbury/
Dover road turn south on B2065
Elham road. After passing under A2
take second turn left; car park on left
before The Mermaid PH on right.

BOOKBINDING
Canterbury Bookbinders
(Christopher Paveley)
60 Northgate, Canterbury CT1 1BB
☎ 0227 452371
Fine leather and cloth bookbinding,
restoration and paper conservation,
gold finishing etc. Small Dickensian
workshop employing three; shop
frontage showing work in progress.
C/R&R/E.
Mon-Sat 9.30am-5pm, lunchtime closing.

Chatham

WOODTURNING
Hartlip Studios
(Colin Abram)
House Carpenters Shop
Ordnance Mews Craft Workshops
Chatham Historic Dockyard
Chatham ME4 4TE
☎ 0634 817721
Specialist woodwork based on
modern classical form with 30
years' experience in three-dimen-
sional design and craftwork:
turning, routing, carving, sculpture,
casting and metalwork. Lecturer,
tuition, talks/demonstrations.
C/R&R/W/E. Credit cards.
Open daily 9am-5pm.
Directions: see under Craft Centres.

ARTIST
Valerie Hayman — Original Oil Paintings
Studio 8
Ordnance Mews Craft Workshops
Chatham Historic Dockyard
Chatham ME4 4TE
☎ *0634 401512 (evenings)*
Oil painting and pen and ink; commissions on any subject. Paintings in private collections, including Royal Family's. C/W/E
Wed-Sun 10am-4.30pm
Directions: see under Craft Centres.

STAINED GLASS
Rochester Stained Glass
(Elayne O'Neill Dracocardos)
Ordnance Mews Craft Workshops,
Chatham Historic Dockyard
Chatham ME4 4TE
☎ *0634 406017*
Stained glass work including designing, making, restoring and relocating windows. Member of guilds and societies. Tuitio. C/R&R/E.
Mon-Fri 10am-4pm, Sat-Sun by appointment.
Directions: see under Craft Centres.

CERAMICS
S S Fitzgerald — Ceramic Artist
Forge II
Ordnance Mews Craft Workshops
Chatham Historic Dockyard
Chatham ME4 4TE
☎ *0634 818530*
All types of ceramic items from functional pieces through to sculptural forms, each item individually hand-crafted. Work widely exhibited. Also sold from workshop. C/W/E. Credit cards.
Tues 1.30-6pm, Wed-Fri 10am-6pm. Open most Sat-Sun 10.30am-5pm.
Directions: see under Craft Centres.

CHANDELIER RESTORATION/ GILDING
Antiquities International
Workshop 5
Ordnance Mews Craft Workshops
Chatham Historic Dockyard
Chatham ME4 4TE
☎ *0634 818866*
Restoration of antique chandeliers and gilding work. C/R&R/W/E.
Open most days, variable times. Advisable to telephone first.
Directions: see under Craft Centres.

ROCKING HORSES & JOINERY
L A Beckley & Sons (Medway) Ltd
(Mr and Mrs L A Beckley)
Unit A, Jenkins Dale, Chatham ME4 5RD
☎ *0634 408099*
Joinery specialists making rocking horses to their unique design, allowing a rocking action in a horizontal plane, as opposed to the normal 'bobbing' motion. Ideal for 2-5-year-old children. Also all types of joinery. C/R&R/W
Mon-Fri 9am-4pm, lunchtime closing.

Cranbrook

THATCHING
Wealden Thatching Services
(Barry Fisher)
9 Brookside, Cranbrook TN17 3BU
☎ *0580 712747*
Thatching contractors employing six; members of National Society of Master Thatchers and Vice-Chairman of regional thatching association.

Dover

CERAMICS
Michael Bayley
Beechcroft Cottage, Green Lane
Temple Ewell CT16 3AS

☎ 0304 822624
Mainly decorative stoneware including wall plaques and bowls, all hand-built. Work exhibited widely in galleries in UK and abroad. Member of Craftsman Potters Association. C/E
Open during normal working hours.
Directions: fromA2, take A256 at Whitfield roundabout to Temple Ewell. At mini-roundabout turn right onto B2060. After Jet Garage 200yd turn right up Wellington Road. Workshop at 5th house on right.

CERAMICS
Nonington Pottery
(David Peacock)
Farthingales, Old Court Hill
Nonington CT15 4LQ
☎ 0304 840174
Small studio pottery producing unique terracotta, stoneware and porcelain for house or garden, domestic or business. Garden pots, cascades, unusual finials for rooftop ridges. C/W/E. Credit cards.
Open daily, advisable to telephone first. Closed 1-2 weeks June/Sept.
Directions: from A2, take B2046 to Aylesham, left at T junc over railway, take 2nd turn right to Nonington. Pottery on right opposite church.

Maidstone

CERAMICS
Peter & Julie Phillips Pottery
Ivy Cottage, Taylors Lane
Trottiscliffe ME19 5DS
☎ 0732 822901
Pottery making stoneware and porcelain. Domestic ware and decorative/sculptural hand-built ware. Work exhibited in UK and Europe. C/W/E
Variable times and days worked;

please telephone first.
Directions: near junction of A20/A25.

SADDLERY
Boots & Saddles
(Mr and Mrs M G Evans)
90 High Street
West Malling ME19 6NE
☎ 0732 870474
Saddlery making and repairing all quality saddles, bridlework and leather goods.
C/R&R/E. Credit cards.
Mon-Fri 9.15am-5.30pm, Sat 9.15am-5pm.

Sandwich

CERAMICS
Summerfield Pottery and Potters' Supplies
(D and A Barnes)
Summerfield Farm, Summerfield Woodnesborough CT13 0EW
☎ 0304 611937
Three employed making domestic stoneware, house name plates, garden pots and bonzai pots, one-off pieces. Pottery materials also supplied and agents for Reward Clay Glaze kilns. C/R&R/W/E.
Mon-Sat 9am-6pm, Sun 10am-4pm. Advisable to telephone for Sat/Sun visiting.
Directions: pottery is situated near Staple, 8 miles east of Canterbury, near Staple Vineyard.

Sidcup

BEAD JEWELLERY/RAG RUGS/KNITWEAR
Twinset
(Debbie Siniska and Vicky Salter)
159 Station Road, Sidcup DA15 7AA
☎ 081 308 0504

Unusual Rocaille beadwork jewellery. Necklace re-stringing and repair service. Traditionally prodded rag rugs and wall hangings. Hand-framed knitwear; Fair Isle, cables, etc. Also yarns available. Tuition: Sunday workshops in beadwork, £15 (inc materials) 10.30am-3.30pm; telephone for details. C/R&R
Mon-Fri (Thurs half-day closing) 9.30am-5.30pm, Sat 9.30-5pm.
Directions: near station, (next to Lamorbey Baths) in Sidcup.

Sittingbourne

ART RESTORATION & PRINTING
Periwinkle Press
(E R, A, A L and J R Swain)
23 East Street, Sittingbourne
☎ *0795 426242*
The restoration of oil paintings, picture-framing. Workshop and gallery. Books old and new. Engravings for sale.
C/R&R/W. Credit cards.
Mon-Sat 9am-5pm.

CERAMICS
Syndale Valley Pottery (Ltd)
(Alison and Bob St Clair Baker)
The Old Dairy Workshop, Forge Farm, Newnham Lane
Nr Newnham ME13 0AT
☎ *0795 890211*
Small country pottery, making hand-thrown craft pottery with a wide range of shapes, designs and Coloured glazes. Associate Members of Craftsman Potters Association. C/W/E
Mon & Fri 9.30am-5.30pm. Other days by appointment.

Tunbridge Wells

CERAMICS
Hook Green Pottery
(Don and Ruth Morgan)
Hook Green TN3 8LR
☎ *0892 890504*
Porcelain and stoneware; vases, bowls, plates, bottles plus a range of oven and tableware. C/W/E
Open normal business hours and Sat.

LANCASHIRE CRAFT WORKSHOP CENTRES

Clifton House Craft Group
(City of Salford Arts & Leisure Dept)
Clifton House Farm, Clifton, Swinton
Manchester M5 4NZ
☎ 061 737 1040
Developing working farm with rare breeds, shire horses etc. Developing rural crafts willow production. Regular meetings at the farm to learn about willow basketry, spinning, weaving, potting, besom making etc; demonstrations, workshops, outings and craft fairs. *At present only Sun 12noon-5pm (to be extended).*

Eccles Farm Needlecraft Centre
(Mr and Mrs Eric Burton)
Eccles Lane, Bispham Green L40 3SD
☎ 0257 463113
Award winning conversion of 17th century barn housing supplies for all types of needlecraft. Also jewellery, pottery, giftware and designer-led crafts. Licensed restaurant; lunch/teas.
Wed-Sun 10am-5pm
Directions: on B5246 between Rufford and Parbold.

The Old Post Office Craft Centre
(K and R Cunliffe)
57 School Lane, Haskayne
Downholland, Nr Ormskirk L39 7JE
☎ 0704 841066
Workshops including wooden toy making and woodturning, hand-painted silks, pen and ink drawing on wood, floral work. Craft shop selling locally made quality crafts. *Shop opening times: Thurs/ Fri/ Sat 9.30am-4.30pm. Workshops with separate opening times; see entries below.*

LANCASHIRE CRAFT WORKSHOPS

Burnley

SLATE GIFTWARE
Slate Age (Fence) Ltd
(P A and K M Rawlinson)
Fence Gate, Fence BB12 9EG
☎ 0282 616952
High quality slate giftware. Hand-crafted green or black slate; range includes clocks, barometers, thermometers, pen stands. Selection of marble. Craft shop. C/W/E
Mon-Fri 8am-4.30pm, Sat 9am-4pm.

FORGEWORK
Barley Forge
(Stephen Marshall)
Barley BB12 9JZ
☎ 0282 603919
Traditional village blacksmith, also producing a variety of decorative wrought ironwork: hay racks, lamp posts and iron brackets with copper lanterns, gates, garden archways etc, specialising in weather vanes. C/R&R/E
Mon-Fri 8am-5pm, Sun 9am-4pm, lunchtime closing.
Directions: Barley Forge can be found at the foot of Pendle Hill.

DRY STONE WALLING
E & D Greenwood
(Eric Greenwood)
Moorside, 33 Hurstwood Lane
Worsthorne BB10 3LF

☎ 0282 37704
Master Craftsman; dry stone walling. DSWA Instructor. Also archaeological restorations.

ANTIQUE FURNITURE
RESTORATION
Walter Aspinall Antiques
(Walter and Beryl Aspinall)
Pendle Antiques Centre, Union Mill,
Watt Street, Sabden BB6 9ED
☎ 0282 76311
Established business employing ten; furniture restoration, furniture-making and woodworking for retail and wholesale markets. Wholesalers and exporters of antiques and older second-hand furniture. C/R&R/W/E. Credit cards.
Mon-Fri 9am-5pm, Sat-Sun 10.30am-4.30pm. Visiting charge:50p.

Bury

FURNITURE RESTORATION
B Baron
11 Linden Avenue
Ramsbottom BL0 0AW
☎ 0706 821681
Restoration work on furniture and antiques, house interiors and colour work on any item. 35 years' experience; specialist in French polishing, varnishing and marbling. Lecturer and examiner and author on woodfinishing. C/R&R
Mon-Fri 9am-5.30pm, Sat 9am-12.30pm.

Carnforth

WOODTURNING
The Wood Revolution
(Malcolm Cobb)
Thie-ne-Shee, Moor Close Lane
Over Kellet LA6 1DF

☎ 0524 735882
Professional woodturner registered with The Worshipful Company of Turners, principally making turned components for new furniture, antique furniture repairs and internal decorative joinery eg newel posts. Medium production runs considered. C/R&R/W/E
Mon-Sat 9am-9pm, Sun 9am-1pm.
Visitors welcome by appointment.
Directions: east of Carnforth, off B6254.

MINIATURE BRASS LIGHTS
Wood 'n Wool Miniatures
(Ken, Joan and Andrew Manwaring)
Yew Tree House, 3 Stankelt Road
Silverdale LA5 0RB
☎ 0524 701532
Family business making specialist brass miniature period lights, lamps and chandeliers in one-twelfth scale for dolls' houses. Also (including transformers) Christmas tree lights, flickering fires and cast resin fireplaces and ceiling sconces. Members of Guild of Master Craftsmen. Shop stocking other miniature items, some by local craftspeople. C/W/E. Credit cards.
Thurs & Sat 10am-4pm, other times by appointment.

GLASS ENGRAVING
Silverdale Engraving
(Barbara Winkfield ARCA)
10 Elmslack Lane
Silverdale LA5 0RX
☎ 0524 701525
Designer working to commission; from simple tumblers for play-groups to commemorative items, windows, decorative mirrors etc. C
Open during normal working hours; advisable to telephone first.

Failsworth

DRY STONE WALLING
Heritage walls
(Robert Arthur Pegler)
61 Cambridge Road
Failsworth M35 OGG
☎ *061 682 8184*
Master craftsman/instructor/
examiner offering courses in
drystone walling. Also ornamental
and full-scale stonework, bridges,
wells, walls, ponds etc. C/R&R
Visitors welcomed to watch work at
any time; telephone beforehand.

Kirkby Lonsdale

PINE PRODUCTS
David H Willan Ltd
Burrow Rural Workshops
Nether Burrow LA6 2RJ
☎ *0468 34328*
Small company specialising in
unusual pine products hand waxed
and finished. Carved bookcases,
dressers and other furniture.
Carved animals: pigs, cats, dogs etc.
Rocking horses. Also painted pub
signs. C/R&R/W/E. Credit cards.
Mon-Fri 9am-5.30pm. Sat opening
varied.

Lancaster

MUSICAL INSTRUMENT MAKING
Robert Deegan Harpsichords
Tonnage Warehouse, St Georges Quay
Lancaster LA1 1RB
☎ *0524 60186*
Fine keyboard musical instruments;
harpsichords, virginals, spinets and
clavichords; high standard of
performance and finish. C/R&R/E.
Mon-Fri 9am-5pm, lunchtime closing.
Visit by appointment only.

UPHOLSTERY
Royal Upholstery
(Roy Goodwin), Greenfields Yard
off Copy Lane, Caton LA2 9QU
☎ *0524 771148*
Personal re-upholstery service and
restoration of soft furnishings. R&R
Mon-Fri 9am-5pm, lunchtime closing
12.30-1.30pm. Sat am by appointment.

DESIGN & SIGNS
Lancaster Fine Arts
(Frank Perkins), Belle Vue Studio
Mewith Lane, Bentham LA2 7DQ
☎ *0524 262219*
Painters, artists, signwriters and
illustrators led by designer Frank
Perkins. General and brewery sign
work, pub signs, pictorials, applied
letters, carved products, fascias,
brackets, marbling, graining etc.
Special projects. C/R&R/W/E
Mon-Fri 9am-5pm.

Oldham

CERAMICS
Artisan
(Anne Hamlett)
3 King Street, Delph OL3 5DL
☎ *0457 874506*
Handmade ceramics and ceramic
sculpture; fantasy castles, lamps,
dragons and figures. Craft shop.
C/W/E. Credit cards.
Wed-Sat 10.30am-5.30pm, Sun 2-5pm.
(Jan-Apr advisable to telephone first).

Ormskirk

WOODEN TOYS & TURNERY
K & R Cunliffe
(Mr and Mrs R Cunliffe)
The Old Post Office, 57 School Lane
Haskayne L39 7JE
☎ *0704 841066*

Workshop making name and stand-up jigsaws, pull-alongs, spinning tops and other toys. Also clocks/barometers, S&P mills, bowls and boxes, unusual gifts. Shop selling craft work. Members of regional craft guild. C/W. Credit cards.
Sat 9.30am-4.30pm, or by appointment.

PEN & INK DRAWINGS ON WOOD & EMBROIDERED PICTURES
Close Connections
(Cathy Brooke)
Workshop 4, The Old Post Office
57 School Lane, Haskayne L39 7JE
☎ *0704 841066*
Pen and ink drawings, pyrography; member of regional craft guild. C. Credit cards.
Sat 9.30am-4.30pm, or by appointment.

DRIED & FABRIC FLOWERS
Floral Studio
(Ruth Brooke)
Workshop 4, The Old Post Office
57 School Lane, Haskayne L39 7JE
☎ *0704 841066*
Fresh, dried and silk flower arrangements. Member of regional craft guild. C. Credit cards.
Sat 9.30am-4.30pm, or by appointment.

Preston

PICTURE FRAMING
Bespoke Artistic Framing Services
(A J Woodruff)
Jasmin House, Hollins Lane
Forton PR3 0AB
☎ *0524 791353*
Contract picture framers, manufacturers of photo-frames, dry mounting and heat sealing.
C/R&R/W. Credit cards.
Mon-Fri 9am-5pm.

DRIED FLOWERS
Grass Roots Craft Centre
(Harold, Marion and Johnny Rigby)
Finney Barr's Farm, Drinkhouse Lane
Croston PR5 7JE
☎ *0772 600221*
Growers and arrangers of dried flowers. Demonstrations by arrangement. Church flowers for hire; wedding flower. Also hat hire; top quality hats. Other crafts for sale. Coffee shop. C/R&R
Wed-Sun 10am-5pm.
Directions: in Croston between Preston and Southport, near Ruffold Old Hall, 10 mins drive from Park Hall and Camelot.

STAINED GLASS
Harlequin Glass
(Deborah Wareing)
Unit 4, Ashley Hall Farm
Inglewhite Road, Longridge PR3 2EA
☎ *0772 785045*
Stained glass artist designing and making windows, doors, decorative mirrors and double glazing. Individual one-off designs a speciality, ecclesiastical work and pub contracts. Member of Guild of Master Craftsmen. C/R&R/E.
Mon-Fri 9am-5.30pm, Sat 9.30am-12noon.

PATCHWORK
Quilters Cottage
(Mrs C M Rigby)
60 Bridge Street, Garstang PR3 1YB
☎ *0995 603929*
Patchwork quilts and cushions made to order. Classes in patchwork, quilting, and needlework crafts. Supplies available. Antique quilt showroom. C. Credit cards.
Mon-Fri 9.30am-5pm (Wed half-day closing). Sat 9.30am-4.30pm.

Rossendale

GOLD & SILVER SMITHING
C Whiting
540 Burnley Road
Crawshawbooth BB4 8NE
☎ *0706 830979*
Designer jewellery and all silver-smithing work undertaken, particularly ecclesiastical plate. Also restoration and repair of antique jewellery and silver, smelting and electroplating. C/R&R/E.
Mon-Fri 8am-5pm, lunchtime closing. Sat-Sun by appointment.

SADDLERY
M Miller
(Michelle Miller)
624 Burnley Road East
Whitewell Bottom BB4 9NT
☎ *0706 226983*
Established Master Saddler's business making and repairing equestrian leather equipment and some general leather goods. Mainly work to specialist requirements. Workshop and shop. Demonstrations given including evenings and weekends, by arrangement. C/R&R/W. Credit cards.
Mon-Sat 9am-6pm (and Suns during Nov/Dec 10am-5pm), lunchtime closing. Telephone first.
Directions: 7 miles from Burnley, 2 miles from Rawtenstall (end of M66) on B6238.

Saddleworth

CERAMICS
Dovestone Pottery
(Les and Karen Roberts)
2&3 Waterside Mill
Greenfield OL3 7NH
☎ *0457 871590*
Distinctive ceramic pottery includ-ing lamps, bowls, candle cups, pomanders, essential oil burners etc. W/E. Credit cards.
Mon-Fri 9am-4pm, lunchtime closing.
Directions: in Greenfield village on A635 Ashton/Holmfirth road, turn off opposite church, take first right, turn left into car park with pottery at end.

Southport

THATCHING
B D Milne
15 Fleetwood Crescent, Banks PR9 8HF
☎ *0704 231510*
Award winning Master Thatcher working in water reed and wheat straw. Roof restoration, listed building work, new work and extensions, recoating work. Plain and decorative ridgework in straw and sedge. Mini timber-framed thatched houses, summer houses and gazebos for sale.

Todmorden

FURNITURE MAKING & RESTORATION
Benchmark Designs
(David Neil Farnworth)
Unit 5, Nanholme Workshops
Shaw Wood Road
Todmorden OL14 6DA
☎ *0422 843853*
Traditional cabinet making business specialising in furniture making and restoration; French polishing etc. Also some contemporary painted pieces undertaken. C/R&R
Mon-Fri 8.30am-6pm, Sat by appoint-ment.
Directions: from Hebden Bridge/Todmorden road A646 turn off at Eastwood, signposted to Mankinholes/Lambutts. Cross narrow bridge and straight on to car park for workshops.

LEICESTERSHIRE CRAFT WORKSHOP CENTRES

East Carlton Countryside Park
East Carlton
Nr Market Harborough LE16 8YD
☎ *0536 770977*

Craft workshops on upper floor of the Heritage Centre, a converted 17th century coach house. Crafts include: glass workshop making figures of birds, animals and flowers, gift items, beads; lacemaking accessories a speciality. Pastel portraits and old master reproductions. Dried floral arrangements. Ceramics; wheel thrown and hand carved pierced work. Blacksmith's forge. Industrial Heritage Centre illustrating Corby's steelmaking past and Ranger Service. Special events arranged in summer months. Café open all year.

Open all year round Mon-Fri 9.30am-4.45pm, Sun 10.30am-6pm.

Directions: 4.5 miles west of Corby off A427.

Ferrers Centre for Arts and Crafts
Staunton Harold
Ashby-de-la-Zouch LE6 5RW
☎ *0332 863337*

Craft workshops converted from old stabling. Visitors can see handcrafted copperware, guitar maker, knitting and weaving of many different textiles, tapestry and embroidery design, childrens' wear, silversmithing, signwriting, potting, mechanical and musical Victorian style automata.

Other workshops only visited by appointment: artist and designer (tuition given), china surgery. Prestigious gallery with art and craft exhibitions throughout the year, with a strong local content. Tea rooms. Workshops have individual opening times. Craft Centre open daily (except Mon) all year including Bank Holidays.

Directions: three entrances to grounds which can be approached on B587 from Melbourne to A453.

LEICESTERSHIRE CRAFT WORKSHOPS

Ashby-de-la-Zouch

PICTURE FRAMING
The Stables
(Mr and Mrs Hampson)
South Street
Ashby-de-la-Zouch LE6 5BR
☎ *0530 414246*

Picture framing workshop for all types of paintings, embroideries, tapestries, mirrors and prints, with gallery attached.
C/R&R. Credit cards.
Tue-Sat 9.30am-5.30pm, closed for lunch 1-2.30pm.

FURNITURE MAKING
The Old Forge Cabinet Workshop
(Tadeusz Rucinski)
North Street
Ashby-de-la-Zouch LE6 5HS
☎ *0530 411400*

Work to commission only making solid wood bespoke furniture, modern and traditional in English hardwoods. Customers include private, small corporate bodies and local authorities. C/R&R/E.
Mon-Fri 9am-5.30pm, Sat 9am-5pm, lunchtime closing.

NEEDLECRAFT & TAPESTRY
The Colour Orchard
(Patricia Stevenson and Sally Harman)
Staunton Harold Nurseries Workshops
Staunton Harold LE6 5RW
☎ 0332 864828
Needlepoint and embroidery
designers for two large manufacturers of tapestry kits and own new
range of kits. Also available: oil
paintings, Batik paintings, watercolours and Raku ware.
C/R&R/wholesale. Credit cards.
Open daily (except Fri) 10.30am-5.15pm,
lunchtime closing 12noon-12.45pm.
Directions: opposite Ferrers Centre
(see under Craft Centres) within
gardens of Staunton Harold Nurseries.

FURNITURE & RESTORATION
M R Clark
Staunton Harold Nurseries Workshops
Staunton Harold LE6 5RW
☎ 0332 792461
Furniture restoration, traditional
hand French polishing, gilding and
woodturning. Workshop and
showrooms. C/R&R
Open daily 2-5pm.
Directions: opposite Ferrers Centre
(see under Craft Centres) within
gardens of Staunton Harold Nurseries.

Coalville

KNITWEAR & TEXTILES
Sharon J Webb
Gladstone Villas, 56 North Street
Whitwick LE6 4EA
☎ 0530 811708
Experienced knitwear and textile
designer producing own knitwear
range (hand/machine knitted) for
retail outlets. Teacher of knitwear/
textiles, art and design at local
college. Also teaching workshops

with crèche facilities; telephone for
details. C/W
Varied working hours; advisable to
telephone first.
Directions: from M1 junc 22 take A50
to Coalville. Follow signs for
Whitwick (4 miles from M1).

Leicester

CABINET MAKING & UPHOLSTERY
S J Nott & Co Ltd
(Stephen Nott)
15a Dorothy Avenue
Thurmaston LE4 8AB
☎ 0533 695127
Established business restoring and
re-upholstering antique and
modern furniture, French polishing,
constructing and re-constructing
bespoke upholstered furniture. Also
soft furnishings. Member of Guild
of Master Craftsmen and Association of Master Upholsterers.
C/R&R/E. Credit cards.
Mon-Fri 9.30am-5pm, Sat 9.30am-1pm.
Directions: in old Thurmaston village on
A607 Leicester-Melton Mowbray road
turn off opposite British Legion club.

FORGEWORK
Walter Allen & Son
(L S Allen and W J Allen)
12 High Street, Great Glen LE8 0FJ
☎ 0533 592225
Ornamental ironwork and general
blacksmithing; also plumbing. C/R&R
Mon-Fri 8am-5pm, lunchtime closing.

SADDLERY
Hurst Saddlers
(Mrs S M Hurst)
52/54 Brabazon Road
Oadby LE2 5HD
☎ 0533 713741
Saddlery and equipment made,

altered and repaired. C/R&R
Mon-Fri (closed Wed) 11am-6pm, Sat 9.30am-6pm.
Directions: from A6 to Leicester, after traffic lights/crossroads in Oadby turn left and fork left into Brabazon Road; saddlery on left.

Loughborough

COPPER WORK
F M Fisher Metalworks
Lime Kiln Farm
Normanton-on-Soar LE12 5EH
☎ 0509 853657
Victorian style copper lamps, metal spinning and general metal work. Engineering workshop with ability to manufacture anything in metal, copper etc. C/R&R/E
Mon-Fri 8am-6pm, Sat-Sun 9am-6pm.

FURNITURE RESTORATION
Wymeswold Country Furniture
(Bill and Jenny McBean)
17 Far Street, Wymeswold LE12 6TZ
☎ 0509 880309
Furniture restoration. Also selling antique and reproduction pine furniture and knitwear. C/R&R. Credit cards.
Open daily 8.30am-5.30pm.
Directions: on A6006. From Loughborough take A60 to Nottingham. At Hoton turn right to Wymeswold. In village follow road round to left, take next left (Clay Street); workshop at top.

CANE & RUSH SEATING
Abbey Cane & Rush Seaters
(M R Abbey)
Moat House, Bramcote Road
Loughborough LE11 2SA
☎ 0509 214154
Experienced re-seaters and importers of high quality rush. C/R&R

Open during normal working hours, please telephone first.

FURNITURE RESTORATION
Quorn Pine
(Stephen Yates and Steven Parker)
Unit 3, 75 Barrow Road
Quorn LE12 8DH
☎ 0509 416031
Workshops employing seven craftsmen, members of Guild of Master Craftsmen. Specialists in restoration of antique pine furniture, both for customers and trade. C/R&R/W/E. Credit cards.
Mon-Fri 9am-6pm, Sat 9.30am-5.30pm
Directions: From A6 Loughborough-Leicester road turn into Barrow Road (garage on corner); workshop on right.

JEWELLERY
Wendy Greene Handmade Fashion Jewellery
94 Maplewell Road
Woodhouse Eaves LE12 8RA
☎ 0509 890403
Fashion jewellery designed and made up in beads, metal, flowers etc. C/R&R/W/E
Visitors welcome at any time by appointment.

LACEWORK
Hand Knitted Lace
(Mrs Maria Dalmar)
37 Atherstone Road
Loughborough LE11 2SH
☎ 0509 266302
Fine hand-knitted lace; work exhibited in UK and abroad. Talks given. C/W
Mon-Fri after 5pm, Sat-Sun any time. Please ring for appointment.
Directions: take ring road off A6 along Shelthorpe Road, over roundabout to Park Road. Atherstone Road 4th on left.

Market Bosworth

LEATHERWORK
Bosworth Crafts
(R Thorley)
23 Main Street
Market Bosworth CV13 0JN
☎ *0455 292061*
Plain and hand-carved leatherwork;
workshop area in general craft
shop. C/W/E. Credit cards
Mon-Fri 9am-5.15pm, Sat 9am-5pm.

BASKET WORK
Country Crafts
(B T and M Sturgess)
Main Street, Market Bosworth
☎ *0530 72469 (evenings)*
English country baskets and babies'
cradles handmade from willow.
Rush and cane seating. Any kind of
wickerwork restored. C/R&R/W/E
Mon-Fri 10am-4pm, Sat-Sun 10am-
5pm, advisable to telephone first.

Market Harborough

CERAMICS
Quorn Pottery
(J W and Mrs A E Brookes)
46-48 Scotland Road
Little Bowden LE16 8AX
☎ *0858 431537*
Unusual slip-cast earthernware
including collectable novelty
teapots. Wall clocks, cruets,
magnets and brooches. C/W/E
Mon-Fri 9am-5pm, Sat 10am-4pm.
Advisable to telephone first.
Directions: off A6 out of Market
Harborough towards Northampton.

CERAMICS
Frank Haynes Gallery
50 Station Road
Great Bowden LE16 7HN

☎ *0858 464862*
Teacher and craftsman, throwing
pots and 'figure' bowls in workshop
at weekends. Also gallery selling art
and work by local craftspeople
(displays changed regularly). C
Tues/Thurs/Fri/Sat 2pm-5pm, Sun
10am-5pm, lunchtime closing. Closed
July/August.

THATCHING
Stephen Duffin
147 Logan Street
Market Harborough LE16 9AP
☎ *0858 434834*
Roof thatching contractor.

Melton Mowbray

SPINNING WHEELS
Timbertops
(Anne and James Williamson)
Wheel Lodge, 159 Main Street
Asfordby LE14 3TS
☎ *0664 812320*
Quality spinning wheels made in
workshop — selection of wheels for
spinners to see and use. Also
accessories. W/E.
Visitors welcome by appointment only.
Directions: in Asfordby 3 miles west
of Melton Mowbray on A6006, almost
opposite The Blue Bell PH.

FURNITURE
Hollies Farm Handcrafts
(J J Atton)
Hollies Farm
Little Dalby LE14 2UQ
☎ *0664 77553*
Workshop employing two crafts-
men making fine furniture and
restoring furniture. Also craft shop
selling a wide variety of goods
handmade by selected crafts people.
Tea room. C/R&R. Credit cards.

Workshop:Mon-Fri 9am-5pm, Sat-Sun 11am-4.30pm. Craft shop open Sat-Sun 11am-4.30pm or by appointment.
Directions: from Melton Mowbray-Oakham road turn south to Little Dalby.

GOLD & SILVER JEWELLERY
Vipa Designs Ltd
(Peter Crump and Vivia Bremer-Goldie)
Freeby View
Waltham Road
Thorpe Arnold LE14 4SD
☎ *0664 78444/78423*
Manufacturers of gold and silver jewellery and silver giftware. Mainly lost-wax casting. Six staff making a wide range of products with exclusive rights to produce Beatrix Potter jewellery. Commissions from companies and associations wanting corporate or incentive gifts. Products sold world-wide, particularly in Japan. C/R&R/W/E.
Mon-Fri 8.30am-5pm, Sat-Sun by appointment only. Closed Xmas-New Year.
Directions: off A607 between Thorpe Arnold and Waltham-on-the-Wolds (1.5 miles), turn right before Freeby View Farmhouse.

CERAMICS
Grange Farm Pottery
(Elaine Pell)
Grange Farm
Plungar NG13 0JJ
☎ *0949 60630*
Pottery workshop and showroom. Handmade pottery, domestic and individual pieces all made on the premises. C
Sat-Sun 11am-6pm (closed Xmas & Easter).

Rutland

FURNITURE
Rutland Cabinet Makers Ltd
(S Brophy, M King and P Baker)
Unit A3
Thistleton Road Industrial Estate
Market Overton, Oakham LE15 7PP
☎ *0572 722166*
Manufacturers of yewtree, mahogany and walnut Regency style furniture for dining rooms, bedrooms, living rooms and offices. Also the Rutland Oak range. Bespoke cabinet making a speciality; repairs and restoration work. Showrooms at The Table Place, 74 Station Road, Oakham. C/R&R/W/E. Credit cards.
Workshop open Mon-Fri 10am-4pm. Visitors welcome by appointment only.
Showrooms: Mon 10am-1.30pm, Tues-Fri 10am-5pm, Sat 10am-4pm, Sun 11am-4pm.

Swadlincote

FURNITURE MAKING
Phil Dennis Woodcraft
Unit 2, Moira Furnace Workshops
Furnace Lane , Moira DE12 6DY
☎ *0283 551577*
Furniture and cabinet making. Also gifts and presentation pieces, turnery, clocks and tantalizing executive toys. Workshop with display area. C
Mon-Sat 8.30am-5.30pm, lunchtime closing.

CERAMICS
Furnace Lane Pottery
(Louise Field)
Unit 7, Moira Furnace Workshops
Furnace Lane, Moira DE12 6AT

☎ *0283 552218*
Hand-thrown wood-fired pots for
domestic use. C/R&R/W/E.
Mon-Fri 10am-5.30pm, Sat 10am-4pm.

Uppingham

FORGEWORK
J F Spence & Son
(Derek C and John C Spence)
The New Forge, Station Road
Uppingham LE15 9TX

☎ *0572 822758*
Ornamental ironwork including
hand-forged items of all descriptions; fire baskets and furnishings,
gates, rose arches, garden furniture,
hanging basket brackets, boot
scrapers etc. C/R&R/W
*Mon-Fri 8am-5pm, lunchtime closing
12.30-1.30pm. Sat 9.30am-12.30pm.
Closed 1 week in July and Bank
Holidays.*

LINCOLNSHIRE CRAFT WORKSHOP CENTRES

Manor Stables Craft Workshops
(Anne Wood)
Grantham NG32 7JN
☎ *0400 72779*
Stone-built stables converted into craft workshops; weaving, spinning, needlework, jewellery, saddlery, furniture making.

Showroom with a variety of local craftwork. Coffee shop with home-made refreshments and produce.
Tues-Sun 10.30am-4.30pm
Directions: from Lincoln on A607 to Grantham, go straight over Leadenham traffic lights at crossroads with Newark-Sleaford A17; workshops shortly after on right.

LINCOLNSHIRE CRAFT WORKSHOPS

Alford

CERAMICS
The Pottery
(Heather & Michel Ducos)
Commercial Road
Alford LN13 9EY
☎ *0507 463342*
Domestic and decorative stoneware pottery specialising in oven-to-table ware. Pierced work (pomanders, lanterns). Commemorative pieces undertaken. Access for wheelchairs. C/W/E
Mon-Fri 9am-5pm, lunch-time closing. Sat by appointment.

Gainsborough

CERAMICS
Kirton Pottery
(Peter and Christine Hawes)
36 High Street
Kirton-in-Lindsey DN21 4LX
☎ *0652 648867*
Pottery and ceramic sculpture; functional and imaginative pottery both in bright colours and subtle stoneware for the garden and tableware. Demonstrations given by arrangement. C/W/E

Mon-Sat (except Tues) 9.30am-5.30pm, Sun 2.30-5pm
Directions: 2 miles east of A15 south of M180.

Grantham

ANIMAL PORTRAITS & PAINTING RESTORATION
Roger Heaton
2 Park Cottages, Lenton NG33 4HQ
☎ *0476 85467*
Painting of animal portraits in oils, watercolours and pastels. Also picture framing and restoration of paintings and frames. Member of artists' societies; work selected for Royal Academy Exhibition. C/R&R/W/E. Credit cards.
Mon-Fri 9am-5.30pm, please telephone first. Sat-Sun by appointment.

STAINED GLASS
A H Associates Ltd
(Tony Hollaway)
Home Farm House, Bottesford Road
Allington NG32 2DH
☎ *0400 81754*
Designers and fabricators of stained and painted glass to commission. Recent commission for Manchester Cathedral. Approved by English

Heritage for stained glass conservation. C/R&R/E.
Mon-Fri 9am-5pm.

SADDLERY
Mark Bushell Saddlers
Manor Stables Craft Workshops
Fulbeck NG32 7JN
☎ *0400 73711*
Working saddler registered with Society of Master Saddlers. Bespoke work, repairs and servicing. C/R&R.
Tues-Sun 10.30am-4.30pm.
Directions: see under Craft Centres.

SPINNING
Sheepshades Spinning Gallery
(Kathie Jackson)
Manor Stables Craft Workshops
Fulbeck NG32 7JN
☎ *0400 72779*
Hand-spinning; tuition given (full/half-day and longer courses for all levels), also equipment and supplies. Fleeces and fibres, yarns and beads. Hand-knitted sweaters. Commissions. Credit cards.
Tues-Sun 10.30am-4.30pm.
Directions: see under Craft Centres.

WEAVING & TEXTILES
Anne Wood
Manor Stables Craft Workshops
Fulbeck NG32 7JN
☎ *0400 72779*
Weaving and textile producing; appliqué work, quilting and patchwork. C/R&R. Credit cards.
Tues-Sun 10.30am-4.30pm.
Directions: see under Craft Centres.

TOYS & CERAMICS
Fulbeck Heath Craft Centre
(Mr and Mrs C Lemmon)
Ryland Grange Cottage
Fulbeck Heath NG32 3JH

☎ *0400 61563*
Hand-carved wooden rocking horses decorated in a wide range of colours; also repairs. Ceramics and soft toys. All crafts made on the premises. Refreshments. C/R&R/E.
Open daily.
Directions: near RAF Cranwell on A17.

Lincoln

FLOWERS & POT POURRI
Romance Flowers
(Karen and Barry Thomas)
9 Bawtry Close, Birchwood Estate
Lincoln LN6 0HS
☎ *0522 689244*
Dried, silk and fresh flower arrangements for all occasions. Pot pourri loose and in arrangements, and accessories. Demonstrations given at rural and London shows. Small groups welcome. C/R&R/W
Mon-Fri 10am-7pm, Sat-Sun 11am-7pm, other times by arrangement. Advisable to telephone first.

NEEDLECRAFT
Stitch Witchery
(Isabel Neale)
43 Chiltern Road, Brant Road
Lincoln LN5 8SB
☎ *0522 540299*
Specialists in custom-designed cross-stitch pictures, clocks, cards and gifts. Unique presents for births and anniversaries. Kits to order. C
Open daily 10am-5pm, advisable to telephone first.
Directions: from A46 Lincoln relief road follow A1434 (approx 3.5 miles) to traffic lights with 'RAF Waddington' signed to right. Turn right at lights (Brant Road) for half mile. Turn first right (Calder Road) immediately left Chiltern Road.

PRESSED FLOWERS
Linda Makin
Unit 10, Cobb Hall Craft Centre
St Paul's Lane, Bailgate
Lincoln LN1 3AL
☎ *0522 510805*
All pressed flower work undertaken; pictures, lampshades, greeting cards. Pressed wedding bouquets etc. Bone china decorated with pressed flowers. Member of regional and craft guilds. C/W. Credit cards.
Tues-Sat 10.30am-4.30pm (closed Jan)

FURNITURE
Cobweb Crafts
(Kevin Burks)
The Old School, Cadney Road
Howsham LN7 6LA
☎ *0652 678761*
Bespoke furniture made in English hardwoods; free standing and fitted furniture including bathrooms, bedrooms and kitchens. C/W/E
Mon-Sat 8am-5pm, Sun by appointment.

PRINTED PICTURES
Left Bank Arts
(Liz and Leonard Read)
63 Little-Bar-Gate Street
Lincoln LN5 8JL
☎ *0522 512390*
Original prints made of scenes, animals etc. Also picture framing and sign writing. C
Mon-Fri 11am-4pm, Sat-Sun 2-4pm.
Please telephone for appointment.

TEXTILES, PRINTING & PAPERMAKING
Timberland Art & Design
(Janet Crafer & Jonathan Korejko)
12 Church Lane, Timberland LN4 3SB
☎ *05267 222*
Art and craft workshop in converted Methodist chapel. Weaving, print-making, paper-making, marbled paper, tapestries, rugs, picture frames. Demonstrations and tuition (for all ages) in weaving, print-making, paper-making. Also materials available. C/R&R/W/E
Mon-Fri 9am-5pm, Sun 1-4pm.
(Closed fortnight afterXmas)

WOODTURNING
Tamcraft Woodworks
(Angela Fleming and Paul Carroll)
2 Gatehouse Cottage
Caenby Corner, Lincoln LN2 3EE
☎ *06737 634*
Wooden toys, country chairs, turned giftware etc. C/W/E
Mon-Fri 10am-8pm, Sat-Sun 10am-5pm.

WOODWORK & WOODTURNING
Pig & Whistle Rocking Horses
(Jenny and Roger Sanderson)
Grove Cottage, Church End
North Somercotes LN11 7PZ
☎ *0507 358648*
Small workshop making hand-carved rocking horses, bespoke furniture, tables, Windsor chairs, clocks and turned goods. C/R&R/W/E. Credit cards.
Mon-Fri (closed Wed) 9am-5pm, Sat-Sun 10am-5pm.

Louth

HAND-PAINTED BONE CHINA
Liz Butterfield — China Painter
Hedgehog Corner, Wragholme Road
Grainthorpe, Louth LN11 7JD
☎ *0472 388179*
English bone china plates hand-painted to commissions for collectors and some limited editions. C
Variable working hours; advisable to telephone first.

FORGEWORK
Alvingham Forge and Striking Designs Gallery
(R F Oakes)
Yarburgh Road
Alvingham LN11 0AG
☎ 0507 327017
Traditional and contemporary ironwork; architectural and decorative, general blacksmithing, restoration and repairs. Two smiths, members of British Artist Blacksmiths' Association working to commission for individuals, civic buildings and companies. Winners of numerous awards. C/R&R.
Mon-Sat 9am-6pm, Sun 10am-5pm. Please telephone first.

HAND-PAINTED JEWELLERY
Blue Heron Designer Jewellery
(Elizabeth Jack)
Muckton LN11 8NX
☎ 0507 480630
Hand-painted wooden bead jewellery, mostly one-off designs. Earrings, brooches, cufflinks and necklaces. C/W/E
Open at any time by arrangement.

CERAMICS
Jackpots
(Pauline Baskcomb)
The Old Stables, Queen Street Place (rear of public car park)
Louth LN11 9BD
☎ 0507 604656 *(evenings)*
Hand-thrown earthenware made in studio pottery producing a large range of functional domestic ware. C
Mon-Sat 10am-4pm, advisable to telephone first.

CERAMICS
Ceramix
(Anna Walsh)
44 Queen Street, Louth LN11 9BL
☎ 0507 604297
Pottery with workshop and gallery making wheel-thrown domestic ware, slipcase vases and sculptural pieces. Work widely exhibited. C
Mon-Sat 9.30am-6pm. (Workshop closed Jan-Mar).

Mablethorpe

ROCKING HORSES & WOODEN TOYS
Western Horseman
(S F Cook)
12 Millfield, Trusthorpe LN12 2PG
☎ 0507 473721
Hand-carved rocking horses (varying sizes), hobby horses, other wooden toys made to order. Visitors welcome to workshop. C/R&R. Credit cards acepted
Open daily 10am-6pm, advisable to telephone first.
Directions: on road from Mablethorpe to Sutton-on-Sea, turn right at bakery (opposite Trustville Holiday Camp).

Market Rasen

WOODEN TOYS
The Stable Workshop
(Dick Watson)
Well Cottage, Main Road
Snarford LN8 3SW
☎ 06735 295
Workshop in old stable with craftsman hand-crafting toys in wood to traditional designs. C/R&R/E.
Open daily (except Fri) 9am-6pm, advisable to telephone first.
Directions: on A46 from Lincoln (8 miles) past Snareford cross roads towards Market Rasen (6 miles). Workshop on left opposite right turn.

Spilsby

DRIED FLOWERS, HERBS & POT POURRI
Candlesby Herbs
(John and Jane Stafford Allen)
Cross Keys Cottage
Candlesby PE23 5SF
☎ 0754 85211
Small cottage industry, growing and utilizing herbs and flowers for all aspects of living. C/R&R/W/E
Tues-Sun (also open Bank Holidays Mons) 10am-5pm.

CERAMICS & SOFT FURNISHINGS
John and Jane Snowden
Thirtytales Cottage, The Cul-de-Sac
Stickford PE22 8EY
☎ 0205 480848
Established pottery producing hand-thrown domestic pottery; free-form sculptures produced on commission. Also appliqué/patchwork cushions, teapot cosies and soft furnishings. High quality work in East Midlands shops and galleries. C/R&R/W/E
Open daily 9am-5pm, lunchtime closing. Visit by appointment only. (Closed Jan).

Stamford

FORGEWORK
Tinwell Forge
(David O'Regan)
Tinwell PE9 2UD
☎ 0780 56341
Hot- and cold-forged decorative ironwork. Stock includes plant-troughs, hay-racks, door-knockers, candlesticks, fencing, hanging baskets. Barbecues, weathervanes, street lights etc to order.
C/R&R/W/E. Credit cards.
Mon-Fri 8.30am-5.30pm, lunchtime closing. Sat-Sun 9am-5.30pm.
Directions: 2 miles west of Stamford on A6121.

Woodhall Spa

CABINET MAKING
Edmund Czajkowski & Son
(Michael Czajkowski)
96 Tor-o-Moor Road
Woodall Spa LN10 6SB
☎ 0526 352895
Cabinet makers and specialists in antique furniture, clock and barometer restoration. French polishing, carving, gilding, marquetry, boule work and upholstery. Reproduction or modern furniture designed and made, bearing oak-leaf trademark. C/R&R.
Mon-Fri 9am-5pm, other times by appointment.

NORFOLK CRAFT WORKSHOP CENTRES

The Raveningham Centre
Beccles Road
Raveningham, Norwich NR14 6NU
☎ 050 846 441/688
Victorian farm buildings renovated
to provide craft workshops, art and
antiques gallery and coffee shop.
Workshops include: furniture
making and restoration by member
of BAFRA, interior design and
piano restoration.
Open daily 10am-6pm.
Directions: from A146 Norwich-Lowe-
stoft road or A143 Great Yarmouth
road, turn onto B1140 near Beccles;
centre on left travelling north.

Alby Crafts
(Valerie Allston)
Cromer Road
Erpingham, Norwich NR11 7QE
☎ 0263 761590 (workshops 0263 761702)
Craft centre housed in old farm
buildings with craft workshops
including furniture making,
knitwear, picture framing, pottery,
stained glass and lace making and
museum. Also bee keeping with
observation hive in covered
accommodation; honey and honey
making. Bottle museum, gift shop,
tea room. Four acres of grounds with
large ponds.

NORFOLK CRAFT WORKSHOPS

Cromer

ARTIST & PICTURE FRAMING
Jane Cort — Picture Framer/Artist
Workshop 11, Alby Crafts
Cromer Road, Erpingham NR11 7QE
☎ 0263 761702
Water colour artist and picture
framer; ready-made frames,
framing commissions, mirrors,
prints and paintings. C/R&R
Tues-Sun 10am-5pm, lunchtime
closing. (Closed Xmas-Apr).
Directions: On A10 Cromer to
Norwich Road.

Dereham

CLOCK MAKING
Clockspares
(R F Charman)
The Yard, Wellington Road
Dereham NR19 2BP
☎ 0362 694165

Workshop employing four craftsmen making spare parts for clocks,
old and new including Norfolk
turret clocks, quartz clocks. Wheel
cutting etc. Mail order. C/R&R/W/E.
Mon-Fri 8.30am-5pm, Sat-Sun 9am-
12noon.

SPINNING & WEAVING/
CANE & RUSH SEATING
Twists & Turns — Spinners
(Vic and Eileen Ringwood)
Spinners, Fakenham Road
Beetley NR20 4BT
☎ 0362 860194
Husband and wife team, spinning
and weaving; mostly passementerie
(fringes, bullion, braids, ruches,
tassels etc). Small colour matching a
speciality. Weaving and spinning
equipment available. Also cane and
rush seating. C/R&R
Fri-Sat 10am-5pm, or by appointment.

Diss

FURNITURE
David Gregson Furniture
Bridge Green Farm, Gissing Road
Burston IP22 3ND
☎ *0379 740528*
Distinctive one-off furniture for domestic and corporate environments. Many awards and important commissions. C/W. Credit cards
Mon-Fri 9am-5pm, lunchtime closing.
Advisable to telephone first.

ANTIQUE FURNITURE
RESTORATION & CLOCK REPAIRS
Brian Harris Antique Restoration
Workshop 3, Gables Yard, The Green
Pulham Market IP21 4SU
☎ *0379 608379*
Furniture restoration including polishing, upholstery etc. Clock repairs. Antiques bought and sold. C/R&R/W/E
Mon-Fri 8.30am-5.30pm, Sat mornings only.

Downham Market

DRIED FLOWERS
Bexwell Hall Dried Flower Centre
(Mrs B A Daniels)
Bexwell Hall
Downham Market PE38 9LZ
☎ *0366 382208*
Growing and retailing dried flowers and arrangements. Workshops in converted stables in grounds of historic hall with Saxon church nearby. Groups welcome by arrangement. Flower arranging classes. C/W.
Mon/Thurs/Fri/Sat 10am-4pm.
Directions: 1.5 miles from Downham Market on Swaffham Road A1122.

MODEL MAKING
Guild Master Models
(Martin Raymond and Christine Field)
Willow Cottage, High Street
Nordelph PE38 0BL
☎ *0366 8351*
Model making; pattern making in brass for collectors and gift industry. One-off model making for architects, museums and collectors. C/R&R/E. Credit cards.
Mon-Fri 9am-10pm, Sat-Sun 9am-6pm.

East Dereham

PUPPETS/MASKS
Doreen James Parsons
Lucky Chance Cottage, Bittering Street
Gressenhall NR20 4EB
☎ *0362 860491*
Sole artist supplying puppets, masks, animal costumes etc. for national theatres, companies, TV and advertising. C/E.
Tues/Thurs/Sat 10am-4pm by appointment.

BASKET MAKING/
CANE & RUSH SEATING
Rob King — English Willow Basketworks
Dick Fool's Lane
Wendling NR19 2NF
☎ *0362 87569*
Traditional and contemporary baskets, cradles, fencing panels etc made in peeled and unpeeled willow. Chair caning and rush seating. Demonstrations. Member of regional craft guild. C/R&R/E
Open most days, any time; please telephone first.
Directions: from A47 Dereham-Norwich road take turning to Wendling, at Rose Cottage PH turn left over A47, Dick Fool's Lane on left.

BRASS FURNITURE FITTINGS
Marshall Brass
(Mr and Mrs Marshall)
Long Ground Cottage, Keeling Hall Road
Foulsham NR20 5PR
☎ *0362 844105*
Brass and steel fittings made for antique and reproduction furniture. Polishing and restoration of metalwork. C/R&R/W/E. Credit cards.
Mon-Fri 8am-6pm, please telephone first.

Fakenham

BASKET MAKING/
CANE AND RUSH SEATING
Mona Leckie
Manor House, Toftrees
Fakenham NR21 7DZ
☎ *0328 863598*
Traditional handmade baskets and restoration of chair seating. Qualified demonstrator. Courses by arrangement. C/R&R/E. Credit cards
Open daily 10am-4pm, other times by appointment.

CANDLEMAKING
Candlemas Cottage Enterprises
(Valerie Goodsell)
The Candle Shop, Guild Street
Little Walsingham NR22 6BU
☎ *0382 820748*
Ornamental candles; carved, sculptured, hand-painted. Sold in shop premises. C/W/E
Open daily 10am-5.30pm (-5pm winter months), lunchtime closing. (Jan/Feb fewer opening hours).

TEXTILE PRINTING
Sheila Rowse Ltd
(S F and J C Rowse)
The Textile Centre, Hindringham Road
Great Walsingham NR22 6DR

☎ *0328 820009*
Housed in traditional Norfolk barns, printing well-known Sheila Rowse designs on aprons, tea towels, oven gloves etc. Craftshop. C/W/E. Credit cards.
Mon-Fri 9.30am-5.30pm, Sat-Sun 10am-5pm, (Closed Jan-mid March).
Directions: on Fakenham to Wells road turn right to Great Walsingham.

CERAMICS
Ryburgh Pottery
(Stephen Parry)
1 May Green, Little Ryburgh NR21 0LP
☎ *0328 78543*
Studio pottery; wood-fired stoneware, domestic and individual pots.
Open daily 11am-6pm.

FINE ART/DRIED FLOWERS
Old Barn Studio
(Frank and Jane Jarvis)
Kettlestone NR21 0JB
☎ *0328 878762*
Qualified artists, specialising in natural history subjects, and illustrators. Also unusual dried flower arrangements. C/W/E
Open daily 2-5pm (closed 6 Jan-6 Mar).

Great Dunham

CERAMICS
Lyn Sandford Pottery
Hill Farm Workshops, Castle Acre Road
Great Dunham PE32 2LP
☎ *0760 755345*
Finely glazed hand-thrown stoneware decorated with Chinese brushwork. C/W/E
Mon-Fri 10am-4pm, Sat-Sun by arrangement.
Directions: from Swaffham to Fakenham on A1065, over A47 crossroads, pass George & Dragon PH on right, turn right to workshops.

Great Yarmouth

CARRIAGE BUILDING
Gt Yarmouth Carriage Co Ltd
(Alan Godbold)
Croft Farm Horse Driving Centre,
Thrigey Road, Filby NR29 3DP
☎ *0493 368275*
All types of horse-drawn carriages
built (all metal and wood). Steel
wheels made to order. C/R&R/E
Mon-Sat 8.30am-5.30pm.

STRAWCRAFT
Poppyland
(Christine and John Webb)
Delph Farm, Horsey NR29 4EQ
☎ *0493 393393*
Corn dolly makers. Also craft shop
and art gallery. Cream teas. C/W/E
Open daily Spring Bank Holiday-
Oct10am-6pm. Winter: Sun only.

CERAMICS
The Willows Pottery
(Beverley Pacey)
Court Road, Rollesby NR29 5ET
☎ *0493 740420*
Pottery making domestic and
functional earthenware. Regional
craft guild member. C/W.
Open daily 10am-5pm. Please
telephone first.

Harleston

CERAMICS
Millhouse Pottery
(Alan and Ann Frewin)
1 Station Road, Harleston IP20 9ES
☎ *0379 852556*
Established pottery producing
domestic slipware and garden pots.
Decorated pie dishes, mugs, jugs
etc, large bowls and plates. Also
large range of terracotta tin-glazed

and slipware garden pots including
frost-resistant bird baths and
fountains. C/W.
Open daily 10am-5.30pm.

Holt

FURNITURE MAKING
P H Roberts & Co
The Tithe Barn, Leatheringsett Hill
Holt NR25 6RY
☎ *0263 713803*
Furniture makers, turners and
antique furniture restorers offering
a complete service from advice to
delivery. Both modern and fine
reproduction furniture in oak,
mahogany and walnut.
C/R&R/E. Credit cards.
Mon-Fri 9am-5pm, lunchtime closing.
Sat 9am-1pm.

ANTIQUE FURNITURE
RESTORATION
Michael Dolling
Church Farm Barns
Glandford NR25 7JP
☎ *0263 741115*
Structural repairs to furniture and
veneering, marquetry, polishing,
ebonising etc. R&R
Mon-Sat 9am-5pm, lunchtime closing.

CERAMICS/JEWELLERY
Made in Cley
High Street
Cley-next-the-Sea NR25 7RF
☎ *0263 740134*
Five self-employed potters and one
jeweller; hand-thrown stone-ware
pottery and contemporary jewellery.
C/W/E. Credit cards.
Showroom: Mon-Sat 10am-6pm
(closed Wed Oct-Jun), Sundays 11am-
5pm. Visitors to workshop by
appointment.

GLASSBLOWING
Langham Glass
(Paul Miller and Diana Mitchell)
The Long Barn, Langham
Holt NR25 7DG
☎ *0328 830511*
Glassmaking factory in converted 18th century barn with viewing gallery. Crystal glass blowing and hand-sculpturing. Also engraving service. Work collected and exported world-wide. Shop, children's play area and tea room. C/R&R/W/E. Credit cards.
Mon-Fri 10am-5pm, Sun & Bank Holidays May-Sept. Charges: adults £2, children £1 & OAPs.

King's Lynn

FAN-STICKS & BEADS
Flint Studio
(John and Pippa Brooker)
East Rudham PE31 8RB
☎ *0485 528303*
Individually designed and traditionally made fan-sticks in English holly. Also lace bobbins, pincushions and over 1000 different beads. C/R&R
Flexible hours; visitors welcome by appointment.

CERAMICS
Friars' Pottery
(David Charles Moore)
17A Tuesday Market Place
King's Lynn PE30 1JN
☎ *0553 771977 (evenings)*
Handmade terracotta and stoneware pottery. Garden and wall pots. Hand-pierced work a speciality. Very small workshop; work best seen at local 'Tuesday Market'. C/W
Mon/Thur/Fri 11am-5.30pm, Wednesday 2pm-5.30pm.

FORGEWORK
Cranwell & Son
(Rodney Cranwell)
The Forge, Stow Bridge PE34 3NL
☎ *0366 382600*
Highly trained forgeworkers providing commissioned work for architects, builders, companies and private customers in all kinds of metals. Past commissions for English Heritage, National Trust, DoE, churches etc.
C/R&R/W/E. Credit cards.
Mon-Fri 7.30am-7pm, lunchtime closing. Sat-Sun by appointment.

FORGEWORK
Gemini Forge
(Mr and Mrs A F Keeble)
The Old Forge, Little Dunham PE32 2DP
☎ *0760 721645*
Decorative wrought-iron work. Livery member of Worshipful Company of Blacksmiths; Diploma of Merit; several awards.
C/R&R/W/E
Open daily 9am-4pm, lunchtime closing.

FORGEWORK
Tony Hodgson & Partners
The Forge, 2 Wesley Road
Terrington St Clement PE34 4NG
☎ *0553 828637*
All types of decorative ironwork, fire and kitchen canopies etc. Winner of Worshipful Company of Blacksmiths Bronze and Silver medals. C/R&R/E
Mon-Fri 8am-5pm.
Directions: from King's Lynn on A17 to Sutton Bridge, after 6 miles turn right signed Terrington St Clement. Go to end of road, turn left, after 200yd turn right, after 100yd turn left into Wesley Road; forge on right.

SOFT TOYS
Mrs Ann Guilbert
3 Jubilee Drive, Dersingham PE31 6YA
☎ *0485 541507*
Wide range of soft toys; traditional
teddy bears a speciality. Member of
rural craftsmen's guild. C/R&R/E
Visit at any time by appointment.

CLOCKS & BAROMETERS
A J Guilbert
3 Jubilee Drive, Dersingham PE31 6YA
☎ *0485 541507*
Chartered engineer making clocks
and barometers. Member of rural
craftsmen's guild. C/R&R/E
*Open daily 9am-6pm, please tel-
ephone for appointment.*

HERALDRY
Dudley Bateman — Heraldic Artist
Revonah, Bourne Close
South Wootton PE30 3LZ
☎ *0553 672381*
Handmade and hand-painted
heraldic wall shields and paintings
(no copper pressing).
C/R&R/E. Credit cards.
Open daily 8.30am-7.30pm.

North Walsham

CERAMICS
Cat Pottery
(Ken Allen)
1 Grammar School Road
North Walsham NR28 9JH
☎ *0692 402962*
Up to life-size life-like pottery cats
and dogs made in traditional
workshop. Also classical heads and
columns. Collection of 'railwayana'
and other curiosities can also be
seen. C/W/E
Mon-Fri 9am-6pm, Sat 11am-1pm.

GLASS ENGRAVING
Michael Virden Engraved Glass
Folgate Road
Laundry Loke Industrial Estate
North Walsham NR28 0AJ
☎ *0692 404417*
Quality glass engraved with a wide
variety of designs; drinking glasses,
bells, plates, ashtrays etc. Personal-
ised engraving. C/W/E. Credit cards.
Mon-Fri 9am-5pm.

CERAMICS
Belaugh Pottery
(George Simmons and Bridget Graver)
Church Road, Felmingham NR28 0LQ
☎ *0692 403967*
Husband and wife partnership
producing hand-thrown stoneware,
both domestic ware and individual
pieces for trade and private
customers. C/W/E
Mon-Fri 9am-4pm, Sat 9am-12noon.

Norwich

CERAMICS
Lenham Pottery
(Mrs A G Funnell)
215 Wroxham Road
Norwich NR7 8AQ
☎ *0603 419065*
Dolls' house china made to one-
twelfth scale. C/W/E
Mon-Fri 9am-12noon.
Directions: on A1151 to Wroxham.

CABINET MAKING
Simon Simpson Cabinet Maker
Cotenham Barns
Panxworth NR13 6JG
☎ *060 549 270*
Country workshop concentrating
on high-quality traditionally styled
furniture in a variety of hardwoods.
Anything made to order. Specialist

joinery and antique furniture restoration also undertaken. Other crafts on sale. C/R&R/E
Mon-Fri 9am-5pm, Sat 9.30am-12.30pm.
Directions: 7 miles east of Norwich on B1140.

SILK SCREEN PRINTING
Fidgen Design
(Ken and Karen Fidgen)
The Old Surgery, Aldborough NR11 7NR
☎ *0263 768007*
Husband and wife producing prints for framing and greeting cards from own original designs. C/E
Mon-Fri 10am-6pm, Sat-Sun variable.

CERAMICS
Sutton Windmill Pottery
(Malcolm Flatman)
Church Road, Sutton NR12 9SG
☎ *0692 580595*
Small workshop producing thrown, reduction-fired stoneware in a variety of glazes; particularly tableware, decorative lamps and individual pieces. Visitors welcome to view and buy in workshop. C
Mon-Fri 9am-6pm, Sat-Sun advisable to telephone first.
Directions: off A149 Cromer-Great Yarmouth road, just south of Stalham, Norfolk Broads.

FORGEWORK
Jonathan Skipper — Designer and Blacksmith
(Jonathan Skipper)
Surlingham Forge, Cross Lane Surlingham NR14 7DE
☎ *05088 8152*
Contemporary forged ironwork and restoration work. Sculpture, gates, furniture, console tables, lighting, fireplace furniture, forged latches etc. Work includes Art Nouveau

ironwork for Sotherbys, London; restoration of Georgian ironwork in Norwich; contemporary gates for Swiss bank. C/R&R/W/E
Mon-Fri 8.30am-6pm, Sat 9am-5pm.

SCIENTIFIC & ARTISTIC GLASSWARE
Emivack Glassware
(Kevin Hughes)
Unit 4, Hall Farm Workshops Morningthorpe NR15 2LJ
☎ *0508 498209*
Manufacture and repair of glassware; historic ships (Viking, long boats, *Santa Maria* etc), hot air balloons, oil rigs, animals, sporting models. C/R&R/W/E. Credit cards.
Mon-Fri 8.30am-5.45pm, Sat-Sun 10am-4pm, lunchtime closing.
Directions: 10 miles south of Norwich. From A140, take B1135 to Hempnall. At rear of Foundry Garden Centre.

FURNITURE MAKING & RESTORATION
S P Long
Rowencroft, Kenninghall Road Banham NR16 2HE
☎ *095 387 8488*
Specialissts in antique furniture restoration. Furniture making and general woodwork. C/R&R
Mon-Fri 9am-5pm, lunchtime closing. Sat by appointment.

CERAMICS
The Particular Pottery
(David and Michelle Walters, Peter and Gillian Anderson)
Church Street
Kenninghall NR16 2EN
☎ *095 387 8476*
Cottage industry in old chapel building. Hand-thrown reduction porcelain and stoneware; also some hand-built pots and ceramic

sculptures. Workshop and showroom. C/W/E. Credit cards.
Open daily 10am-4.30pm.

THATCHING
Master Thatchers (Broadland)
(Alan Wotherspoon)
Montpellier House, 20 Clarendon Road
Norwich NR2 2PW
☎ 0603 618943
Thatching contractors for a wide variety of properties, both commercial and domestic.

SADDLERY
C P Hunt — Saddlers
Malthouse Farm, Scottow NR10 5DB
☎ 0692 69687
Saddlery and harness making; manufacture and repair of saddlery and all leather goods. C/R&R/W/E
Mon-Fri 9am-5pm, Sat-Sun by appointment.

FORGEWORK
Bill Cordaroy
Old Farm Forge, East Ruston NR12 9JE
☎ 0692 650724
Wrought iron; sculptural, architectural, domestic and industrial forgings. Commissions for restoration of period houses and ecclesiastic work including gates in St Johns College, Cambridge. C/R&R/E
Open from 8am daily, advisable to telephone first.

FORGEWORK
Capricorn Architectural Ironwork — Ironage
(David Townsend and Wendy Alford)
Haddiscoe Forge, Church Lane
Haddiscoe NR14 6PB
☎ 050 277 519
Recently restored forge opened June 1992. Second workshop and

showroom in London. Designers and consultants; architectural ironwork, restorers of period ironwork, etc.
C/R&R/W/E. Credit cards.
Mon-Fri 8am-6pm, Sat 9am-1pm.

KNITWEAR
Fun Nits
(Janet Daynes)
10 Mousehold Lane, Norwich NR7 8HF
☎ 0603 419048
Exclusive knitwear decorated with appliqué flowers and beaded insects, butterflies, dragonflies, ladybirds etc. C/W. Credit cards.
Visitors welcome by appointment only.

FURNITURE MAKING
Charles Matts Furniture
(Charles and Marianne Matts)
Manor Farm, School Road
Thurgarton NR11 7PG
☎ 0263 768060/713088
Furniture making workshop employing ten craftsmen making individual and traditional furniture from local wood. Difficult access to workshops, therefore prospective clients should visit by appointment. Work on display at Alby Crafts (see under Craft Centres) and at 11 Fish Hill, Holt. C/E.
Mon-Sat 10am-5pm. Visit by appointment.

SPINNING & WEAVING/
FELTWORK & BRAIDS
Leanda
(Daphne Crisp)
39 Borrowdale Drive
Norwich NR1 4LY
☎ 0603 34707
Manufacturing and tuition workshops; also supplies and repairs of equipment. Mail order service.

Hands-on demonstrations by arrangement. C/R&R/W.
Open at any time, by arrangement.
Directions: off main A47 to Great Yarmouth, Norwich Inner Link Road.

CERAMICS/CHILDREN'S CLOTHES
O & B Licencing
(Paul and Penny Jackson)
Corner House, Aldborough Road
Calthorpe NR11 7QP
☎ 0263 768346
Porcelain fantasy sculptures. Also hand-painted children's clothes. Oscar and Bertie teddy bear range (cut-outs, tins, cards etc). London exhibitions and press coverage. C/E
Winter months: Wed/Fri/Sat 10am-5pm, also Mon in summer. (Closed Jan-Feb).
Directions: south of Cromer-Norwich A140, road to Erpingham.

TEXTILES/SCREEN PRINTING
Raindrop
(Helen Howes)
9A St Mary's Works, St Mary's Plain
Duke Street, Norwich NR3 3AF
☎ 0603 767653
Artist of multiple skills working from own designs; primarily textiles (particularly kites), knitwear (hats, waistcoats), jewellery, woodwork etc. Noted for bright colours and eccentric objects. C/W/E
Mon-Fri 10am-6pm.

INTERIOR DESIGN
B Design
(Brenda Korn)
Unit 8, The Raveningham Centre,
Castell Farm, Beccles Road
Raveningham NR14 6NU
☎ 050 846 721
Interior design service from simple re-arrangement of furniture to re-creation of a room, house, office or hotel. Styles from Baroque to Bauhaus. Special design of wallpaper and fabrics; garden design. C/W/E
Mon-Fri 8.30am-6pm, Sat-Sun 10am-6pm.
Directions: see under Craft Centres.

FURNITURE MAKING & RESTORATION
David Bartram Furniture
Unit 8, The Raveningham Centre,
Castell Farm, Beccles Road
Raveningham NR14 6NU
☎ 050 846 721
Fine furniture and restoration work by highly qualified member of BAFRA, listed on Museums & Galleries Conservation Register. C/R&R/W/E. Credit cards.
Mon-Fri 8.30am-6pm, Sat-Sun 10am-6pm.
Directions: see under Craft Centres.

PIANO RESTORATION
Piano Sales & Restoration
(Andrew & Jill Giller)
Unit 10, The Raveningham Centre,
Castell Farm, Beccles Road
Raveningham NR14 6NU
☎ 050 846 8914
Complete restoration of pianos including action re-building, re-stringing, cabinet work and French polishing. Also sales of restored pianos, piano stools etc. R&R
Mon-Fri 9am-5.30pm, Sat-Sun 10am-2pm.
Directions: see under Craft Centres.

CERAMICS
Ken & Ivy Hooker
Ceramic Workshop, Alby Crafts,
Cromer Road
Erpingham NR11 7QE
☎ 0263 761702
Hand-built porcelain and stoneware; small sculptured ceramics.

C/R&R/W/E
Tues-Sun 10am-5pm (closed Dec-Mar).
Directions: see under Craft Centres.

KNITWEAR
2 Jays
(Jim and Betty Jay)
Unit 12, Alby Crafts, Cromer Road
Erpingham NR11 7QE
☎ 0263 761702
Shetland wool knitwear designed,
hand-framed and finished. Members
of regional craft guild. C/W/E
*Tues-Sun 10am-5pm (closed mid Dec-
mid Mar).*
Directions: see under Craft Centres.

LACE MAKING
Alby Lace Museum & Study Centre
(Mrs Lesley Thomas)
Alby Crafts, Cromer Road
Erpingham NR11 7QE
☎ 0263 768002
Bobbin and needle lace made.
Primarily a museum; lace-making
supplies. Demonstrations by
arrangement. Courses and overseas
teaching tours. C/R&R/E. Credit cards.
*Tues-Fri & Sun 10am-5pm, lunchtime
closing. (Closed Xmas-mid Mar).*
Entry charge: adults 30p, children 15p.
Directions: see under Craft Centres.

MUSICAL INSTRUMENTS
David Van Edwards
102 Earlham Road
Norwich NR2 3HB
☎ 0603 629899
Lutes, archlutes, theorboes and
chitarrones made to authentic
medieval, renaissance and baroque
designs. Renaissance and baroque
bows. All instruments made to
players' requirements. C/R&R/E.
Visitors welcome by appointment.

FORGEWORK
The Forge
(Brian Reynolds)
Mundesley Road, Paston, Mundesley
☎ 0263 721871
General forgework, fabrication and
ornamental ironwork. C/R&R/W/E
Mon-Fri 8am-6pm, lunchtime closing.
Sat-Sun 9am-1pm.

Sheringham

CERAMICS
Sheringham Pottery
(H L and I Farncombe)
30 Church Street
Sheringham NR26 8QS
☎ 0263 823552
Award-winning pottery making a
wide range of table and functional
ware; coffee and tea sets, lamps,
bowls, plates etc. Carved work cut
free-hand with tree, leaf and flower
designs. C/W/E. Credit cards
Open daily 9am-5pm.

UPHOLSTERY
Tariq Sharif Uphostery
Church Farm Barn
Glandford, Holt NR25 7JP
☎ 0263 741115
Re-upholstery service and antique
furniture restoration. C/R&R/W/E
Mon-Sat 9am-5pm.

Swaffham

WOODTURNING
Breckland Woodcrafts
(Brian Elliott)
4 Castleacre Rd
Swaffham PE37 7HS
☎ 0760 24282
Specialists in fine woodturned work
using English and exotic timbers.
Hand-made wall and mantle clocks,
small items of furniture. C/R&R/W

Mon-Fri 9.30am-5.0pm
Directions: rear of Paynes Garage in Castleacre Road.

FURNITURE
CY Woodrow
The Goodshed
Little Dunham PE32 2DJ
☎ *0760 22348*
Fine furniture made to order in English hardwoods. All types of work, from individual designs to replicas of period pieces. C
Open daily 9am-5pm, advisable to telephone first.
Directions: 6 miles north east of Swaffham.

CERAMICS
The Pottery
(John & Kate Turner)
Narborough
☎ *0760 337208*
Domestic pottery, original sculpture. Also designer woollen clothes and paintings. Work exhibited widely abroad. C/W/E
Open daily 9am-9pm.

Thetford

ANTIQUE FURNITURE RESTORATION
M J Antiques
(Michael Chapman)
22 Gallants Lane
East Harling NR16 2NQ
☎ *0953 717784*
Family business restoring antiques and manufacturing furniture. C/R&R/W/E. Credit cards
Mon-Fri 8am-6pm, Sat 8am-4pm.

SILVERSMITHING
Philip Isern Jewellery
(Philip and Carol Isern)
Crosshill House, 1 Old Feltwell Road
Methwold IP26 4PW
☎ *0366 728573*
Figurative silver jewellery incorporating stones in intricate designs of birds, insects, animals, flowers etc. Mail order service. Credit cards.
Mon-Fri 9.30am-5.30pm, lunchtime closing.
Directions: off B1106 between Brandon and Stoke Ferry.

Wells-next-the-Sea

CERAMICS
Burnham Pottery
(Thom and Jan Borthwick)
Old Railway Station, 2/4 Maryland
Wells-next-the-Sea NR231LX
☎ *0328 710847*
Pottery making cast ceramics and hand-thrown domestic pottery, specialising in cats and pigs. 400 retail outlets supplied nationwide. Mail order service. C/W/E
Mon-Fri 9am-5pm, Sat 10am-5pm (lunchtime closing).

Wymondham

PICTURE FRAMING
Locality Arts Ltd
(Gwyn Jones)
Coign House, Hackford Road
Wicklewood NR18 9PZ
☎ *0953 602186*
Picture framing and extensive melamine laminating service. Also manufacturers of picture framing equipment. Member of Fine Art Trade Guild. C/R&R/W/E
Mon-Fri 9am-6pm, Sat-Sun 9am-4pm, lunchtime closing.

NORTHAMPTONSHIRE CRAFT WORKSHOP CENTRES

Woodnewton Craft Studios
(Rob Bibby)
The Old Chapel
43 Main Street
Woodnewton, Oundle PE8 5EB
☎ *0780 470866*
Four craft studios housed in a converted Methodist Chapel; pottery, woodcarving, stencilling, leatherwork, puppet-making. Group visits by arrangement; talks and demonstrations given for a small charge.
Mon-Fri 9am-5pm, Sat-Sun by appointment.

The Old Dairy Farm Centre
(Mrs Helen Brodie)
Upper Stowe, NrWeedon NN7 4SH
☎ *0327 40525*
An award-winning conversion of 19th century farm buildings. Craft workshops include woodturning, forgework, picture framing, woodworking, Batik and silk screen printing, stained glass and pottery. Shops, demonstration/conference room and restaurant.
Open daily 10am-5.30pm, closed Xmas-1st Mon after New Year.
Directions: Upper Stowe lies west of A5 Weedon-Towcester road.

NORTHAMPTONSHIRE CRAFT WORKSHOPS

Kettering

THATCHING
W A Shiells
41 Benefield RD, Brigstock NN14 3ES
☎ *0536 373292*
Thatching contractor with 47 years' experience; house thatching in long straw, combed wheat reed and water reed. Spars, liggers and wooden sways made for thatching.

FORGEWORK
George James & Sons (Black-smiths) Ltd
(David and Tim James)
22 Cransley Hill
Broughton NN14 1NB
☎ *0536 790295*
Fifth generation of family business employing several blacksmiths. Commissions for all types of ironwork and restoration. Royal Show 1st and Championship Prizewinner, 1991. C/R&R/W/E
Mon-Fri 8.30am-5.30pm, Sat by appointment.

Northampton

MINIATURE FIGURINES
Phoenix Model Developments Ltd
(B L and S G Marlow)
The Square, Earls Barton NN6 0NA
☎ *0604 810612*
English pewter or 'white metal' miniature figurines and other scale models produced in kit form ranging from the Georgian period to fantasy figures. Free quotations for completed models, special trophies and commissioned work. C/R&R/W/E. Credit cards.
Mon-Fri 9am-5pm by appointment.
Directions: 6 miles east of Northampton.

CERAMICS
Brixworth Pottery
(Evelyn Campbell and Dorothy Watkins)
Beech Hill, Church Street
Brixworth NN6 9BZ
☎ 0604 880758
Workshop selling hand-painted pottery direct to customers, both off the shelf and to order; individually designed house name plaques, anniversary plates, portraits of houses/pets from photos. All types of ceramic restoration. C/R&R
Mon-Sat 10am-5pm, lunchtime closing. (Closed Xmas-6 Jan).

CERAMICS
Design Ceramics
(Verity Rosas)
Unit 1, Dovecote Pottery
Upper Harlestone NN7 4EL
☎ 0604 402697
Potter producing quality stoneware table ware; mugs, jugs, casseroles etc. Decorated white earthenware. Work widely exhibited. C/E.
Visitors welcome by appointment only.

STAINED GLASS
Anthony MacRae
1 South Street, Lower Weedon NN7 4QP
☎ 0327 40313
Stained glass artist and designer working from studio workshop. C/R&R/W/E
Variable working hours; advisable to telephone first.

SILK/BATIK WORK
Heather Trotter Silk Batik
The Old Dairy Farm Centre
Upper Stowe, Nr Weedon NN7 4SH
☎ 0327 40525
Silk Batik scarves, cotton Batik pictures/wallwork, hats and textiles. Tuition given; member of Batik and regional Designer Craftsmen Guilds. C/W/E. Credit cards.
Tues-Fri 1.30-5pm, Sat-Sun 10.30am-5pm.
Directions: see under Craft Centres.

FORGEWORK
Derek Hawkins Ornamental Ironwork
The Old Dairy Farm Centre
Upper Stowe, Nr Weedon NN7 4SH
☎ 0327 40525
Ornamental wrought ironwork designed and manufactured to order. C/R&R/W/E
Open daily 10am-5pm, advisable to telephone first.
Directions: see under Craft Centres.

CABINET MAKING
G Linnell — Woodworker
The Old Dairy Farm Centre
Upper Stowe, Nr Weedon NN7 4SH
☎ 0327 40525
Joinery, cabinet making and furniture making. C/R&R
Tues-Fri 7.30am-5.30pm, Sat-Sun 10am-5.30pm, lunchtime closing.
Directions: see under Craft Centres.

PICTURE FRAMING
Village Workshop
(Michael and Valerie Benton)
The Old Dairy Farm Centre
Upper Stowe, Nr Weedon NN7 4SH
☎ 0604 830628
Picture framing service and restoration. Also gallery exhibiting a selection of prints. C/R&R
Wed-Sun & Bank Holidays 11.30am-5.30pm.
Directions: see under Craft Centres.

WOODTURNING

Guy Ravine - Woodturner
The Old Dairy Farm Centre
Upper Stowe, Nr Weedon NN7 4SH
☎ 0327 40416
Fine woodturning and treen. Mostly small ornamental work but some larger exhibition work. Member of regional Designer Craftsmen Guild. Work widely exhibited and exported worldwide. C/W/E
Mon-Sat 10am-5pm.
Directions: see under Craft Centres.

HAND-PAINTED AND PRINTED TEXTILES

Lorna Hardingham
The Old Dairy Farm Centre
Upper Stowe, Nr Weedon NN7 4SH
☎ 0327 40525
Hand-painted and screen printed silk; one-off silk scarves and material. C/W/E. Credit cards.
Mon-Sun 2-5pm.
Directions: see under Craft Centres.

THATCHING

E Dunkley & Son
Highview, Little Street
Yardley Hastings NN7 1EZ
☎ 0604 696293
Long established family business. Master Thatcher; consultant specialising in all types of thatching. Member of national and regional thatching associations.

Oundle

CERAMICS

Glazed Expressions
(Rob Bibby)
Woodnewton Craft Studios
The Old Chapel, 43 Main Street
Woodnewton PE8 5EB
☎ 0780 470866
Hand-thrown pottery, glazed and hand-painted; pottery house name plaques. Work sold through shops and at craft fairs. C/W/E. Credit cards.
Mon-Fri 9am-5pm, Sat-Sun by appointment.

WOOD CARVING

Ken Barnard
Woodnewton Craft Studios
The Old Chapel, 43 Main Street
Woodnewton PE8 5EB
☎ 0780 470866
Wooden sculptured figures for fair organ decoration. 20 years' teaching at college of furniture. C/R&R/E.
Mon-Fri 9am-5pm, Sat-Sun by appointment.

CARVED SIGNS AND HERALDRY

Glyn Mould
Woodnewton Craft Studios
The Old Chapel, 43 Main Street
Woodnewton PE8 5EB
☎ 0780 470866
Carved wooden coats of arms, village signs etc. Woodcarving courses held at workshops. Award winner; recent commission for House of Commons. C/R&R/E
Mon-Fri 9am-5pm, Sat-Sun by appointment.

FORGEWORK

Fotheringhay Forge
(Barry Keightley)
Fotherinhgay PE8 5HZ
☎ 08326 323
Family business producing customers' requirements in an 18th century smithy. Showroom. Stock includes a cast iron products. C/R&R/W/E.
Tues-Fri 9am-5pm, Sat 9am-4pm, lunchtime closing.
Directions: Fotheringay lies north east of A605 Oundle-Peterborough road.

Rushden

LEATHERWORK
Harmatan Leather Ltd
(M Lamb)
Westfield Avenue
Higham Ferrers NN9 8AX
☎ *0933 312471*
Specialist manufacture of northern
Nigerian goatskin and calf leather
for bookbinding, used worldwide.
W/E
Mon-Fri 8.30am-5.30pm, Sat-Sun by
appointment.

Yelvertoft

ALABASTER & MARBLE WORK
Nigel Owen
(Nigel and Michael Owen)
42 High Street, Yelvertoft NN6 7LQ
☎ *0788 822281*
Craftsmen making alabaster and
marble table lamps, clocks, barometers and other small items in
English alabaster; marble hearths,
chess tables, shelves, bases etc. to
order. C/R&R/E.
Mon-Fri 9am-5.30pm, Sat 9am-5pm.
Directions: 2 miles from M1 junc 18.

Alnwick

FURNITURE MAKING &
RESTORATION
John Smith of Alnwick Ltd
West Cawledge Park, Alnwick NE66 2HJ
☎ 0665 604363
Fine furniture designed and made
by a small team of craftsmen.
Antique furniture restoration
tuition. C/R&R/W/E. Credit cards.
Daily 10am-5pm
Directions: 1m S of Alnwick on A1
(signposted).

STAINED GLASS
Iona Art Glass
(Chris and Sarah Chesney)
The Woodlands, Warkworth NE65 0JY
☎ 0665 711533
Stained glass artists working in
English antique glass; restorers of
stained glass. New works in local
churches. C/R&R/W/E. Credit cards
*Mon-Fri 9am-5pm, Sat 9am-4pm,
lunchtime closing.*

CERAMICS
Harehope Forge Pottery
(Richard Charters)
Harehope, Eglingham NE66 2DW
☎ 06687 347
Diverse range of traditional wheel-
thrown, wood-fired terracotta pots.
Impressed and roulleted relief
designs, natural 'smoke-flashed'
and 'blushing' colourings. C/W/E
Mon-Sat 10am-5pm.

CERAMICS
Breamish Valley Pottery
(Alastair Hardie)
Branton, Powburn, Alnwick NE66 4LW
☎ 066 578 263

Hand-thrown kitchen and domestic
stoneware, fired with wood. Work
can be viewed from showroom.
C. Credit cards.
*April-Oct Tue-Sat 9.30am-5.30pm,
lunchtime closing. Opening times
variable during winter months.
Advisable to telephone first.*

CERAMICS
Narrowgate Pottery
(Andy Chilcott)
*22 (Back) Narrowgate
Alnwick NE66 1HJ*
☎ 0665 604744
Domestic ware and ornamental
pieces in earthenware; also some of
stoneware. C/W

ARROW-SASH WEAVING
Dilys Anderson
*East Winds, Newton Barns
Alnwick NE66 3DY*
☎ 0665 576675
Demonstrations and tuition of
Arrow-sash weaving, an old
French-Canadian craft with an
interesting history.
Visitors welcome by appointment.

CHILDRENS' KNITWEAR
Little Knits
*(Mrs Kay Moody)
Kay's Kabin, Christon Bank
Alnwick NE66 3ES*
☎ 0665 576767
Chidrens' picture jumpers at
affordable prices. C

KNITWEAR
Socks For All
*(Ada Straughan)
20 North Side, Shilbottle
Alnwick NE66 2YE*

☎ 0665 575412
Hand-framed socks made to order; wool and wool/nylon mix. C

PUPPETS & TOYS
Pat Thompson Designs
No 1 The Shambles
Alnwick NE66 1TN
Hessian bags, glove puppets, birthday cards and finger puppets. C

ARTIST
Poppies
(Jacqui Chilcott)
22 (Back) Narrowgate
Alnwick NE66 1HJ
☎ 0665 604744
Watercolours of local wildlife and landscapes. C

KNITWEAR
Elizabeth Ann Knitwear
(Ann Widdows)
Warden's Cottage
Chillingham NE66 5NP
☎ 06685 250
Jackets and jumpers made from pure new wool; original designs incorporating Celtic art and fretwork patterns from the Lindisfarne Gospels, also bird plumage designs. C/W. Credit cards.
Apr-Oct open daily (except Tues) 10am-5pm, by appointment.

Ashington

CERAMICS
Butterflies
(Dawn Hopper)
1 Cheviot View
Ashington NE63 9ER
☎ 0670 812716
Ceramic masks, clocks, mirrors, brooches, butterflies etc. C/W

GLASS ENGRAVING
Corrib Crafts
(Colin Prow)
53 Beverley Drive, Wansbeck Estate
Choppington NE62 5YA
☎ 0670 815534
Diamond engraving on glass. C/W

FORGEWORK & FARRIERY
Cuthbert Wilkinson — Blacksmith
Stables Block
Woodhorn Colliery Museum
Woodhorn Road, Ashington
☎ 0670 856968
Ornamental wrought ironwork and farriery. Agricultural repairs. C/R&R

ANTIQUE FURNITURE RESTORATION
Woodhorn Furniture
(Ray Blake)
Woodhorn Colliery Museum
Queen Elizabeth Country Park
Ashington NE63 9YF
☎ 0670 856592
Restoration of antique furniture and furniture made in all woods. C/R&R

PINE STRIPPING
Ashington Stripping Centre
(Matthew Bambrough)
Old Co-Op Buildings
Back Lintonville Terrace, Ashington
☎ 0670 814816
Antique restoration and stripped pine furniture. C/R&R

GARDEN ORNAMENTS
British Gnome Stores
(Michael Peary and Colin Neary)
Unit 9, Senet Workshops, Green Lane
Ashington NE63 0EF
☎ 0670 521646
Manufacture of garden ornaments.

Sean Fitzgerald making furniture at Cranleigh, Surrey

Semi-elliptical console table in maple, pear and rosewood by Nicholas Dyson, Wantage, Oxfordshire

Re-rushing an antique ladder-back chair (photo: Rural Development Commission)

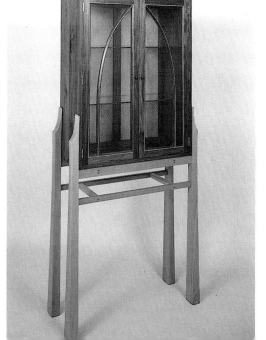

Contemporary display cabinet by Mark Ripley, South Moreton, Didcot, Oxfordshire

Hand engraved glass
by Tony Gilliam,
Alresford, Hampshire

Coloured glass bowl
decorated by
sandblasting, Ruth
Dresman, Tisbury,
Salisbury, Wiltshire

Blacksmith at his forge, producing a wide range of wrought iron objects
(photo: Rural Development Commission)

'Love Goblets' turned from British grown woods by Maurice Mullins, Hesket Newmarket, Wigton, Cumbria

Turned wooden vessel by Melvyn Firmager, Stourton Cross, Wedmore, Somerset, in *Eucalyptus Gunni* rescued from a storm damaged tree

Silverwork by Patricia Hamilton, Hardwick, Witney, Oxfordshire

Silver boxes decorated with niobium flowers by Pauline Gainsbury, Chartham, Canterbury, Kent

Handpainted and carved wooden screen by Mark Houlding, Emsworth, Hampshire

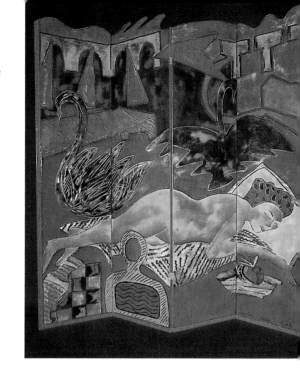

Canework by John Haywood, Beaulieu, Hampshire

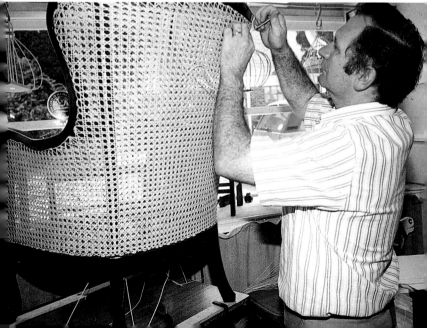

Papier maché pots by Miriam Troth, Southbourne, Bournemouth, Dorset

Carved sycamore relief panel by Jeremy Turner, Gawcott, Buckingham

Potter at work making a large earthenware container
(photo: Rural Development Commission)

Finely painted pottery by Michael and Barbara Hawkins, Stroud, Gloucestershire

Raku globe in white crackle and dry metal glaze, by Richard Capstick, Shifnal, Telford, Shropshire

Terracotta garden ware decorated with a black vitreous slip by Belford Pottery, Belford, Northumberland

Earthenware lidded jar, decorated with slips and underglazed colours, by Jan Bunyan, Butlers Marston, Warwick

Hand-carved carousel-style rocking horse in English limewood by Stan Greer, Whittington, Worcester

Wooden Noah's Ark set by David Plagerson, Totnes, Devon

Handwoven rug by Jacqueline James, York

Wheelwright finishing a traditional cart wheel
(photo: Rural Development Commission)

Thatching, a traditional rural craft which is enjoying a revival
(photo: Rural Development Commission)

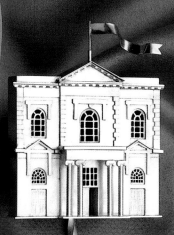

Crafts Council
4a Pentonville Road
Islington
London N1 9BY
(Angel Tube)
Telephone 071 278 7700

Open
Tuesdays to Saturdays 11 - 6
Sundays 2 - 6
Closed Mondays

Crafts Council
national centre for the crafts

Free entry

The Crafts Council Gallery

Gallery Shop

Information centre

Picture library of 33,000 slides

Reference library

Cafe

The Worshipful Company of Blacksmiths
London

BY·HAMMER·AND·HAND·ALL·ARTS·DO·STAND

For further information contact the Clerk:
Raymond C. Jorden, F.Inst.D., F.Inst.AM.
27 Cheyne Walk
Grange Park, London N21 1DB
Telephone: 081-364 1522

LETTERING
J Ferguson Lettering
15 Prospect Place
Newbigin-by-the-Sea NE64 6DN
☎ *0670 811968*
Beautiful lettering in a variety of
scripts for any occasion. C

LANDSCAPE ARTIST
Robert Ritchie
56 Kenilworth Road
Ashington NE63 8DD
☎ *0670 850460*
Northumberland landscapes in oil
on canvas. C

Belford

KNITWEAR AND TWEED
Hazon Mill Knitwear
(Christine Moffat)
6 High Street
Belford NE70 7NM
☎ *0668 213808*
Hand-framed knitwear in natural
yarns individually designed to
commission, also small range of
ready-to-wear garments including
matching tweed skirts. C. Credit cards.
Mon-Sat (closed Thurs) 9.30am-4.30pm
Directions: In High Street, opposite
Blue Bell Hotel.

CERAMIC FIGURES
Norselands Gallery & Studio
   Two Workshop
(Barrie and Veronica Rawlinson)
The Old School, Warenford NE70 7HY
☎ *0668 213465*
Ceramic figures featuring ladies,
lovers, children, toddlers, fairies
and fungi; glazed stoneware in matt
cream and brown. Other crafts on
sale. C/W/E. Credit cards.
*Open daily 9am-5pm (-9pm in
summer).*

CERAMICS
Belford Pottery
(John and Sareth Atkin)
2 Market Place, Belford NE70 7ND
☎ *0668 213044*
Small pottery making terracotta
gardenware; sgraffito decoration on
black vitreous (glossy) slip. Also
white stoneware domestic pottery.
Gallery next door with crafts, cards,
exhibitions, books, maps etc and tea
room. C/W/E. Credit cards.
Visitors to pottery by appointment.

GOLD AND SILVERSMITHING
Silverfoot Design
(Susanne Spence)
27 West Street, Belford NE70 7QB
☎ *0668 213230*
Unique handmade pieces in silver
and gold using precious and semi-
precious stones and woods. C/W

Berwick-upon-Tweed

FARRIERY & FORGEWORK
Michael J Cross
Primrose Cottage & Smithy
28 Main Street, Lowick TD15 2UA
☎ *0289 88403*
Mainly farriery work; shoes made
in the forge and hot shoeing carried
out. Also blacksmithing work.
Note: the smithy is only open to the
public when the farrier is at work
there — much of his work is out of
the workshop. C/R&R/W
*Mon-Fri 8.30am-5.30pm, Sat-Sun
variable times. Advisable to telephone first.*

CERAMICS
S D Binns Ceramics
Unit 2, The Chandlery
Berwick-upon-Tweed TD15 1HE
☎ *0289 330273*

Pottery producing porcelain models of mainly locally found animals and birds, sculpted by Stephen Binns. Bronze finish animals also brooches and earrings. Gallery with paintings and prints. C/W/E. Credit cards.
Mon-Sat 9.30am-4.30pm.

WOODWORKING & FORGEWORK
Errol Hut Smithy & Woodworkshop
(P A Smith and J Hanson)
Errol Hut Smithy, Letham Hill
Etal, Cornhill-on-Tweed TD12 4TP
☎ 089 082 317
Woodturning; hard wood furniture and turned items including spinning wheels. Also general blacksmithing work. Small souvenirs; spinning demonstrations given. C/R&R/W/E. Credit cards.
Open daily 11am-5pm.
Directions: from B6354 Berwick road, between Heatherslaw Mill and Etal Manor, turn right to Slainsfield; Errol Hut on left.

CERAMICS
Lindisfarne Limited
(Ronald Tait)
Palace Green
Berwick-upon-Tweed TD15 1HR
☎ 0289 305153
Hand-thrown stoneware, particularly goblets, flagons for drink industry etc. Dried flowers and pot pourri. Original Victorian chemist and craft shop. Free sampling of the famous Lindisfarne Mead. C/W/E
Mon-Fri 8.30am-5pm, Sat 9pm-5pm.

CERAMICS
Tower House Pottery
(Peter and Margaret Thomas)
Tower Road, Tweedmouth
Berwick-upon-Tweed TD15 2BD
☎ 0289 307314

Three potters making hand-thrown decorated domestic slipware. C/W/E
Mon-Fri 10am-4pm, lunchtime closing.

BESPOKE RIDING WEAR
Clothes-horse
(Mrs Valerie Allen)
Branton, Cornhill-on-Tweed TD12 4SW
☎ 0890 82219
Bespoke clothing for horse and rider. C

KNITWEAR
Me Knitwear & Crafts
(Mr and Mrs R T P Brown)
Oxford Farm Shop, Oxford Farm
Ancroft TD15 2TA
☎ 0289 87253
Knitwear manufactured on the premises. Also collection of crafts. C

LEATHERWORK
Exley Brothers
(Graham and Michael Exley)
6 Bridge Street
Berwick-upon-Tweed TD15 1AQ
☎ 0289 330292
All types of handmade leatherwork, repairs and alterations. C/R&R

LEATHERWORK
Dave Downie Enterprises
Letham Hill, Nr. Etal
Cornhill-on-Tweed TD12 4TP
☎ 0890 82227
High quality tooled belts and leatherwork, also fancy buckles. C

PICTURE FRAMING
John Mallen Woodcarvings &
  Picture Frames
Westfield Cottage, Ford TD15 2PY
☎ 0890 82559
Small picture frames, woodcarvings, cigarette cards framed etc. C

FURNITURE
S W Taylor & A Green
The Old Power House
Etal Village, Cornhill-on-Tweed
TD12 4TW
☎ *0890 82376/82566*
Handmade furniture in a variety of
styles and timbers. C/W

ARTIST
The Quayside Gallery
(Peter Knox)
5 The Chandlery, Quayside
Berwick-upon-Tweed TD15 1AH
☎ *0289 330165*
Painter of Northumbrian coastal
views and other marine subjects,
past and present. C

ARTIST
Sally Port Gallery
(Derek Jones)
48 Bridge Street
Berwick-upon-Tweed TD15 1AQ
☎ *0289 307749*
Figure and landscape painting in
acrylic and watercolour. C

Blyth

ARTIST
Edwin Blackburn
15 Newlands Road, Blyth NE24 2QJ
☎ *0670 369904*
Landscape/marine artist. Watercolour and pastel paintings, limited
edition prints. C
Visitors welcome by appointment.

DRIED & SILK FLOWERS
Floral Design
(Mrs Joan Campbell)
60-62 Maddison Street, Blyth NE24 1EY
☎ *0670 366679*
Silk and dried flower arrangements. C

Chathill

SPINNING, WEAVING AND
DYEING COURSES
Margaret Blackett Hunter
Pinewood, Chathill NE67 5JP
☎ *0665 89218*
Teaching workshop running courses
(including one-day) in textiles.
Visitors welcome by appointment.
Summer months only.
Directions: 11m N of Alnwick.

Haltwhistle

PINE FURNITURE
Border Pine
(Barrie Irving)
Unit 1, Park Road
Haltwhistle NE4 99AR
☎ *0434 321825*
Small workshop making solid pine
reproduction furniture.
C/W/E. Credit cards.
Mon-Fri 9am-5pm, Sat 9am-12noon.

FORGEWORK
Stan Pike
Waylands Forge, Sook Hill
Haltwhistle NE49 9PS
☎ *0434 344309*
Decorative ironwork including
indoor and outdoor furniture made
and restored. C/R&R/W/E
Mon-Fri 8.30am-4.30pm, Sat 9am-
12noon.

SOFT FURNISHINGS
Fabrications
(Rachel Phillimore)
Matties Green
Kellah, Haltwhistle NE49 0JN
☎ *0434 320706*
Bedspreads and cushions (feather
filled) in a variety of materials. C
Visitors welcome by appointment.

Hexham

STRINGED INSTRUMENTS
Nial Cain Violins
(Nial Cain and David Mann)
The Violin Shop, 31a Hencotes
Hexham NE46 2EQ
☎ 0434 607897
Violins, violas and 'cellos made and
restored. Instruments in any
condition purchased. C/R&R
Mon-Sat 10am-5pm.

LEATHERWORK
Austin Winstanley
19b Priestpopple, Hexham NE46 1PH
☎ 0434 602267
Leatherwork and case-making,
attaché and gun cases, musical
instrument cases. All work handmade in best quality hide and aircraft ply, padded and lined. C/R&R/E
Mon-Fri 9am-6pm, lunchtime closing.
Visit by appointment.

CERAMICS
Bardon Mill Pottery
(Errington Reay & Co Ltd)
Tyneside Pottery Works
Bardon Mill NE47 7HU
☎ 0434 344245
Seven employed in pottery making
large salt-glazed garden and
domestic clay pots. Many commis-
sions in UK and overseas. C/W/E
Mon-Fri 9am-4.45pm, all year. Sat-Sun
10am-4.45pm Easter-Sept only.

CERAMICS
Shire Pottery
(I. Mackay, D. Barrow and F. Wedel)
Unit 3, Station Yard
Corbridge NE45 5AZ
☎ 0434 633503
Hand-thrown and hand-finished
earthenware garden pottery. C/W

CERAMICS
Thornton Pottery
(Pip and William Thornton)
Humshaugh NE46 4AG
☎ 0434 681406
Earthenware pottery — original
designs, colourful, modest prices. C/W

STAINED GLASS
Jenny Hammond
High Greenleycleugh, Ninebanks
Hexham NE47 8DE
☎ 0434 345307
Stained glass windows and panels
designed, made and repaired.
Commissions/repairs,restoration
Mon-Fri 9am-5.30pm, by appointment.

KNITWEAR
Packman Crafts
(Mrs S J Finney)
1 King Street , Bellingham NE48 2BW
☎ 0434 220006
Knitwear made on the premises and
other Northumbrian crafts. C

FURNITURE
Fine Grain
(Derek Dellow)
Unit 8, Matfen Hall Workshops
Matfen NE20 0RH
☎ 0661 886269
High quality furniture of all styles
handmade to order. C
Mon-Fri 8.15am-6.30pm, Sat 9.30am-
6.30pm.

FURNITURE RESTORATION
Milverton Restorations
(Ian Brown)
The Stables, Station Road
Hexham NE45 5DN
☎ 0434 606828
Comprehensive furniture restora-
tion service. C/R&R

HERBAL PRODUCTS
Hexham Herbs
(Kevin and Susie White)
Chesters Walled Garden
Chollerford NE46 4BQ
☎ 0434 681483
Herbal products, pot pourri, herb plants and extensive display gardens. Winners of Large Gold Medal, National Garden Festival, Gateshead 1990 for display herb garden sponsored by Barclays Bank. *Easter-Oct open daily 10am-5pm; shorter hours out of season.*

PRESSED FLOWERS
Summer Days
(Katherine Wilkinson)
Mill House Farm Cottage
Bardon Mill NE47 7JA
☎ 0434 344468
Pressed flowers, mainly collage and miniatures; especially light switch surrounds. C

SCULPTURE
Wildlife Studio
(Eric Heron)
Old Foundry Yard
Bellingham NE48 2DA
☎ 0434 240317
Sculptures of animals and birds.

CHINA DECORATION
Nexus Design
(Phil and Anne Bowyer)
Unit 11, Matfen Hall Workshops
Matfen NE20 0RP
☎ 0661 886412
Limited edition china; plates and mugs decorated with unique Celtic designs based upon Lindisfarne Gospels etc. C/W/E
Mon-Fri 9.30am-4.30pm, advisable to telephone first.

CERAMICS
Sourdust Pottery
(John Scott)
Unit 10, Matfen Hall Workshops
Matfen NE20 0RP
☎ 0661 886412
Handmade pottery, particularly pottery wellingtons, agate ware, casserole dishes etc. C/W

Morpeth

NORTHUMBRIAN SMALLPIPES
D G Burleigh
Rothbury Road
Longframlington NE65 8HU
☎ 0665 570635/0670 760430 *(evenings)*
Maker of traditional Northumbrian smallpipes. Visitors can see pipes being made and hear them being played. Also music for sale. C/R&R
Mon-Fri 8am-4pm, lunchtime closing. Sat-Sun by appointment.

SILVERSMITHING
Chantry Silver
(Alan le Chard)
The Chantry, Chantry Place
Morpeth NE61 1PJ
☎ 0670 58584
Silversmith producing hand-crafted jewellery and silverware in modern designs; some in gold. Gallery includes work of local artists and glass engravers who engraving on presentation items. C/R&R/W
Mon-Sat 9.30am-5pm.
Directions: in town centre at south end of Bridge Street behind The Chantry.

FURNITURE MAKING
Karva Furniture
(Eric and Jean Tollett)
Widdrington Station
Morpeth NE61 5DW
☎ 0670 790325

Furniture making business working in solid wood seasoned in own kiln, using traditional methods with attention to detail. C. Credit cards.
Mon-Fri 9am-6pm, Sat-Sun 12noon-5pm.

GOLDSMITHING
Leif Design
(Anthony John Hope)
Keepers Cottage
Clennel, Nr Alwinton NE65 7BG
☎ *0669 50352*
Workshop making fine modern jewellery in gold, using precious stones and diamonds. Work supplied to jewellery shops throughout UK. C/W/E
Visitors welcome by appointment.

FORGEWORK
J S Lunn & Sons
(John Smith/John and Stephen 'Lunn')
The Forge, Red Row
Morpeth NE61 5AU
☎ *0670 760246*
Forge employing five craftsmen designing and manufacturing ironwork. Winners of awards. C/R&R/W/E
Mon-Fri 8.30am-5.30pm, lunchtime closing 12.30-1.30pm. Sat 8.30am-12.30pm.

STAINED GLASS
Sarah Richardson
2 Bullocks Hall Farm Cottages
West Chevington, Morpeth NE61 5BE
Not on the telephone
Stained and leaded glasswork using traditional techniques to produce contemporary work for private, public and church commissions. Work exhibited internationally and winner of awards. C/R&R/W/E.
Daily 9am-5pm; visitors welcome by appointment only.

CERAMICS
Linda Sher
East Molesden House
Molesden NE61 3QF
☎ *0670 75380*
Hand-thrown domestic and garden stoneware pottery. C/W
Open daily 9am-6pm, advisable to telephone for appointment.

CRAFT JEWELLERY
Chameleon Jewellery
(C. Greenhaugh and F. Swailes)
Bolland House, Pottery Bank
Morpeth NE61 1BH
☎ *0670 516513*
Unique hand-crafted jewellery in wire, bead and papier maché. C/W

WOOD CARVING
Bauern Kunst
8 Fifth Avenue, Stobhill Gate
Morpeth NE61 2HH
☎ *0670 519240*
Artist-craftsman in wood; icon making and restoration, sculpting and wood carving. C
Visitors welcome by appointment.

LACE MAKING
Anne Evens
26 Embleton Terrace, The Lee
Longframlington NE65 8JJ
☎ *0665 570289*
Handmade bobbin lace. C
Weekends only.

FURNITURE
Stephen Down Furniture
Unit 5, Whitehouse Centre
Stannington NE61 6AW
☎ *0670 789727*
Free standing and fitted furniture. Architectural joinery. C/R&R

FURNITURE
Morpeth Handmade Furniture
(Graham Willis)
Unit 3D, Coopies Field
Coopies Lane Industrial Estate
Morpeth NE61 6JT
☎ *0670 511159*
Furniture traditionally made to
individual requirements using solid
woods. Member of Guild of Master
Craftsmen. C/R&R. Credit cards.
Mon-Fri 9am-5pm, Sat 9am-3pm.

FURNITURE RESTORATION
Richard Pattison
Unit 3E, Coopies Field
Coopies Lane Industrial Estate
Morpeth NE61 6JT
☎ *0670 511582*
Traditional furniture restoration
and French polishing; old pine
kitchens a speciality. C/R&R

FURNITURE RESTORATION
Town & Country Antique
   Furniture Restorers
(David Hicks)
Herald Office Yard, Bridge Street
Morpeth NE61 1NT
☎ *0670 511287*
Cabinet makers, antique furniture
restorers, French polishing. C/R&R

STRAWCRAFT
Cornucopia
(Dennis David Parker)
4 Lorbottle Cottages
Lorbottle, Thropton NE65 7JS
☎ *0665 74651*
Corn dolly maker — traditional and
modern designs. C

PORCELAIN PAINTING
Louise Liddell
Meadow View, Red Row, Morpeth

☎ *0670 760373*
Lithographed fine bone china and
hand-painted china.

HORN WORK
Northumbria Horncraft
(Terry Cuthbert)
The Old Telephone Exchange
Manor House Dairy, Whalton Village
☎ *0670 519174*
Horn handle walking sticks, spoons,
shoehorns etc; antiques. C/R&R

Newcastle-upon-Tyne

GLASS ENGRAVING
Margaret Gamlin
32 Dunsgreen, Ponteland NE20 9EH
☎ *0661 23833*
Engraving on glass. Silver jewellery
and enamelled copper. C/W

FIRE SURROUNDS
The Fireplace
(B Fitzgerald and R Elrington)
Berwick Hill, Ponteland NE20 0JZ
☎ *0661 860154*
Makers of fine period fire surrounds in
pine, mahogony and oak. C/R&R
Mon-Fri 8.30am-5pm, Sat 9.30am-4pm.

FURNITURE RESTORATION
Country Ways
(Nigel Oswald)
Unit 2, West End Farm, Berwick Hill
Ponteland NE20 0JZ
☎ *0661 860076*
Renovation of antique pine furniture,
manufacture of kitchen tables. C/R&R

DECORATIVE PAINT FINISHES
The Stencil Library
(Helen Morris and Michael Chippendale)
Nesbitt Hill Head
Stamfordham NE18 0LG
☎ *0661 886529*

Huge range of stencils and equipment available. Day courses. Credit cards.
Mon-Sat 9.30am-5.30pm.

Otterburn

CERAMIC SCULPTURE
Keith Maddison
Hill View, Elsdon NE19 1AA
☎ *0830 20538*
Highly detailed single edition figures and animals in English porcelain. Demonstrations given. Work widely exhibited. C/W

SCULPTURE
Julia Barton
Redesdale Dairy, Soppitt Farm Otterburn NE19 1AF
☎ *0830 20764*
Sculptor and environmental artist. C
Visitors welcome by appointment.

FORGEWORK
S Farrell & Co Ltd
Otterburn NE19 1NP
☎ *0830 20431*
Wrought ironwork, copper and brass. C/R&R

Prudhoe

DRIED FLOWERS
Froggatts
(Ann and John Froggatt)
6a Earls Court, Princess Way Low Prudhoe NE5 1NB
☎ *0661 832587*
Dried flower arrangements, collages, spice garlands, pot pourri etc for department stores, interior design, hotels and private commissions. Components and floral sundries available. Tuition. C/W/E
Open daily 9am-4pm.

Seaton Burn

SADDLERY & CANVAS WORK
Bart J Snowball
(K Lyndon-Dykes)
4/7 Milkhope Centre Blagdon NE13 6DE
☎ *0670 789732/3*
All saddlery work, harness making and casemaking.
C/R&R/E. Credit cards.
Mon-Fri 9am-5.30pm, Sat 9am-5pm, Sun 10am-2pm

Stocksfield

PRINTING
Bacon & Bacon
(Christopher Bacon)
Cherryburn Museum Mickley NE43 7DB
☎ *0661 842368/843276*
Traditional copperplate and wood block printing. C/W
Visitors welcome by appointment

Wooler

CERAMICS
The Pottery
(Vanessa Taylor)
High Humbleton House High Humbleton NE71 6SU
☎ *0668 81623*
Specialist in Mocha-ware. The patterns result from an acidic solution on the slip, producing a resemblance to moss-agate stone. C/W/E
Mon-Fri 10am-5pm, Sat 10am-4pm.
Directions: take Burnhouse Road in Wooler, then 1st left.

NOTTINGHAMSHIRE CRAFT WORKSHOP CENTRES

Church Farm Craft Workshops

Mansfield Rd, Edwinstowe NG21 9NJ
☎ 0623 823767
Craft workshops in former farm next to parish church. Dolls' houses and miniatures, brass and bronze casting, knitwear, landscape painting, canal art and sign writing, stained glass, toys, coppersmithing, leathercraft, fossils, minerals and natural stone jewellery.
Wed-Sun 11am-5pm.

Longdale Rural Craft Centre

Manager: Janet Purcell
Longdale Lane, Ravenshead NG15 9AH
☎ 0623 794858/796952
Founded by wood carver Gordon Brown the craft centre has over fifty craftspeople and an international reputation. Disabled visitors welcomed. Lecture tours and demonstrations by arrangement.

Displayed as a mid-1800s village street with authentic period workshops, the craft centre has many full-time artists and craftsmen including wood carving, sculpting, stained glass, toy making, potting, print making, metalwork and jewellery. The Craft Museum incorporates a collection of traditional craftsmen's tools and equipment.
Licensed restaurant (Egon Ronay recommended) and coffee lounge, open daily 9am-10pm.
Open daily all year 9am-6pm
Charges: Free entry to main gallery selling arts and crafts. Entry charge for museum and workshops: £1.20 adults, £1 children, students and OAPs. Groups by arrangement.
Directions: between Nottingham-Mansfield A60 and Nottingham-Ollerton A614.

NOTTINGHAMSHIRE CRAFT WORKSHOPS

Mansfield

BRASS & BRONZE CASTING
Manor Brass
(B G and H M Bradley)
Units 5 & 8
Church Farm Craft Workshops
Mansfield Road
Edwinstowe NG21 9NJ
☎ 0623 823767
Cast aluminium house nameplates, ornamental brass, industrial castings (brass, bronze and gunmetal). C. Credit cards.
Open daily 10am-6pm.

Newark

ANTIQUE FURNITURE RESTORATION
T S Barrows & Son
(Norman and John S Barrows)
Hamlyn Lodge, Station Road
Ollerton NG22 9BN
☎ 0623 823600
Cabinet making and antique furniture restorers; French polishing etc. Specialist commissioned pieces. Visitors welcome in the workshop. C/R&R. Credit cards.
Tue-Sat 10am-5pm, Sun 12noon-5pm, -4pm in winter. Open on Bank Holidays.

CERAMICS/STONE CARVING
Harvey Wood (Heron Rock)
The Wharf, Trent Lane
Collingham NG23 7LZ
☎ 0636 892627
Pottery and stone carving; work widely exhibited and pieces in private and commercial collections. C/R&R/W/E. Credit cards.
Visitors welcome at any reasonable time.

METALSMITHING
T Marshall & Son
(R A and A Marshall)
Barnby Hall, Barnby NG24 2SA
☎ 0636 626332
Father and son metalsmiths working in solid silver, bronze, brass, copper etc. Mainly presentation awards and gifts (business promotion, sports, civil/military functions, celebrations etc). Designing, etching, engraving, hand-beaten copper, brass, pewter.
C/R&R/E. Credit cards.
Mon-Fri 8am-5.30pm, Sat-Sun 9am-12.30pm, lunchtime closing. Advisable to telephone first.

DOLLS' HOUSE/MODEL MAKING & WOODWORK
Heritage Concern Ltd
(Barry Atkinson FRSA)
Woodside House, Norwell NG23 6JN
☎ 0636 86377
Sole craftsmen enlisting expertise from specialist associates, making dolls' houses, models and toys. Also furniture making, woodworking and sign writing.
C/R&R/E. Credit cards.
Open daily.
Directions: 7 miles north of Newark, 1.5 miles from Al.

Newthorpe

FORGEWORK & FARRIERY
Terry Martin — Blacksmith & Farrier
The Old Mission Forge, Main Street
Newthorpe NG16 2DH
☎ 0773 713868
General smithing and farriery; registered shoeing smith and Diploma of Worshipful Company of Farriers. Forgework: wrought ironwork; gates, fencing, fire baskets etc. Tools sharpened, electric welding. Site work.
C/R&R/W/E
Open daily 8am-5pm.

Nottingham

CERAMICS
Carrington Pottery
(Mrs Judith Firmin)
15 Church Drive, Carrington NG5 2AS
☎ 0602 603479
The pottery features the wide ranging work of the proprietor and also sells the work of two other local potters. Childrens' and adults' ceramic workshops. C/R&R/W/E
Mon-Sat (closed Thur) 10am-4pm
Directions: Carrington is just north of Nottingham on A60 Mansfield road. Turn left at St John's Church.

RUSH & CANE WORK
Antique Cane Restoration
(C R Lathlane)
28 Mensing Avenue
Cotgrave Village NG12 3HY
☎ 0602 892459
Small workshop established in 1986 carrying out repair work to cane and rush seating; machine woven rattan cane, also seagrass. R&R
Mon-Sat 9am-6pm.

CAST METAL NAMEPLATES
Classic Names
(M Moreland)
14 Magdalen Drive
East Bridgford NG13 8NB
☎ *0949 21036*
Workshop making plain and decorated cast metal nameplates in a large variety of colours and designs. Awards won at county shows etc. C/R&R/W/E
Mon-Fri 10am-4pm, Sat-Sun by appointment.
Telephone for directions or ask at local Post Office.

PICTURE RESTORATION AND FRAMING
The Bart Luckhurst Gallery
(Bart Luckhurst)
9 Union Street, Bingham NG13 8AD
☎ *0949 837668*
Full art restoration and picture cleaning service. Bespoke framing and frame restoration. 'Picture Framer of the Year' award won several times in succession. C/R&R
Thur 9am-5.30pm, Sat 9am-1pm.

WOODEN TOYS & GIFTWARE
Geppetto's Woodcrafts
John Elliott — 'Mr Geppetto')
The Old Forge, Gonalston
Nr Lowdham NG14 7JA
☎ *0602 663499*
Traditional woodcrafts, specialist in fretwork. Also antique and modern restoration and repairs undertaken. Business awards; large design commissions. C/R&R/W/E
Mon-Fri (closed Thur) 12noon-6pm, Sat 11.30am-6pm, Sun 1-6pm. (Bank Holidays as for Sun timings)
Directions: from A612 Nottingham-Southwell road turn west to Gonalston at crossroad with Hoveringham road.

SADDLERY
Saddlecraft
(Mr W J and Mrs L P Cliff)
3 Cropwell Road
Radcliffe-on-Trent NG12 2FJ
☎ *0602 332800*
Husband and wife team established for 18 years, members of BETA. Saddlery and leatherwork. C/R&R
Mon-Fri (closed Wed) 9am-5pm, Sat 9am-4.30pm, lunchtime closing.
Directions: in the centre of Radcliffe village, close to A52 and A46 (10 mins from Nottingham).

FORGEWORK
O J Blood
45 Church Street
Ruddington NG11 6HD
☎ *0602 211842*
Wrought ironwork and general blacksmithing work. C/R&R
Mon-Fri 10am-4.30pm, lunchtime closing. Sat 10am-12noon. (Closed Xmas fortnight).

LETTERING & HERALDRY
The Calligraphers
(Deborah Hammond and Denise Hagon)
The Calligraphers Studio
The Lace Hall, High Pavement
Nottingham NG1 1HN
☎ *0602 483157*
Calligraphy, lettering and design specialising in fine lettering, illumination and heraldry for presentation awards, charters and formal documents. Framed poetry, family trees, greeting cards and stationery designed for all occasions (weddings etc). Many notable clients, work for BBC television etc. C/W/E
Mon-Sat 9.30am-5pm, Sun by appointment.

TEXTILES
Workshop Gallery
(Paula Linklater & Freda Raphael)
Bucks Lane
Sutton Bonington LE12 5PF
☎ *0509 673543*
Spinning and weaving workshop; handwoven rugs, hand-laid and embroidered felt. Also gallery, in early 15th century timber framed building, selling art and craft work. C. Credit cards.
Tues/Fri/Sat 10am-4pm, Wed 10am-12noon, Thur 10am-7pm. Sat 10am-4pm. (Closed 2 weeks in Aug).

Ravenshead

CERAMICS
Ravenshead Pottery
(Brian and Joan Marris)
23 Milton Drive
Ravenshead NG15 9BE
☎ *0623 793178*
Husband and wife business making domestic and decorative stoneware: soup, wine and coffee sets, bowls, jugs and casseroles.
C/W. Credit cards.
Visitors welcome at any time, but advisable to telephone first.

BOOKBINDING
Elaine Saxton — Bookbinding & Restoration
Longdale Craft Centre, Longdale Lane
Ravenshead NG15 9AH
☎ *0623 794858*
Bookbinding and book and paper restoration. C/R&R
Open daily 9am-6pm.
Entry charges and directions: see under Craft Centres.

STAINED GLASS
Michael Stokes — Stained Glass Artist
Longdale Craft Centre, Longdale Lane
Ravenshead NG15 9AH
☎ *0623 794858*
Restoration of ecclesiastical and modern domestic windows; design of contemporary and traditional windows. C/R&R/E. Credit cards.
Open daily 9am-6pm
Entry charges and directions: see under Craft Centres.

SILVERSMITHING
Joanne Pond — Metalsmith & Jeweller
Longdale Craft Centre, Longdale Lane
Ravenshead NG15 9AH
☎ *0623 794858*
BA (Hons) degree graduate of Art & Design in silversmithing and jewellery; design and manufacture of jewellery working in base metals and precious metals.
C/R&R/W/E. Credit cards.
Open daily 9am-6pm
Entry charges and directions: see under Craft Centres.

CRAFT PRINTING
The Crafty Printmaker
(Rachel Morley)
Longdale Craft Centre, Longdale Lane
Ravenshead NG15 9AH
☎ *0623 794858*
Lino and woodcut printing; production of handmade greetings cards, notepads and a range of gifts by BA(Hons) degree graduate in Art & Design. C/W/E. Credit cards.
Open daily 9am-6pm
Entry charges and directions: see under Craft Centres.

172

CERAMICS
Francesca Hollings Ceramics
Longdale Craft Centre, Longdale Lane
Ravenshead NG15 9AH
☎ *0623 794858*
One-off hand-built vessels and
thrown domestic and decorative
work by BA(Hons) degree graduate
in ceramics. C/E. Credit cards.
Open daily 9am-6pm
Entry charges and directions: see
under Craft Centres.

Retford

FURNITURE MAKING &
RESTORATION
Simon Denton
Woodside Cottage, Low Holland
Sturton le Steeple DN22 9HH
☎ *0427 880600*
Small family business restoring and
making furniture and other
woodwork. C/R&R/E.
Mon-Fri 10am-3pm, advisable to
telephone first.
Directions: between Gainsborough
and Retford take A620, through North
Wheatley. Sturton le Steeple after 1.5
miles; just before village church turn
left towards Littleborough. Workshop
is first house past school on right.

FURNITURE
Lee Sinclair Furniture
Endon House, Main Street
Laneham DN22 0NA
☎ *0777 228303*
Small, versatile workshop employing four skilled craftsmen making
individual pieces of furniture and
batch-produced items. A design

and prototyping facility is available
to large companies and interior
designers. Work extensively
exhibited. C
Mon-Fri 9am-5pm, Sat-Sun by
appointment.

CERAMICS
Chris Aston Pottery
(Chris and Phillippa Aston)
Yew Tree Cottage, 4 High Street
Elkesley DN22 8AJ
☎ *0777 83391*
Established pottery in traditional
brick and pantile village; husband
and wife business making an
extensive range of hand-thrown
stoneware, and screen-printed
mugs, tankards, plates etc.
C/W/E. Credit cards.
Generally open daily, 10am-6pm,
lunchtime closing.
Directions: Turn off A1 (20 miles
north of Newark) into old village;
pottery is near the church.

Worksop

STAINED GLASS
Reg and Dorothy Pritchard
Stained Glass Studio
Bawtry Road, Blyth S81 8HJ
☎ *0909 591205*
Design, making and restoration of
stained glass windows; also
lampshades, mirrors etc. and etched
glass panels. Many domestic and
ecclesiastical commissions.
C/R&R/W
Mon-Thur (closed Fri) 9am-5pm, Sat-
Sun 10am-4pm. Advisable to tel-
ephone first.

OXFORDSHIRE CRAFT WORKSHOP CENTRES

Cross Tree Centre
(Sir John Cripps)
Cross Tree, Filkins
Nr Lechlade GL7 2JL
☎ *0367 86491*
Woollen mill with craft workshops. Spinning and weaving, stone masonry, woodturning, cane and rush seating, antique furniture restoration. Gallery exhibiting paintings and crafts. Tea rooms.
Mill: Mon-Sat 10am-6pm, Sun 2pm-6pm. Gallery: Tue-Sat 10am-5.30pm, lunchtime closing.

Middleway Workshops
(The Old Bakehouse Trust Ltd)
Middleway
Summertown
Oxford OX2 7LG
A group of craft workshops including: specialists in sliding sash windows, paper print recycling, printing, upholstery, appliquéd clothing.
Workshops have independent working hours.

OXFORDSHIRE CRAFT WORKSHOPS

Abingdon

GOLD & SILVERSMITHING
Geoffrey Harding
31 The Green, Steventon OX13 6RR
☎ *0235 831371*
Handmade silver and gold articles of all kinds made to commission; domestic, ecclesiastical and ceremonial. Some jewellery. Pieces of original design made for local churches, Oxford colleges and private customers. C/R&R/E
Available at any reasonable time; advisable to telephone first.

GOLD & SILVERSMITHING
J & L Poole
(Dr M J Poole and Mrs E L Poole)
38 Wootton Road, Abingdon OX14 1JD
☎ *0235 520338*
Designer-makers of domestic and ecclesiastical silverware. Jewellery in gold and silver with precious stones. Small workshop and showroom selling products. Work also exhibited and sold at major craft fairs. Member of regional craft guild. C/R&R. Credit cards.
Visitors welcome by appointment any day, 10am-8pm.

ENGRAVING
Allsports Trophies
(Raymond Ledger)
Trophy Lodge, Sugworth Lane
Abingdon OX14 2HX
☎ *0865 736028*
Trophies assembled and engraved; also glass engraving. C/R&R/W/E
Mon-Fri 9am-5pm, Sat 9am-12noon.

FURNITURE
Mark Ripley — Furniture Designer/Maker
Unit 8 Hall Farm Workshops
High Street, Southmoreton OX11 9AG
☎ *0235 819545*
Contemporary and traditional furniture in temperate hardwoods. Built-in work also undertaken. Member of Guild of Master

Craftsmen. Work widely exhibited (Philips Auction Rooms and 'Antiques of the Future' Exhibition). C
Mon-Fri 9.30-5pm, some Sat 10am-4pm. Advisable to telephone first.
Directions: 2.5 miles east of Wallingford (take Hithercroft road). Hall Farm Workshops marked on red brick wall on left-hand side when entering village.

FURNITURE
Andrew J Kirkland — Fine Furniture Maker
The Old Dairy Workshop, Hill Farm Steventon OX13 6SW
☎ 0235 820176
Small specialist woodworking workshop. All work to commission only, furniture designed and made using traditional skills. Also feature staircases in hardwoods, doors, shop counters and displays. C/E.
Mon-Fri 8.30am-5.30pm, some Sats 9am-2pm. Advisable to telephone first.

FORGEWORK
R & M Turnpike Forge
(Roy Hanson)
Clifton Hampden OX14 3DE
☎ 086 730 7755
All types of ornamental metalwork and small steel fabrication contracts. Fire baskets, tools, canopies, gates, railings, decorative grilles. Member of BABA. C/R&R/W/E. Credit cards.
Mon-Fri 8am-5pm, Sat 8am-3pm. Some Suns 8am-12noon; advisable to telephone first.

Banbury

CERAMICS
Hook Norton Pottery
(Russell Collins)
East End Farm House
Hook Norton OX15 5LG

☎ 0608 737414
Studio pottery producing a wide range of work including garden pottery, lamps and general domestic ware. C/W/E. Credit cards.
Mon-Fri 9.30am-5.30pm, Sat 9am-5pm. Visiting charge for groups £5.

FORGEWORK & WOODTURNING
P Giannasi
The Close
South Newington OX15 4JN
☎ 0295 720703
Craftsmen in metal and wood. Hand-forged wrought-ironwork from garden furniture to fire-furnishings, mostly to commission. Turned work in British woods, some incorporating metalwork; bowls, complete lamps, goblets, skittles and croquet sets. C/R&R
Open at all reasonable times.

PICTURE FRAMING
Castle Farmhouse Framing
(Richard Barnden)
The Barn, Castle Farm House
Overthorpe OX17 2AD
☎ 0295 710450
Picture framing service; exhibition standard work by Member of Guild of Craftsmen. C/W/E
Mon-Fri 9.30am-5.30pm, lunchtime closing. Sat 9.30am-1pm.

WOODWORK
Fairmitre Windows & Joinery Ltd
(N D Carter and P Grossi)
The Potteries
Barford St Michael OX15 0RF
☎ 0869 38804
Workshop manufacturing hand-made joinery using English and imported timbers. Restoration of old/listed properties. C/R&R/W
Mon-Fri 8am-5pm, lunchtime closing.

UPHOLSTERY
Goldford Furnishings
Hudson Street
Deddington OX15 0SW
☎ *0869 38165*
Traditional hand upholstery. Range of 'in house' designed sofas and chairs (nation-wide delivery). C/R&R/W/E. Credit cards.
Mon-Fri 9am-5pm, Sat 10am-1pm, Sun by appointment. (Closed Xmas-New Year and Bank Holidays).

Bicester

WOODWORK
Hayward, Carey & Vinden
The Stables, Garth Park
Launton Road, Bicester OX6 0JB
☎ *0869 249654*
Commissioned cabinet work, woodturning and architectural joinery including staircases and traditional sash windows. C/E. Credit cards.
Mon-Fri 9am-6pm, Sat 9am-1pm by appointment.

CERAMICS
Chesterton Pottery
(Tony Smythe)
Chesterton OX6 8UW
☎ *0869 241455*
Small rural pottery producing handmade earthenware with traditional slip decoration, high fired for domestic use. Commemorative work. Showroom and workshop open to the public. C/W/E
Mon-Fri 9am-5pm and most week-ends; advisable to telephone first.

DRIED & FABRIC FLOWERS
Lark Rise Flowers
(Mrs Sandra Green)
Green Hill Cottage, Main Street
Hethe OX6 9ES

☎ *0869 278166*
Floristry and flower arranging; designer of bridal head-dresses made to order. Floral decorations for interior design. Associate member of The Society of Floristry. Work exhibited at large shows and featured in bridal publications. C/W
Visitors welcome by appointment.

Brize Norton

WOODWORK
R Griffiths Woodwear Ltd
(Ray and Ann Griffiths & Clive Austin)
Viscount Court
Brize Norton OX8 3QQ
☎ *0993 851435*
Designers and makers of giftware, furniture and all forms of joinery work. C/W/E. Credit cards.
Mon-Fri 8.30am-5.30pm, Sat 8.30am-1pm.

DRY STONE WALLING
Adrian Caunter
10 Elm Grove, Brize Norton OX8 3NE
☎ *0993 844300/(mobile)0831 295033*
Dry stone walling and repair work. Fencing, patios and turfing.

Burford

FURNITURE
Robert H Lewin
Overgreen, Lower End
Alvescot, Bampton OX18 2QA
☎ *0993 842435*
Furniture designed and handmade from hardwoods, from small boxes and bowls to fitted kitchens; coffee tables a speciality. C
Mon-Fri 9am-5.30pm, Sat-Sun by appointment only
Directions: 6 miles south of Burford on B4020.

FORGEWORK
F C Harriss & Sons
(M I Harriss and F C Harriss)
The Forge, Sturt Farm Industrial Estate
Sturt Farm, Burford OX8 4ET
☎ *0993 822122*
Several forgeworkers producing
traditional wrought ironwork,
specialising in restoration work and
work to commission. C/R&R
Mon-Fri 9am-5pm, Sat 9am-12noon.

ANTIQUE FURNITURE
RESTORATION
Colin Piper Restoration
Highfield House, The Greens
Leafield OX8 5NP
☎ *0993 87593*
Small workshop specialising in the
restoration and repair of fine
antique furniture and clock cases,
using traditional cabinet making
and polishing techniques. C/R&R
Mon-Fri 8.30am-5pm, Sat 8.30am-1pm.

Chipping Norton

CLASSICAL GUITARS
Paul Fischer — Luthier
West End Studio, West End
Chipping Norton OX7 5EY
☎ *0608 642792*
Baroque, 19th century and modern
concert guitars made by luthier
with an international reputation.
Instruments exported worldwide
and played by leading guitarists.
Lecture tours. Winner of awards
and fellowships. C/W/E
Mon-Fri 10am-5pm, lunchtime
closing. Sat 10am-12noon.

WOODTURNING
John Sparling
Unit 12, Langston Priory Workshops
Kingham OX7 6UP

☎ *0608 658872*
Decorative woodturning; hand and
copy-turning, architectural turning
and pattern making. Also traditional rush-seated ladder-back
chairs. Member of regional craft
guild. C/W/E
Mon-Fri 8.30am-5.30pm.
Directions: next to Kingham Station
(well signposted).

ANTIQUE FURNITURE RESTORA-
TION/FURNITURE MAKING
John Hulme
11a High Street
Chipping Norton OX7 5AD
☎ *0608 641692*
Restoration work and woodwork
for private and trade customers.
Furniture, modern and traditional,
designed and made. C/R&R/E.
Mon-Fri 8.30am-6pm, lunchtime
closing. Sat-Sun by appointment.

BAROMETERS & ANTIQUE
INSTRUMENTS
Peter Wiggins
Raffles Farm, Southcombe
Chipping Norton OX7 5QH
☎ *0608 642652*
Workshop making high quality
mercurial barometers. Also repair
and restoration of antique instru-
ments, scientific instruments and
clocks. C/R&R/W/E. Credit cards.
Mon-Sat 9am-5.30pm.

Didcot

CERAMICS
Sidney Hardwick
Cedarwood, Stream Road
Upton OX11 9JG
☎ *0235 850263*
Stoneware and porcelain pottery for
house and garden including bowls,
wash basins, lamps. C

Open daily at any time up to 6pm; advisable to telephone first.

GOLD & SILVERSMITHING
Ruth Harris
Post Office Studio, Post Office Gallery, 24 Main Road East Hagbourne OX11 9LN
☎ 0235 510280
Gold and silver jewellery incorporating precious and semi-precious stones. Also silver chasing; boxes, bowls etc. Workshop and studio with a range of work, also shown on Open Days in conjunction with the gallery exhibiting paintings. C
Mon-Fri (Tues early closing) 9am-5.30pm, lunchtime closing. Sat 9am-12.30pm by arrangement.

Faringdon

VINTAGE CAR RESTORATION
Alpine Eagle
(Roy Partridge and John Hodson) The Mill, Mill Lane, Clanfield OX8 2RX
☎ 0367 81401
Full restoration of vintage and post-vintage motorcars including bodywork repairs, paintwork, mechanical and interior renovation. Specialists in Rolls-Royce and Bentley cars.
Mon-Fri 9am-5.30pm, telephone for appointment.

WOODTURNING
Michael Bradley
Unit 4, Regal Way, Faringdon SN7 7BX
☎ 0367 22316
All kinds of specialist woodturning for domestic or trade use. Work ranges from spinning wheels to staircase balustrades and newel posts. Large front door columns. C/R&R/W
Mon-Fri 9am-4pm, lunchtime closing. Visit by appointment.

Henley-on-Thames

ANTIQUE FURNITURE RESTORATION
Manor Farm Furniture Restorations
(Paul Kelaart and Nicola Shreeve) Manor Farm Workshop Nettlebed RG9 5DA
☎ 0491 641186
Small workshop with 40 years' experience in restoration; repairs to marquetry, mouldings, decorated and turned furniture. C/R&R
Mon-Fri 8.30am-5.30pm.

CERAMICS
Cookie Scottorn — Potter
8 Swedish Houses, Park Corner Nettlebed RG9 6DT
☎ 0491 641889
Domestic pottery decorated with images from the countryside; also tiles. Sold through shops and galleries; also private sales. C/W/E
Visitors welcome by appointment.

BOATBUILDING
Henwood and Dean — Boatbuilders
(Colin Henwood) Greenlands Farm, Dairy Lane Hambleden RG9 3AS
☎ 0491 571692
Traditional boatbuilding and restoration work on early Thames craft. Several times winner of Best Restored Boat at Thames Traditional Boat Festival, Henley.
C/R&R/W/E
Mon-Fri 8am-6pm, Sat 9am-5pm.

FURNITURE
The Old Pine Workshops
(André Lambelin) Northend Workshops Northend RG9 6LE
☎ 0491 63216

Workshop employing four crafts-men making furniture in old pine, mainly fitted kitchens; other work undertaken. C/R&R/E
Mon-Fri 8.30am-5.30pm, Sat-Sun by arrangement.

Oxford

WOOD CARVING & CABINET MAKING
John Bye, Woodworker
Springfield, 42 North Hinksey Lane North Hinksey OX2 0LY
☎ *0865 721814*
Wood carver and cabinet maker specialising in church carving and furnishings, letter-cutting and heraldry. Also upholstery and reproduction furniture. C/R&R/E
Mon-Sat 8am-5pm, advisable to telephone first.

GOLD & SILVERSMITHING
Cyndy Silver — Goldsmith
19 Cumnor Rise Road Oxford OX2 9HD
☎ *0865 862295*
Gold and silver jewellery incorpo-rating precious and semi-precious stones. Clients' own stones can be used. Any item made to commis-sion and specification. Pearls and bead necklaces inventively restrung. C/R&R/W/E
Visitors welcome by appointment only.

FURNITURE MAKING & RESTORATION
Matthew Collins Furniture
Units 7/8 Lakeside Industrial Estate Stanton Harcourt OX8 6EQ
☎ *0865 883443*
Bespoke cabinet making and furniture restoration; all types of furniture, fine reproductions and

repolishing undertaken. C/R&R/W/E. Credit cards.
Mon-Fri 8am-5pm, Sat 9am-1pm.

BRAIDS & TRIMMINGS
Rodrick Owen — Braidmaker
38 Argyle Street, Oxford OX4 1SS
☎ *0865 245681*
Maker of braids and trimmings; narrow fabrics for fashion accesso-ries, interior design and wall hangings. C/E.
Visitors welcome by appointment.

APPLIQUÉD CLOTHING
Raija Medley
Middleway Workshops, Middleway Summertown, Oxford OX2 7LG
☎ *0865 52320*
Silk and cotton clothes with patchwork details, marketed from the workshop and craft shows. C
Tues-Fri 10am-5pm.

UPHOLSTERY
Chair Flair
(Christine Turley)
Middleway Workshops, Middleway, Summertown, Oxford OX2 7LG
☎ *0865 514586*
An open workshop employing two people upholstering and repairing antique and modern furniture. Also advice on fabrics, curtains and loose covers. C/R&R
Mon-Fri 8.30am-4pm.

WOODWORK
Oxford Sash Window Company
(Nick Rogers MSE(Eng))
Middleway Workshops, Middleway, Summertown, Oxford OX2 7LG
☎ *0865 513113*
Specialists in traditional box-framed sliding sash windows made to match original designs. Single or

double glazed. Replacement frames or sashes only. Supply and installation. Renovation service includes fully overhauling existing windows and draught-proofing. C/R&R
Mon-Fri 8.30am-5pm.

CABINET MAKING
Roger Cash, Cabinet-maker
Spareacre Works, Spareacre Lane
Eynsham OX8 1NH
☎ 0865 881214
Fine furniture and architectural joinery designed and made to commission using temperate hardwoods and combining traditional and modern techniques. C/W/E
Mon-Fri 9am-5pm, other times by appointment.

GOLD & SILVERSMITHING
Chorley Works — Designer Jewellery
(Rosamund Chorley)
90 Lonsdale Road, Summertown
Oxford OX2 7ER
☎ 0865 54518
Artist producing individual work often inspired by ancient cultures and medieval architecture. Silver and gold jewellery, small sculpture, mixed media painting on paper and silk, painted boxes. Work in local galleries and overseas. C/W/E
Visitors welcome between 10am-3pm by appointment only.

WOODCARVING/PICTURE FRAMING & GILDING
Geoffrey Butler
71 High Street, Chalgrove OX9 7SS
☎ 0865 890612
High quality hand-carved and gilt picture frames. Also woodcarving for furniture makers. Limewood mirror frames, pine mantel carving,

musical instrument carving and interior design work. C/R&R/E.
Visitors welcome by appointment.

FORGEWORK
Mark Crockett
Rycote Lane Farm
Milton Common OX9 2NZ
☎ 0844 278692
Traditional forge making wrought ironwork and fabrications. C/R&R
Mon-Fri 8am-5pm.

FURNITURE
Bates & Lambourne
(R Bates, D Lambourne and T Smith)
The Camp, Rycote Lane
Milton Common OX9 2NP
☎ 0844 278978
Small workshop specialising in country style furniture, contemporary and traditional designs, and Windsor chairs. Also ecclesiastical work. Members of Guild of Master Craftsmen. C/R&R/W/E.
Mon-Fri 9am-6pm, Sat 9am-1pm.
Directions: just off M40 junc 7, 3 miles from Thame on A329 Wallingford road.

TRADITIONAL CHAIRS & WOODWORK
David Bennett
The Villa, Swinbrook Road
Carterton OX18 1DT
☎ 0993 841158
Windsor, ladder and spindle-back chairs. Also furniture in native timbers, turned domestic and decorative bowls, boxes, lamp bases etc. Member of Worshipful Company of Turners specialising in greenwood furniture using traditional tools, techniques and materials. Tuition. C/R&R/W
Mon-Fri 9am-5pm, Sat-Sun by arrangement.

Directions: on east side of Swinbrook Road beyond Carterton boundary, 15 miles west of Oxford.

FURNITURE
Martin Dodds Country Furniture
Spelsbury Road, Charlbury OX7 3LP
☎ *0608 810944*
Small business designing and making quality country furniture, to commission, using English hardwoods (specialising in elm). C/R&R/E
Mon-Fri 9.30am-7pm, Sat 10am-7pm, closed Bank Holidays.
Directions: north of Charlbury on B4026 road to Chipping Norton.

BOOKBINDING
Chris Hicks — Bookbinder
64 Merewood Av, Sandhills OX3 8EF
☎ *0865 69346*
Specialist bookbinding and restoration service; fine and decorative bindings. C/R&R/E.
Mon-Fri (except Wed) 9am-5pm, Sat-Sun by appointment. Advisable to telephone first.

Reading

CABINET MAKING
Philip Koomen Furniture
Wheelers Barn, Checkendon RG8 0NJ
☎ *0491 681122*
Workshop, established in 1975, with a team of designer makers and cabinet makers producing contemporary furniture, particularly dining room, study and library schemes; also special one-offs. Work widely exhibited in UK and overseas. Notable private and business clients. Work featured in many publications. C
Mon-Fri 9am-5.30pm, other times by appointment. (Closed Xmas fortnight).

Thame

FORGEWORK
The Forge
(David Moss)
71 High Street, Thame OX9 3AE
☎ *0844 215979*
Handmade forged metalwork of any description undertaken, plus all repair and restoration work of iron, steel, copper, brass etc. C/R&R
Open daily 9am-6pm.

GLASS CARVING & ENGRAVING
Gillian Cox
Kestrel Cottage, Spriggs Alley
Nr Chinnor OX9 4BU
☎ *workshop: 0831 328536*
(studio 0494 483374)
Carved and engraved layered crystal of rich colours, also engraved crystal. Work widely exhibited, on display in V&A Museum and notable collections. C/E.
Mon 10am-4pm, lunchtime closing.
All other times by appointment.
Directions: from M40 junc 6, take B4009 north to Princes Risborough. After 3 miles turn right in Chinnor at mini-roundabout, towards High Wycombe. Ignore all left turns. Go straight ahead signed after 1 mile Spriggs Alley. After 1 mile, cottage on right, before Sir Charles Napier PH.

FURNITURE
C S Design
(Christopher Simpson)
54 High Street
Thame OX9 3AH
☎ *084421 2500*
Industrial Design and Design Education consultants. Bespoke modern furniture to commission as part of overall business. C/W/E
Visitors welcome by appointment.

ART RESTORATION
Malcolm H Burt Restorations
(Malcolm and Christine Burt)
15 Thame Park Road, Thame OX9 3PJ
☎ 084 421 2074
Business established for 40 years,
offering full restoration of items of
graphic art — oil painting in
particular. Corporate member of
Guild of Master Craftsmen. C/R&R
Mon-Sat 9am-5pm strictly by
appointment.

Wallingford

CERAMICS
Blenheim Pots & Plaques
(Lucienne de Mauny and Anthony
Fletcher)
Blenheim Farm
Nr Wallingford OX10 6PR
☎ 0491 39707
Ceramic house name and number
plaques, hand-thrown slip decorated pottery for everyday use,
commemorative plates, plaques,
cups etc. Work widely exhibited, also
at Liberty's, Regent Street. C/W/E
Wed-Sat 10am-5pm, Sun 2-5pm (Sat
lunchtime closing). Other times by
appointment.
Directions: 2 miles east from Wallingford. Take turning off Henley road
A423 signed Turners Court; pottery
next to Blenheim Riding Centre.

PICTURE FRAMING
The Studio
(P M and E Smith)
85 Wantage Road
Wallingford OX10 0LT
☎ 0491 37277
Husband and wife business,
established 20 years. All types of
picture framing. C/R&R/E
Mon-Sat 8am-6pm.

ANTIQUE FURNITURE
RESTORATION
Ipsden Woodcraft
(Martin Small)
Post Office , Ipsden OX10 6AG
☎ 0491 680262
Furniture restoration work includ-
ing French polishing. Also repro-
duction and modern furniture
made. C/R&R/E
Mon-Fri 8am-6pm, lunchtime closing.
Sat 8am-1pm.

UPHOLSTERY/SOFT FURNISHINGS/
RUSH & CANE SEATING
Springs & Things
(Elizabeth Wooldridge) Unit 14
Station Road Industrial Estate (West)
Wallingford OX10 0HX
☎ 0491 36529
Small business employing several
craftspeople working on traditional
and modern upholstery and all
types of soft furnishings and sewing
work. Also rush and cane seating
refurbishment.
C/R&R/W. Credit cards.
Mon-Fri 9am-5pm, Sat by appointment.
(Closed Xmas-New Year)
Directions: from Wallingford traffic
lights, head north towards Didcot, go
over crossroads and take 2nd turn left.

ARTIST
Rebecca Hind
8 Manor Farm Road
Dorchester on Thames OX10 7HZ
☎ 0865 340633
Professionally trained artist
working in a studio and the local
landscape painting in oils, water-
colours; landscapes and still life.
C/W/E
Advisable to telephone for appointment.
Directions: studio is located behind
Dorchester Abbey.

Wantage

PICTURE FRAMING
Ardington Gallery
(Charles Escritt)
Home Farm, Ardington OX12 8PN
☎ 0235 833677
Picture framing and restoration and art gallery promoting work of local artists.Oval and multiple-opening mounts, wash and line work, stretching and framing of textiles and box-framing. C/R&R
Mon-Fri 9am-5.30pm, Sat 9am-1pm.

PEWTER WORK
Tom Neal's Pewter Studio
28 Adkin Way, Wantage OX12 9HW
☎ 02357 69292
Craftsman, with training in metallurgy, producing classical and contemporary pewterware, mainly to commission. Domestic, commemorative and ecclesiastical work, trophies etc. Unusual commissions welcomed (pieces not necessarily expensive). C/R&R/W/E. Credit cards.
Available at any reasonable time; please telephone first.
Directions: Take turning opposite Fitzwarren School into Witan Way, continue down to T junc (Adkin Way); turn right, studio immediately on left.

RUSH & CANE SEATING/
FURNITURE RESTORATION
Country Chairmen
(A M and K M Handley)
Home Farm, School Lane
Ardington OX12 8PY
☎ 0235 833614
Furniture restorers specialising in rush and cane seating, using Thames rushes. Traditional dining, rocking and childrens' chairs made to order. Furniture restoration includes re-polishing, turning, veneering, structural repairs and desk/table leather top repairs. Supplies of rush and cane materials. C/R&R/W/E.
Mon-Fri 8.30am-5.30pm, Sat 10am-1pm.
Directions: from A34 Newbury-Oxford road turn west onto A417, take 4th turning left to Ardington, then follow signs to Home Farm.

FURNITURE
Nicholas Dyson Furniture
Unit 2, Home Farm, School Lane
Ardington, Wantage OX12 8PN
☎ 0235 834311
Contemporary furniture for private, professional and corporate clients. Work widely exhibited and found in offices of major companies, colleges and institutions and private houses throughout UK. C/W/E
Mon-Fri 9am-5.30pm, visit by appointment. (Closed Xmas-New Year).
Directions: from A34 Newbury-Oxford road turn left onto A417, take 4th left, then right, workshop on left.

CERAMICS
Ardington Pottery
(Les and Brenda Owens)
15 Home Farm, School Road
Ardington, Wantage OX12 8PN
☎ 0235 833302
Studio pottery in Victorian dairy making a wide range of highly fired stoneware and porcelain, including tableware, kitchen ware, vases, lamps etc. Several TV appearances, 'Incredible Export' award. Throwing demonstrations by arrangement. C/W/E. Credit cards.
Mon-Sat 10am-5.30pm, Sun 11.30am-5.30pm. (Closed Xmas-New Year)
Directions: 1.5 miles east of Wantage off A417. From A34 take A417 towards Wantage, pass East and West Hendred, then follow signs.

PRINTING
The Black Swan Press
(Peter Lord)
28 Bosley's Orchard
Grove, Wantage OX12 7JP
☎ 02357 4517
Design, letterpress printing and hot
metal typesetting of books, book-
lets, brochures, prints, maps etc on
high quality materials. C/W/E
Mon-Fri 9am-5pm, Sat-Sun by arrange-
ment. Please telephone first.

Witney

CERAMICS
The Stable
(Jane and Stephen Baughan)
Kingsway Farm, Aston OX8 2BT
☎ 0993 850960
Colourful, hand-decorated mugs,
jugs, bowls etc. C/W/E
Open daily 9am-6pm.

THATCHING
County Thatchers Ltd
(M J Minch)
Chilbrook Farm
Barnard Gate OX8 6XD
☎ 0865 882693
Thatching contractors established in
1977 working with long straw,
combed wheat straw, water reed.
Member of regional and national
Master Thatchers Associations.

SILVERSMITHING & JEWELLERY
Patricia Hamilton, Bill and Nikki
  Maddocks
The Mill, Hardwick OX8 7QE
☎ 0865 300407
Co-operative of jewellers and silver-
smiths producing designer jewel-
lery, flatware and holloware. Metal-
work. C/R&R/W/E. Credit cards.
Mon-Fri 9am-5pm.

Woodstock

CERAMICS
Bladon Pottery
(Graham & Corri Piggott)
2 Manor Road, Bladon OX20 1RT
☎ 0993 811489
Specialists in fantasy and figurative
stoneware sculpture, dragons,
wizards, Alice in Wonderland
figures, studies of children and
animals. Also unusual pottery and
watercolour paintings. C/W/E
Open daily 10am-6pm (often later
during summer). Other times by
appointment.

PAINTING ON SILK
Sally MacCabe
14 Hensington Rd, Woodstock OX7 1JL
☎ 0993 812051
Textile work; painting on silk
(tuition given), hand embroidery,
textured abstract panels. Workshop
in artist's home, open to the public
with permanent exhibition of work.
Member of regional craft guild. C/E
Visitors welcome by appointment.
Directions: opposite public car park, 5
mins from Blenheim Palace.

DECORATIVE PAINT FINISHES
Decorative Painting Workshop
(Annie Sloan)
Knutsford House, Park Street
Bladon OX20 1RW
☎ 0993 812590
All types of decorative painting
including distressed and scumble,
glaze finishes, woodgraining,
marbling, lined wood etc. Specialist
courses and liming kits available.
Some commissions, particularly in
colour work, with other decorative
painters. C/W/E. Credit cards.
Visitors welcome by appointment.

SHROPSHIRE CRAFT WORKSHOP CENTRES

Dinham House
Ludlow SY8 1EH
☎ 0584 874240
Exhibition and craft centre; working studios, art and craft exhibition, historical displays. Workshops include pottery and bone china jeweller. Music and drama held in the grounds. Regular outdoor events throughout the summer.
Open daily. Admission charge: adults £1, children 50p.

Llangedwyn Mill
Nr Oswestry SY10 9LD
☎ 0691 780618
Old mill with individual workshops (see entries below), situated on the banks of River Tanat in the valley below the Berwyn Mountain Range.

Free admissions and parking.
Directions: between Llynclys and Llanrhaedr on B4396.

Maws Craft Centre
Manager: Geoffrey Clarkson
Ferry Road, Jackfield
Ironbridge TF8 7LS
☎ 0952 882088
Workshops in the buildings of a famous Victorian tile company on the banks of the River Severn . Wide range of craft businesses with individual opening times, mostly open throughout the week and weekend afternoons.
No admission charge.
Directions: situated on the west bank of River Severn (footbridge from Coalport China Museum).

SHROPSHIRE CRAFT WORKSHOPS

Bishop's Castle

CABINET MAKING
Downes Furniture Company
(N B Downes)
Criftin House, Wentnor SY9 5ED
☎ 058 861 270
Fine furniture, mostly reproduction, made to order by two cabinet makers; also fitted furniture. The range includes furniture in oak, mahogany and walnut; Hepple-white, Chippendale etc.
C/R&R/W/E
Mon-Fri 9am-5pm, Sat 9am-12noon. Visitors welcome by appointment only.
Directions: from A489 turn off at Eaton to Wentnor; after 1.5 miles Criftin House on right.

GOLD & SILVERSMITHING
Marian Watson
2 Salop Street, Bishops's Castle SY9 5DB
☎ 0588 638864
Contemporary jewellery using gold with silver, set with precious and semi-precious stones. Also boxes, hatpins etc. Repairs undertaken. Work exhibited nationally and sold from workshop. C/R&R/W
Mon-Sat 10am-4pm, Sun by appointment

Ironbridge

MUSICAL INSTRUMENT MAKING
Paul Tozer — Guitar Maker
Unit 7, Maws Craft Centre, Ferry Road Jackfield, Ironbridge TF8 7LS
☎ 0952 884413

Musical instrument maker producing guitars and fretted instruments. Also repairs of most accoustic instruments, guitar, mandolin, banjos etc. C/R&R/E.
Mon-Fri 9am-5pm
Directions: see under Craft Centres.

FOSSIL PREPARATION
Revelations
(Alf Cawthorn)
Unit B2, Maws Craft Centre, Ferry Road
Jackfield, Ironbridge TF8 7LS
☎ 0952 884416
Fossil preparation service; casting and associated work. Specialist in detail work. C/R&R/W. Credit cards.
Mon-Fri 10am-7pm, Sat-Sun (2 weekends per month) 10am-6pm.
Directions: see under Craft Centres.

FORGEWORK
Anvilcraft
(C E Thorne)
Unit 7, Maws Craft Centre, Ferry Road
Jackfield, Ironbridge TF8 7LS
☎ 0952 882580
Blacksmith forging single items to customers' design and requirements. C/R&R/W/E. Credit cards.
Mon-Fri 10am-6.30pm, Sat-Sun 11am-6pm
Directions: see under Craft Centres.

JIG-SAW PUZZLES
Enigma Puzzles
(David Andrews)
Unit 7, Maws Craft Centre, Ferry Road
Jackfield, Ironbridge TF8 7LS
☎ 0952 883635
Three-dimensional jig-saw puzzles; made to individual design and complexity. Fretwork in wood. C/E.
Mon-Fri 10am-6pm, Sat-Sun advisable to telephone first.
Directions: see under Craft Centres.

PRINTING AND BOOKBINDING
The Orchard Press
(A J Mugridge)
Unit D3, Maws Craft Centre, Ferry Road
Jackfield, Ironbridge TF8 7LS
☎ 0952 883700
Printing using antique presses with a large variety of typefaces, specialising in book and short-run work. Also bookbinding. C/R&R/W/E.
Visitors are welcome by appointment.
Directions: see under Craft Centres.

CABINET MAKING
Dovetail (Fine Quality Woodwork)
(Barry Cheesman and Tony Evans)
Unit B3/1, Maws Craft Centre, Ferry Road
Jackfield, Ironbridge TF8 7LS
☎ 0952 883960
All types of woodwork on commission, specialising in high quality cabinet and joinery, working in English hardwoods. C/R&R/W/E
Mon-Fri 8.30am-6pm, Sat 9am-1pm. (Closed Xmas-New Year).
Directions: see under Craft Centres.

SOFT FURNISHINGS
Cottage Textiles
(Rebekah Cole)
Unit 19, Maws Craft Centre, Ferry Road
Jackfield, Ironbridge TF8 7LS
☎ 0952 883651
Maker of cushions, curtains, bed covers, patchwork and quilting, clothing, gifts. C. Credit cards.
Open daily 10am-5pm.
Directions: see under Craft Centres.

CERAMICS
Leisurecraft Hobby Ceramics
(K E and B F Shell)
Unit 17/18, Maws Craft Centre,
Ferry Road, Jackfield
Ironbridge TF8 7LS
☎ 0952 883815

Ceramics classes, greenware, supplies, firing service and finished pieces. Credit cards.
Open daily (except Tues-Wed) 11am-5pm, Sat-Sun 11am-5pm
Directions: see under Craft Centres.

TILE MAKING
John Burgess Tiles
Unit B25, Maws Craft Centre, Ferry Road Jackfield, Ironbridge TF8 7LS
☎ *0952 884094*
Manufacture of reproduction Victorian and art nouveau tiles. Range of glazed, embossed ceramic tiles using methods patented in 1841. Also dust-pressed encaustic floor tiles. Major restoration projects including Foreign Office and V&A Museum. C/R&R/W/E.
Mon-Fri 8.30am-4.30pm (Fri early closing). Sat-Sun by appointment.
Directions: see under Craft Centres.

ENAMELLING
Sue Daly Enamels
Unit 14, Maws Craft Centre, Ferry Road Jackfield, Ironbridge TF8 7LS
☎ *0952 883873*
Enamelled mirror frames and jewellery working mainly on copper in a variety of colours and designs. C/W/E. Credit cards.
Open daily 10am-7pm, advisable to telephone first.
Directions: see under Craft Centres.

METAL & RESIN FIGURINES
Marshall Arts
(Edward Marshall)
Unit 3.2 Maws Craft Centre, Ferry Road Jackfield, Ironbridge TF8 7LS
☎ *0952 883854*
High quality art deco reproduction figurines, lamps, small boxes and other items. C/R&R/W/E. Credit cards.

Mon-Fri 8am-4.30pm, Sat-Sun variable.
Directions: see under Craft Centres.

GOLDSMITHING/JEWELLERY
W J Bishop
Unit 24, Maws Craft Centre, Ferry Road Jackfield, Ironbridge TF8 7LS
☎ *0952 883914*
Handmade precious metal jewellery. Repairs while you wait. R&R/W/E. Credit cards.
Mon/Wed/Fri 9.30am-4.30pm, Sat (alternate) & Sun 10am-6pm.
Directions: see under Craft Centres.

CERAMICS
Winterwood Pottery
Maws Craft Centre, Ferry Road Jackfield, Ironbridge TF8 7LS
☎ *0952 883643*
Stoneware pottery, hand-thrown and slip cast with sprigged decorations; bas relief a speciality. C/W/E
Mon-Fri 10am-5pm, Sat-Sun 1-5pm.
Directions: see under Craft Centres.

STAINED GLASS
Stained Glass Art
(David and Carol Green)
Unit C15 Maws Craft Centre, Ferry Road Jackfield, Ironbridge TF8 7LS
☎ *0952 884240*
Decorative stained glass; manufacturing, design and repair of stained glass windows, lampshades and mirrors. C/R&R. Credit cards.
Mon-Fri 10am-4pm, Sat 11am-4pm, some Suns 2-5pm, advisable to telephone first.
Directions: see under Craft Centres.

DRIED FLOWERS
Wallflowers
(Jaqui Bagguley)
Unit 21 Maws Craft Centre, Ferry Road Jackfield, Ironbridge TF8 7LS

☎ 0902 788793
Dried flower arrangements and
original wild flower drawings made
to any size and specification. Special
orders undertaken. C/W/E
Open daily 10am-5pm.
Directions: see under Craft Centres.

WOODWORK
Peter Meyrick Designs
Unit 20 Maws Craft Centre, Ferry Road
Jackfield, Ironbridge TF8 7LS
☎ 0952 883873
Woodworker manufacturing
figurative furniture, automata,
carvings, toys. C/W
Open daily 10.30am-5pm, advisable
to telephone first.
Directions: see under Craft Centres.

PICTURE FRAMING
Maws Gallery
(Trevor Davies and Robyn Whitney)
Maws Craft Centre, Ferry Road
Jackfield, Ironbridge TF8 7LS
☎ 0952 883967
Picture framing workshop and
gallery exhibiting a wide selection
of work of artists; prints and photo-
graphs. C/R&R/W/E. Credit cards.
Open daily 10.30am-5.30pm.
Directions: see under Craft Centres.

CABINET MAKING &
WOODTURNING
The Boring Mill
(Raymond Read)
Rose Cottage, Dale Road
Coalbrookdale TF8 7DS
☎ 0952 433080
High quality handmade furniture in
English hardwoods. C/R&R/E
Mon-Fri 9am-5pm, Sat 9am-1pm.
(Closed Xmas-New Year)
Directions: behind Rose Cottage, one of
Ironbridge Gorge Museum buildings.

RECYCLED PRODUCTS
Scrap Scrap
(Jakki Moase), Endurance Works,
High Street, Coalport TF8 7HX
☎ 0952 586754
Unique products made from recycled
materials including fabric, plastic
and tin. Mirrors, candlesticks, hats,
bags, quilts, jewellery, cushions,
rugs etc. W/E. Credit cards.
Mon-Fri 8am-6pm, Sun 10am-6pm.
Directions: see under Craft Centres.

Ludlow

TEXTILES
Woodstock House Craft Centre
(Hugh and Wendy Rulton)
Brimfield SY5 4NY
☎ 0584 72445
Fully equipped fabric centre.
Textiles, work made and sold.
Courses in embroidery, dress-
making and soft furnishing. C/R&R
Mon-Sat 9am-5pm. Closed two weeks
in Jan.
Directions: 5 miles south of Ludlow on
A49.

CERAMICS
Old School Pottery
(Pierre Brayford)
Edgton, Craven Arms SY7 8HN
☎ 05888 208
Pottery making functional and
decorative stoneware — mainly
thrown pottery. C/W
Tues-Sun 10am-6pm.
Directions: 4 miles west of Craven Arms.

CERAMICS & GIFTWARE
Alison Williams — Bone China
Jewellery
Dinham House Craft & Exhibition Centre
Dinham SY8 1EH
☎ 0584 874240

Bone china flowers; floral jewellery, baskets and decorated mirrors. Workshop is one of several studios, for further details see under Craft Centres. C/W/E. Credit cards. *Open daily 10am-5pm.*

FINE ART & PICTURE FRAMING
Forge House Gallery
(R de Sylva and Catherine Odgen)
Forge House, Brimfield SY8 4NG
☎ *0584 711500*
Housed in blacksmith's forge built in 1731, workshop and gallery with special exhibitions. Bespoke craft picture framing and print making (Richard de Sylva). Paintings in oil and watercolour by Catherine Ogden. C/R&R/E. Credit cards. *Mon-Sat 10am-6pm.*
Directions: from A49 Leominster-Ludlow road turn east to Brimfield.

Market Drayton

DRIED FLOWERS
Carolyn's Flowers
Longford Farm
Longford TF9 3PW
☎ *0630 638295*
Dried flower business employing four people; flowers mainly grown on farm; arrangements, bouquets and bunches. Tuition. C/W
Open daily (except Tues) 10am-4pm.
Directions: from Tern Hill roundabout (A53/A41) go north signed Whitchurch. After 50yd turn right, 1 mile to Longford, bear right signposted Market Drayton; workshop at first farm on right.

Much Wenlock

HEDGE LAYING & DRY STONE WALLING
Carl Liebscher
The Row, Easthope TF13 6DW

☎ *074 636 497*
Traditional estate work undertaken within 30 mile radius; hedge laying, dry stone walling, coppice produce, gates and stiles, fencing etc.

FLORAL CERAMICS
Cleehills Crafts
(Mrs Carol Colley & daughters)
The Leasowes, Stanton Long TF13 6LH
☎ *074 634 677*
Floral modelling in terracotta clay dug on the premises. Extensive range available in site shop and through retail outlets. Modelling in progress on Tues, Thur & Fri. Also rare breeds farm and picnic facilities during summer . C/W
Open daily 9am-6pm
Directions: from Stanton Long heading south with church on right, take left fork; pottery on right.

Oswestry

CERAMICS
Tickmore Pottery
(Felicity Cripps), Trefonen SY10 9DZ
☎ *0691 661842*
Pottery with display area where visitors can watch production of hand-thrown stoneware. Jugs, carafes, coffee and tea sets, salt pigs, bowls, mugs etc some decorated with unique sheep scenes. C/W/E
Open daily; Sat-Sun advisable to telephone first.

GOLD & SILVERSMITHING/
CLOCK RESTORATION
Studio Bee
(F G Parker H Dip NEWI, Dip RJ)
Llangedwyn Mill, Oswestry SY10 9LD
☎ *0691 780744*
Jeweller and clock repairer producing varied work including trophies

for 'Best Dressed Man' award and Wool Mark 21st Anniversary. Wood turned clocks and barometers. C/R&R *Tues, Thur, Fri, Sat 11am-5pm. Sun by arrangement.*
Directions: see under Craft Centres.

STAINED GLASS
Mirage Glass
(Mrs Wolfe Van Brussel)
Unit 11, Llangedwyn Mill
Oswestry SY10 9LD
☎ 0691 780618
Stained glass/leaded lights designed and made mainly to commission. Traditional and modern techniques including sandblasting, fired painting etc. C/R&R/W. Credit cards.
Mon-Fri 10am-5pm, Sat-Sun by appointment.
Directions: see under Craft Centres.

ANTIQUE FURNITURE RESTORATION
Abbey Antiques
(Leigh Kellaway)
Unit 3, Llangedwyn Mill
Oswestry SY10 9LD
☎ 0691 780403
Furniture restoration work and handmade custom-built furniture. C/R&R/W/E
Mon-Fri 8.30am-5.30pm, Sat-Sun by appointment.
Directions: see under Craft Centres.

Shrewsbury

SADDLERY
J C Evans
(Miss Jane Evans)
Units 9 & 10, Hardwicke Industrial Estate
Hadnall SY4 4AS
☎ 0939 210588
Master saddlers manufacturing leatherwork for saddlery, including

equestrian rugs, sheepskin saddle covers and waxed cotton products. Full repair service for any leather goods. C/R&R/W/E. Credit cards.
Mon-Fri 8am-4pm, Sat-Sun by appointment.

FURNITURE MAKING
Michael Patrick Niblett
The Lilacs
Bayston Hill Common SY3 0DZ
☎ 074 372 2814
High quality traditional furniture making and woodwork. C/R&R/E
Visitors welcome at any time by appointment.
Directions: 2.5 miles south of Shewsbury off A49 on Bayston Hill Common.

TRADITIONAL CHAIRS
John Porritt Furniture
7 Montague Place, Belle Vue SY3 7NF
☎ 0743 361603
Small workshop making Windsor chairs, English and American; restoration of period furniture. Work in Winchester Cathedral. C/R&R/E
Mon-Fri 9am-6pm, Sat by appointment.

Telford

CARRIAGE BUILDING
The Wellington Carriage Co
(Philip Holder)
Long Lane, Telford TF6 6HD
☎ 0952 242495
Horse-drawn carriages designed and built to customers' requirements including replicas. 25 years' experience. C/R&R/E
Mon-Fri 8.30am-6pm, Sat 9am-5.30pm, lunchtime closing. Sunday by appointment.
Directions: from Telford north 7 mile on A442. Turn left at Long Lane service station, past Bucks Head PH, left along Arlscott road, g workshops on left.

ARTIST
Trinity Marketing
(R W R Evans)
56 King Street, Wellington TF1 1NR
☎ 0952 253991
Painter and sculptor working in
different mediums. National diploma
in design. C/W/E. Credit cards.
*Open daily 9am-5pm, advisable to
telephone first.*

CERAMICS
Richard Capstick Ceramic Design
4 Newhouses, Aston Road
Shifnal TF11 8DX
☎ 0952 462460

Pottery specialising in Raku fired
ceramics. Work sold in galleries and
exhibitions. Member of regional
Potters Association. C/W
Mon-Fri 10am-5pm, Sat 10am-1pm.

Whitchurch

BESPOKE JOINERY
Burleydam Enterprises
(Colin Stockton)
Wilkesley Smithy
Burleydam SY13 4BB
☎ 0948 871245
Four employed in specialist joinery
business. C/R&R/W/E
Mon-Fri 8am-5pm, Sat 8am-3pm.

SOMERSET CRAFT WORKSHOP CENTRES

Black Swan Guild
2 Bridge Street
Frome BA11 1BB
☎ *0373 473980*
Art and craft centre with some craft workshops including pottery, furniture making, glass, stained glass, basket making and knitwear. Also a gallery with regular art/craft exhibitions and craft shop selling top quality British craftwork. Wholefood restaurant.
Mon-Sat 10am-5pm.

Riverden Craft Workshops
(Bill, Margit and Stella Poirrier)
Roadwater, Nr Watchet TA23 0QA
☎ *0984 40648*
Crafts include functional and decorative fire-forged metalwork, handmade ornamental candles, hand-knitted garments made from hand-spun Exmoor and Jacob sheep's wool.
Mon-Fri 9am-5.30pm. Sat-Sun by appointment.
Directions: 400yd from Valiant Soldier public house.

SOMERSET CRAFT WORKSHOPS

Bridgwater

BASKETWORK
The Somerset Willow Company
(A J and D J Hill)
Units 10-12, The Wireworks Estate, Bristol Road, Bridgwater TA6 4AP
☎ *0278 424003*
Traditional family business making willow furniture, balloon baskets and various other basketwork, employing seven craftsmen. Work recently on display in Paris.
C/R&R/W/E
Mon-Fri 8am-5pm. (Closed Xmas-New Year).
Directions: (avoiding town centre) from M5 junc 23, take A38 southbound. Workshop/showroom on left.

BASKETWORK
Somerset Levels Basket & Craft Centre
(S J and R C Loveridge)
Lyng Road, Burrowbridge
Bridgwater TA7 0SG
☎ *0823 698688*
Business started in 1864 making basketware. Other crafts for sale.
C/W/E. Credit cards.
Mon-Sat 9am-5.30pm.

SPINNING/RUG WEAVING/ KNITTING
Quantock Weavers
(Wendy Cobbledick)
The Old Forge, Plainsfield
Over Stowey TA5 1HH
☎ *0278 671687*
Workshop in 17th-century forge building. Hand-spun knitting wool and knitted garments in natural colours and plant dyes, woven rugs and wall coverings. Courses in spinning and weaving. Group visits by arrangement. Member of regional craft guild.
C/W/E. Credit cards
Mon-Fri 11am-5pm, Sat-Sun 2-5pm.
Directions: follow ETB signs from A39 Bridgwater-Minehead road at Nether Stowey.

Burnham-on-Sea

CERAMICS
Haven Pottery
(David & Sylvia Lemon)
West Huntspill, Highbridge TA9 3RQ
☎ 0278 783173
Stoneware and tableware using a
variety of glazes in muted colours.
Cider jars and ciderware. Much
work to customers' specifications.
C/W/E. Credit cards.
Mon-Sat 10am-5pm, lunchtime closing.

Chard

FORGEWORK
Michael Judd
The Forge
Cricket St Thomas Wildlife Park
Chard TA20 4DA
☎ 0460 30502
Wrought iron, from authentic
traditional work to modern
ornamental work. Small forge and
showroom where blacksmith can be
watched at work.
C/R&R/W/E. Credit cards.
Open daily 10.30am-5pm.
Directions: access to Wildlife Park
from A30 Crewkerne-Chard road.
(Free access to forge).

Cheddar

FORGEWORK
Peter Phillips (Blacksmith)
The Forge, Cliff Street
Cheddar BS27 3PL
☎ 0934 742345
Blacksmith trained in the traditional
manner in a village forge. Now
producing wrought ironwork of
high quality mainly to commission.
C/R&R/W/E.
Mon-Sat 9am-6pm.

Crewkerne

CERAMICS
D B Pottery
(David Brown)
Highway Cottage, Church Street
Merriott TA16 5PR
☎ 0460 75655
Stoneware pottery, domestic and
sculptural; teapots a speciality.
Workshop and showroom operated
part-time by lecturer in ceramics. C/W
Open daily 10am-dusk, lunchtime
closing on week days.
Directions: 2 miles south of A303.
Take A356 from Lopen Head signed
to Crewkerne and Merriott; take first
turn left into Church Street.

THATCHING
Richard A Fryer — Reed Thatcher
10 Easthams Road
Crewkerne TA18 7AQ
☎ 0460 74250
House thatching undertaken in
wheat reed and water reed. Also re-
timbering work.

TOYS
Hullabaloo
(Sarah Jane Drew)
14 Abbey Street, Crewkerne TA18 7HY
☎ 0460 75805
Member of British Toymakers'
Guild making mainly wooden toys;
also some papier maché and soft
toys. Framed nursery pictures.
C/W. Credit cards.
Tues-Sat 10am-5pm, lunchtime
closing. Advisable to telephone first.

ANTIQUE FURNITURE
RESTORATION
Christopher Weeks
Jubilee House, 59 Lower Street
Merriott TA16 5NW

☎ 0460 75060
Comprehensive antique furniture restoration service. C/R&R
Mon-Fri 9am-5pm, Sat-Sun by appointment.

PICTURE FRAMING & RESTORATION/GILDING
Webster & Hill Fine Art Studio & Gallery
(David Webster and Timothy Hill)
26B Abbey Street
Crewkerne TA18 7HY
☎ 0460 74665
Master framers, restorers of frames, paintings (oils) and gilding. Mounting and framing of antique and contemporary works, dry mounting, heat sealing, re-lining etc. Gallery and exhibitions. C/R&R/W/E. Credit cards.
Mon-Fri 9.30am-5pm, Sat 9.30am-1pm. (Closed Xmas-New Year).

Frome

FORGEWORK
Somerset Smithy
(Alan Patterson)
Christchurch St West, Frome BA11 1EQ
☎ 0373 62609
Hand-crafted ornamental wrought-ironwork, decorative and functional. Designs to suit individual requirements. C/R&R/W/E
Mon-Fri 8am-5pm.

GLASS/ENAMEL WORK
Gaynor Ringland — Glass Artist
The Black Swan Guild, 2 Bridge Street
Frome BA11 1BB
☎ 0373 452962
Textured bowls and vessels coloured with vibrant enamels and scratch patterning. Slumped bowls, vases, interior flat glass panels and jewellery. Crushed glass and powdered enamels used to form crystalline earrings and brooches. Work exhibited throughout UK and overseas; featured in publications. C. Credit cards.
Mon-Sat 9am-5pm.

STAINED GLASS
Victoria Blight — Stained Glass Artist
The Black Swan Guild, 2 Bridge Street
Frome BA11 1BB
☎ 0373 453153
Stained glass and decorative mirrors designed and made by glass artist who studied with notable tutors. C/R&R/W/E. Credit cards.
Mon-Sat 9am-5pm.

KNITWEAR
Cynthia Rennie Knitwear
The Black Swan Guild, 2 Bridge Street
Frome BA11 1BB
☎ 0373 453153
Machine-made knitwear in wool and cotton to original designs by lecturer in textiles and design with 20 years' experience. C/W/E
Tues-Sat (closed Wed) 10am-4.30pm. (Closed Xmas-New Year, 3 weeks in summer)

BASKETWORK
Rebecca Board
The Black Swan Guild, 2 Bridge Street
Frome BA11 1BB
☎ 0373 453140
Individually designed baskets made in dyed and natural willow, sometimes incorporating metals and other unusual materials. Member of regional crafts guild. C/W/E
Mon-Sat 9am-5pm.

Ilminster

CIDER MAKING
Perry's Cider Mills
(H W Perry)
Dowlish Wake TA19 0NY
☎ *0460 52681*
Ciders made from locally-grown apples; cider making in autumn (free admission). Shop selling cider (taste before buying), stone jars, mugs, corn dollies, pottery etc. 16th-century barn houses collection of country bygones. W
Mon-Fri 9am-1pm, 1.30-5.30pm; Sat 9.30am-1pm, 1.30-4.30pm; Sun 10am-1pm.

Langport

CERAMICS
Aller Pottery
(Bryan and Julia Newman)
Aller TA10 0QN
☎ *0458 250244*
Utilitarian and decorative stoneware pottery made to individual design. C/W. Credit cards.
Mon-Sat 9am-6pm, Sun by appointment.

RUSH AND CANE WORK
The Cane Workshop
(John Excell)
The Gospel Hall, Westport TA10 0BH
☎ *0460 281636*
Specialists in re-caning and rush seating furniture. Also supplies and basket making tools. C/R&R
Variable working hours; best to telephone first.

CERAMICS
Muchelney Pottery
(John Leach)
Langport TA10 0DW
☎ *0458 250324*

'Muchelney' kitchenware made since 1965 by the grandson of Bernard Leach. Showroom selling other John Leach designs including his 'Black Pots'. Work in collections and museums in UK and overseas. W/E. Credit cards.
Mon-Fri 9am-5pm, lunchtime closing. Sat 9am-1pm. Visits to workshop by appointment.
Directions: 2 miles south of Langport.

UPHOLSTERY & SOFT FURNISHINGS
Quality Upholstery
(Dodie Huxter)
South Ham Farm , Muchelney Ham Langport TA10 0DJ
☎ *0458 250816*
Upholstery and soft furnishing (loose covers, curtains, cushions, blinds, tie-bands etc). Member of Assoc of Master Upholsterers. Tuition: 'Fabric Magic' courses at regular intervals. C/R&R/W
Mon-Sat 9am-6pm
Directions: From A303 at Podimore roundabout take Langport exit, turn left at Long Sutton. Turn right signposted Muchelney; 2nd on left after passing Muchelney Ham sign.

UPHOLSTERY & SOFT FURNISHINGS
Sutton Upholsterers
(Alexander Coulter and Jennifer Elizabeth Patton)
Picts Hill, Langport TA10 9AA
☎ *0458 252492*
Small family business employing several craftsmen; re-upholstery and furniture repair service, members of Assoc of Master Upholsterers. Also curtain/blind making, loose covers and office furniture repairs. C/R&R
Mon-Fri 8.30am-5pm, Sat-Sun 9am-12noon.

Midsomer Norton

BOOKBINDING
Downside Abbey Bindery
(Richard and Margaret Norman)
Abbey Road
Stratton-on-the-Fosse BA3 4QW
☎ 0761 233392
Fine binding, antiquarian book
restoration, library and liturgical
bindings, restoration of prints,
documents and maps. C/R&R/W/E.
Mon-Sat 9am-5.30pm, Sun by
appointment.
Directions: on A367 between Radstock
and Shepton Mallet, in the grounds of
a Benedictine monastery.

Minehead

SHEEPSKIN GOODS
John Wood (Exmoor) Ltd
Old Cleeve Tannery
Old Cleeve TA24 6HT
☎ 0984 40291
Factory tours showing the making
of sheepskin products. Coats, rugs,
hats, moccasins, toys etc. Showroom
and seconds' shop and café.
C/W/E. Credit cards.
Showroom open Mon-Fri 9am-5pm,
Sat & Bank Holidays 10am-4pm.
Guided tours Apr-Oct Mon-Fri at
10.45, 11.30, 2.15 & 3pm. Guided tour
charge: Adults 50p, children, OAPs &
students 25p. Group charge 40p.

CERAMICS
Mill Pottery
(Michael Gaitskell)
Wootton Courtenay TA24 8RB
☎ 0643 841297
Well-established pottery producing
high quality thrown stoneware for
use and decoration. Showroom and
workshop with working water-

wheel. C/W/E
Mon-Sat 9am-6pm, lunchtime closing.
Sun by appointment.
Visiting charge: adults 50p.
Directions: 4 miles west of Dunster on
A396 turn right for Wootton
Courtenay. Look out for circular Mill
Pottery sign at top of entrance to lane.

PORCELAIN PAINTING
Sweet Somerset
(Susan Myrick)
1 Tregonwell Rd, Minehead TA24 5JD
☎ 0643 702035 *(evenings)*
Hand-painted china and porcelain;
Somerset wildlife and wild flowers,
childrens' subjects. C/W/E
Mon-Fri 10am-5pm, lunchtime
closing. Sat 10am-1pm.

CERAMICS
The Exmoor Pottery
(Pat Jeffs)
1B Brunel Way, Vulcan Road
Mart Road Business Park
Minehead TA24 5BY
☎ 0643 706188
Small business producing sculpture
house plaques, pots, figurines and
tree sculpture. Demonstrations
arranged. Exhibitions. Classes in
autumn/winter months. C/W/E
Mon-Sat 10am-5pm, lunchtime closing.
Directions: adjacent to car park and
new swimming pool in Brunel Way.

SADDLERY
McCoy Saddlery & Leathercraft
(Mr M and Mrs M V McCoy)
High Street, Porlock TA24 8QD
☎ 0643 862518
Manufacture and repair of saddlery,
riding chaps, handbags and other
leather goods. Also comprehensive
mail order supplies. C/R&R/W/E
Mon-Fri 9am-5.30pm, Sat 9am-5pm.

FORGEWORK & GLASS ENGRAVING
Doverhay Forge Studios
(James Horrobin and Gabrielle Ridler)
Doverhay, Porlock TA24 8QB
☎ 0643 862444
Decorative, domestic and architectural forged ironwork. Recent work for V&A Museum and Canary Wharf. Also hand-engraved glass; individual designs and lettering for commemorative or decorative glass. C/R&R/E. Credit cards.
Tues-Fri 10am-5pm, Sat 10am-2pm. (Closed Jan-Easter).
Directions: situated opposite Doverhay car park.

Shepton Mallet

STAINED GLASS
Unicorn Glass
(Frances Davies)
Cooses Farm, Stoke St Michael BA3 5JJ
☎ 0749 840654
Small workshop making stained glass windows. C/W/E.
Open daily 10am-5pm, advisable to telephone first.

PATCHWORK QUILTS
Judith Gait American Patchwork
8 Bath View
Stratton-on-the-Fosse BA3 4RE
☎ 0761 232564
Work usually to commission, but about a dozen quilts to view or buy. Mentioned in several publications. C
Visitors welcome by appointment.

Shipham

MINERALS & GEMSTONE PRODUCTS
Neopeg Gems
(George and Kate Reed)
Hawthorn Cottage, Hollow Road
Shipham BS25 1TG

☎ 0934 842609
Unusual gifts made from minerals and semi-precious stones; gemstone trees, agate clocks, shell and gemstone jewellery, pen sets, bookends and polished mineral specimens. Engraved nameplates. C/W
Mon-Fri 9am-6pm, Sat-Sun 10am-5pm.

Simonsbath

CERAMICS
Simonsbath Pottery
(R A Billington)
Simonsbath, Exmoor TA24 7SH
☎ 0643 83443
Casseroles, mugs, vases, garden pots and lamp bases made on the premises. Other crafts and paintings for sale, including hand-blown glass and prints. W. Credit cards.
Winter: Nov-Mar Wed-Fri 9am-5pm, Sat-Sun 10am-5pm. (Closed Jan-1st week Mar). Open daily 10am-5pm throughout summer.

Somerton

TAPESTRY SCHOOL & KITS
The Icelandic Tapestry School
(Jóna Sparey)
The Little Lynch, 6 Behind Berry
Somerton TA11 7PD
☎ 0458 73111
Courses on ancient Icelandic embroidery and tapestry kits produced. Lectures given. W/E. Credit cards.
Mon-Fri 9am-5.30pm, visit by appointment. Sat: open to public 10am-4pm.

SADDLERY
Timothy Abbot Saddlery
3 Chapel Yard Workshops
Babcary TA11 7DU
☎ 0458 223846

Manufacture and repair of saddlery and harness work, leatherwork, upholstery, riding chaps etc. C/R&R/W/E. Credit cards. *Mon-Fri 9am-5pm, Sat 9am-1pm.* Directions: 1 mile off A37 south of Lydford-on-the-Fosse. Opposite 'phone box and bus shelter in village.

FORGEWORK & TEXTILES
Angela Osborne Textiles & Dundon Forge
(Angela and Peter Osborne)
Lockyers Farm, Peak Lane
Compton Dundon TA11 6PE
☎ 0458 74130
Contemporary textiles; embroidery, feltmaking, wool plying; conservation and repair of ethnographic and oriental rugs, flatweaves, textiles etc. Some paper making and plaster carving. Conservation, research and analysis. Also blacksmithing and metalwork by Peter Osborne, member of regional Guild of Wrought Iron Craftsmen. C/R&R *Textiles: Mon-Fri 10am-2.30pm, Sat-Sun by arrangement; please telephone first. Forge: varied working hour; advisable to telephone first.* Directions: on B3151 between Street and Somerton. Take Ham Lane or Peak Lane off road; Lockyers Farm is between Dundon Beacon and Lollover Hill near church, up small lane next to Middle Farm.

South Petherton

THATCHING
Jack Lewis
Perrins Cottage
Compton Durville TA13 5ET
☎ 0460 40027
High quality thatching work in water reed and combed wheat reed on all types of domestic and agricultural properties. Commissioned work for National Trust and in USA.

Stalbridge

FORGEWORK
Filleybrook Forge
(Tim Fortune, BA)
11d Marsh Lane
Henstridge Trading Estate
Templecombe BA8 0TG
☎ 0963 63315
Individually designed contemporary ironwork, interior and exterior. Work exhibited in galleries throughout the south of England including the Barbican Centre. C/R&R/W/E. Credit cards. *Mon-Fri 9.30am-5.30pm, Sat-Sun by appointment.*

Street

SHOE MAKING KITS
Simple Way
(David Price)
Unit 5, The Tanyard, Street BA16 0HR
☎ 0458 47275
Manufacturers of 'do-it-yourself' leather shoe and handbag kits containing everything needed to make them up, plus full instructions. A made-to-measure service i available. Mail-order catalogue. W/E. Credit cards. *Mon-Fri 9.30am-4.30pm, Sat 9am-12.30pm. Closed Bank Holidays.*

Taunton

BASKETWORK
Willow Wetlands Visitors Cent
(C B and M A Coate)
Meare Green Court
Stoke St Gregory TA3 6HY

☎ 0823 490249
An old family business employing
33, engaged in willow growing,
processing and basket-making in
the traditional manner. Baskets
available in shop. C/R&R/W/E.
Mon-Fri 9am-5pm, Sat 10am-5pm.
Tours: Mon-Fri every half hour.
Tour charges: adults £1.75, OAPs
£1.50, children 75p.
Directions: 8 miles east of Taunton.

CERAMICS & KNITWEAR
Pavement Workshop
(Nancy Wells and Jane Fairfax)
The Shambles, North Curry TA3 6LD
☎ 0823 491025
Workshops and shop selling pottery
and designer knitwear made on the
premises; also cards, paintings,
prints and unusual gifts. C
Tues-Fri 10am-4pm, Sat 10am-1pm.

BASKETWORK & CHARCOAL
English Basket Centre
(Nigel Hector)
The Willows , Stoke St Gregory TA3 6JD
☎ 0823 69418
Basketware of all kinds including
willow garden furniture, summer
houses, arbours, rose arches.
Traditional willow farm; willow
processed in the workshop. Hurdles
and artists' charcoal also produced.
Commissions for Chelsea, Hampton
Court Garden Festivals also garden
designers and local authorities.
C/W/E. Credit cards.
Mon-Fri 8am-5pm, Sat-Sun 9am-4pm.

CERAMICS & TEXTILE PAINTING
Fitzhead Studio
(John and V A Watt)
Fitzhead TA4 3JW
☎ 0823 400359
Individual ceramic pieces, sculp-

tural and garden items. Also Batik
pictures and panels on cotton and
silk. Silk scarves and clothing by
commission. C
*Visitors welcome at any reasonable
time.*
Directions: 8 miles west of Taunton off
B3227 (A361); studio half-mile east of
village.

CERAMICS & CRAFTS
Quantock Design Pottery
(Mr and Mrs R A Billington)
Chapel Cottages
West Bagborough TA4 3EF
☎ 0823 433057
Decorative stoneware made by a
variety of moulding methods (slab,
pressed, slipped) with some
throwing. Showroom with other
handmade crafts, paintings and
prints. Also licensed restaurant/tea
room. C/W/E. Credit cards.
Mon-Fri 9am-5.30pm, Sat-Sun 11am-
5pm.
Directions: 9 miles from Taunton,
signed from A358.

DRIED FLOWERS
The Flower Bower
(Mrs S Tucker)
Units 4B-4C Williton Industrial Estate
Williton TA4 4RF
☎ 0984 33677
Manufacturers of dried flower
arrangements; suppliers of dried
flowers, baskets and sundries, both
wholesale and retail. C/W/E
Mon-Fri 10am-4pm (workshop closed
Jan-Feb. Cash-and-carry open all year).

CERAMICS & HAND-PAINTED SILKS
Vellow Pottery & Silks
(David Winkley and Sibylle Wex)
Lower Vellow, Williton TA4 4LS
☎ 0984 56458

Handmade stoneware and porcelain pots made by David Winkley, Fellow of Craftsmen Potters Assoc. Hand-painted silk scarves, blouses, dresses, pictures and cards by Sibylle Wex. Members of regional craft guild. C/W/E. Credit cards. *Mon-Sat 8.30am-6pm, Sun (summer months) 10am-6pm.*
Directions: signposted off A358 2 miles south of Williton. Take road to Vellow and Stogumber; workshop is first building in Vellow village.

CERAMICS
Yarde Pottery, Somerset Terracotta
(Tim Conway)
Yarde, Williton TA4 4HW
☎ *0984 40111*
Handmade terracotta garden and patio pots made to unique designs. C/W/E
Open daily 2-6pm.

Washford

FORGEWORK
Bill Poirrier
Riverden Forge, Roadwater TA23 0QH
☎ *0984 40648*
Design and manufacture of wrought ironwork using traditional techniques, ie tennons, rivets etc. C/R&R/W/E
Mon-Sat 8am-5.30pm.
Directions: from A39 at Washford turn right before White Horse PH to Roadwater. Workshop is 400yd past Valiant Soldier PH on Luxborough road on right.

Watchet

FURNITURE
Justin Williams & Jane Cleal
Unit C, South Road Workshops
Watchet TA23 0HF
☎ *0984 33123*

Contemporary furniture designed and made to commission. Also some smaller batch items (jewellery boxes, clocks etc). C/R&R/E.
Mon-Fri 9am-5pm.
Directions: from A39 Minehead-Bridgwater road at junc with A358 turn north onto A3191. Turn right to Watchet. Turn right at station for South Road.

Wedmore

WOODTURNING
Melvyn Firmager — Sculptural Woodturner
Nut Tree Farm
Stoughton Cross BS28 4QP
☎ *0934 712404*
Sculptural woodturning; organic hollow forms in all sizes and classical shaped vases. Wood used is mainly green wood from storm-damaged trees. Work exhibited in UK and USA. Member of regional guilds. Tuition and demonstrations given. B&B accommodation. C/E. Credit cards.
Open daily 10am-5pm; advisable to telephone first.

Wellington

STAINED GLASS
Crystal Lines
(C Alexander)
Unit 66, Fox's Factory, Tonedale Mill
Wellington TA21 0AW
☎ *0823 666671*
Stained glass and leaded lights, decorative glasswork, acid etching and sandblasting. Restoration and commission of period windows; salvage panels bought and sold. Member of regional Guild of Craftsmen. C/R&R/W/E
Mon-Fri 9am-5.30pm, Sat 9am-1pm.

DOLLS' HOUSES
Monument Miniatures
(John Wensley)
Unit 66F8, Tonedale Mills
Wellington TA21 0AW
☎ 0823 665291/665221
Quality dolls' houses; mainly supplying specialist shops, some to customers' specifications. C/R&R/W/E.
Mon-Fri 10am-6pm, some Sats.
Advisable to telephone first.
Directions: from M5 junc 26, at Wellington town centre traffic lights turn right to Milverton on B3187. After 1 mile cross over railway, after which Tonedale Mills is 1st turn left.

Wells

FORGEWORK
James W G Blunt AWCB
Barnaby, Roughmoor Lane
Westbury-sub-Mendip BA5 1HQ
☎ 0749 870666
Hand-forged traditional and contemporary wrought ironwork for private and commercial customers. Master blacksmith working to commission only. Champion blacksmith at county shows. C/R&R/E.
Visitors welcome by appointment only.

CERAMICS
Black Dog of Wells
(Philippa Threlfall and Kennedy Collings)
8 Tor Street, Wells BA5 2US
☎ 0749 672548
5 years' experience in making a wide range of relief work for architectural use. Terracotta miniatures and reliefs sold through cathedral shops and mail order. Seconds' sold from studio. C/W
Open at most reasonable times;
telephone first, expecially if wishing
to view ceramic work in progress.

Weston-super-Mare

LEATHERWORK
Mendip Leathercraft
(David and Doreen Treasure)
Sidcot Lane, Winscombe BS25 1LA
☎ 093484 2783
Handbags, purses, belts, gift items and small leather goods in over 250 different fine leathers. Basic designs can be varied to suit customers' requirements. C/R&R/W/E.
Open Mon-Sat (early closing Wed & Sat).

Wincanton

FURNITURE RESTORATION
Ottery Antique Restorers
(Charles James and Roderick Cole)
Wessex Way, Wincanton Business Park
Wincanton BA9 4RR
☎ 0963 34572
Small firm employing three craftsmen; furniture restoration for BADA and LAPDA dealers, as well as private customers. C/R&R
Mon-Fri 8am-5pm, Sat-Sun by
appointment.

PAINTED & PRINTED TEXTILES
Diane Davies Designs
2 Bratton House
Higher Bratton Seymour BA9 8DA
☎ 0963 34670
Hand-printed textile designs; dress lengths, scarves and shawls, napkins, cushions, accessories. Workshop courses in silk painting and textile screen printing. C/W
Open day Wed 10am-4pm, please
telephone first. (Closed August-mid
Sept).
Directions: turn off A371 Castle Cary-Wincanton road at lay-by signed Bratton Seymour. Take 1st turn left (Church Way) up hill to large house at top.

Wiveliscombe

PORCELAIN SCULPTURE
Helen Maasz
(Helen and Ronald Maasz)
'Whispers', Langley Marsh
Wiveliscombe TA4 2UL
☎ 0984 24414
Studio sculptured porcelain;
original designs and glazes. C/W/E
Tues & Thur 9.30am-7pm, Sat 9am-
1pm, Sun open all day. Other times by
appointment.

Yeovil

CERAMICS
Ridge Pottery
(Douglas & Jennie Phillips)
High Street
Queen Camel BA22 7NF
☎ 0935 850753
Handmade useful pots of all shapes
and sizes in stoneware and porce-
lain, wood-fired and colourfully
decorated. Summer workshop
courses at the pottery. C/W
Mon-Fri 9am-6pm, Sat 9am-1pm.
Visit by appointment.

GOLD & SILVERSMITHING
Michael Burton
Osborne Cottage
Hurst, Martock TA12 6JU
☎ 0935 822362
Silver jewellery and smithing
(teapots, napkin rings, chess sets,
buckles etc) designed and made by
Member of Goldsmiths Hall. Work
exhibited and held in collections
worldwide. C/R&R/E.
Mon-Fri 9am-5.30pm, Sat-Sun by
appointment.

CABINET MAKING
Robinsons
(Keith Robinson)
High Street, Queen Camel BA22 7NH
☎ 0935 850609
Business established in 1977;
traditional furniture made to
authentic designs from old oak,
walnut, mahogany etc. Reference
library and design books available
for consultation. Showroom with
examples of work. C/R&R/E
Mon-Fri 9am-6pm, Sat by appointment.

CERAMICS
Coker Hill Pottery
(Elsa Benattar)
Bridge Cottage, 10 West Coker Hill
West Coker BA22 9DG
☎ 0935 863630
Small pottery specialising in highly
durable kitchenware (freezer/
dishwasher/micro oven proof).
Some earthenware pit-fired pots for
special exhibitions. C/W/E
Visitors welcome at any reasonable
time by appointment.
Directions: from Yeovil on A30 pass
through West Coker, then right just
before bridge over road, turn left
passing over bridge; property on left
after short distance.

DRIED FLOWERS
Pinkster Dried Flowers
(Jane Pinkster)
Church Farm House, Barwick BA22 9T
☎ 0935 23267
Growing, drying and arranging
dried flowers (no artificial colouring
used); three people employed. 10-
year old business supplying
embassies, hotels, schools, ships
and general public. C/W/E
Visitors welcome by appointment only.

STAFFORDSHIRE CRAFT WORKSHOP CENTRES

Longnor Craft Centre
(Mrs S Fox)
The Market Hall, Longnor SK17 0NT
☎ 0298 83587
Craft centre in former market hall in
centre of village, with exhibitions of
local craft work; high quality
furniture, turned wooden items,
paintings, photography, enamel-
ing, patchwork and quilting,
woven work, decorated china.
Gallery with special monthly
exhibitions of craft work.
Open daily 10am-5pm.

Ridware Arts Centre
(Jennifer & Chris Hobbs)
Hamstall Hall , Hamstall Ridware
Rugeley WS15 3RS
☎ 0889 22351
Crafts in converted Tudor and
Victorian buildings include wood-
work, pottery, fine art, contempo-
rary precious jewellery, glassware.
Galleries and craft shop. Restaurant.
*Late March-Xmas open daily (except
Mon) 10.30am-5.30pm. Xmas-late
March only Sat-Sun and Bank
Holidays 10.30am-5pm.
Admission and car parking free.*
Directions: between A513 Rugeley-
King's Bromley road and A515 King's
Bromley-Ashbourne road.

STAFFORDSHIRE CRAFT WORKSHOPS

Alstonefield

FURNITURE
M K Griffin
Wesleyan House, Alstonefield
☎ 0335 27249
Oak furniture of modern and
traditional design handmade to
customers' specifications; other
English hardwoods available.
Gates, garden furniture. C
Mon-Sat 8.30am-6.30pm
Directions: 7m NW of Ashbourne.

Burton-on-Trent

GLASS ENGRAVING
David Whyman Crystal Engraver
7 High Street, Tutbury DE13 9LP
☎ 0283 520368
Engraving to commission. Any
subject — animals, birds, portraits

etc. A wide selection on display.
C. Credit cards.
Mon-Sat 9.30am-5pm (closed Wed).

CRYSTAL GLASS
Tutbury Crystal Glass Ltd
*Tutbury Glassworks, Burton Street
Tutbury DE13 9NG*
☎ 0283 813281
Handmade hand-cut lead crystal
glassware of exceptional quality. A
complete range of table glasses,
plus many lines in fancy pieces such
as bowls and vases, available in a
variety of designs. C. Credit cards.
*Mon-Sat 9am-5pm. Tours by appoint-
ment.*

GLASS BLOWING
Georgian Crystal (Tutbury) Ltd
Silk Mill Lane, Tutbury DE13 9LE
☎ 0283 814534

Manufacturers of glass employing 19 craftspeople to blow and decorate crystal glassware. C/W/E. Credit cards.
Mon-Sat 9am-5pm.

Cheadle

ANIMAL PORTRAITURE
Dot Merry
The Blacksmith's Shop, Churnet Road Oakamoor ST10 3AB
☎ 0538 702744
Artist specialising in equine, horticultural and domestic animal portraiture. C/E. Credit cards.
Open any time, preferably by appointment.

SPINNING & WEAVING
Windmill Spinners
(Alton Towers)
Alton, Stoke-on-Trent ST10 4DB
☎ 0538 702200
A team of spinners and weavers, all members of regional guild of Spinners, Dyers and Weavers, demonstrating at Alton Towers and offering hands-on experience. Some hand-made items for sale.
Open daily 10am-5pm
Directions: located in the windmill at the entrance to Alton Towers Farm.

CERAMICS
Alton Pottery
(Alton Towers)
Alton, Stoke-on-Trent ST10 4DB
☎ 0538 702200
Founded in 1952, Alton Pottery is one of the attractions within the park. All pottery is hand-thrown, stoneware and slipware. Two master potters, Steve Parry-Thomas and Albert Maddox, well known in the area. Featured in TV pro-

grammes *(Generation Game, Blue Peter* etc). C/W/E. Credit cards.
Open daily Mar-Nov, 9am-6pm.

GLASS BLOWING
Kingfisher Glass
(R Adam)
32 Kingfisher Crescent Cheadle ST10 1RZ
☎ 0538 753894
Demonstrations of glassware making at Alton Towers and craft fairs. W. Credit cards.
Open daily 8am-6pm, lunchtime closing.

Leek

FURNITURE
Aspley Antiques
(John Aspley)
Compton Mill, Compton Leek ST13 5NJ
☎ 0538 373396
Workshop making reproduction furniture in all woods. Three outlets also selling antiques, pictures, ceramics and silver. C/W/E. Credit cards.
Mon-Sat 9am-6pm, Sun 10.30am-5.30pm.

Longnor

FURNITURE
Fox Country Furniture
(Peter and Sheila Fox)
The Old Cheese Factory Reapsmoor, Longnor SK17 0LG
☎ 0298 84467
Family business making quality handmade furniture in English hardwoods and turned items. Work displayed at Longnor Craft Centre. C
Mon-Fri 8am-5.30pm (lunchtime closing 12.30-1.30pm).

Rugeley

MINERALS & GEMS
Midland Crafts
(G J Foulkes and D W Fecher)
Cromwell House
Wolseley Bridge ST17 0XS
☎ *0889 882544*
Manufacturers of onyx and alabaster goods; tables, lamps, clocks etc. Mineral specimens, gemstones and gemset jewellery at wholesale prices. Demonstrations for parties by appointment. W/E. Credit cards.
Mon-Fri 10am-5pm, Sat-Sun 1.30-5pm.
Directions: 2 miles north of Rugeley on A51 Stone-Lichfield road.

WOODWORK
The Wood Experience
(F M Wallace)
The Studio, Ridware Arts Centre, Hamstall Hall, Blithbury Road Hamstall Ridware, Rugeley WS15 3RS
☎ *088922 423*
Conservatories, gazebos, porches, arbours and garden furniture in Western Red Cedar, known for its durability. C/W
Late March-Xmas open daily (except Mon) 10.30am-5.30pm. Xmas-late March only Sat-Sun and Bank Holidays 10.30am-5pm.
Directions: see under Craft Centres.

CERAMICS
Ridware Pottery
(Russell Parker and Nicola de Whalley)
Ridware Arts Centre, Hamstall Hall, Blithbury Road, Hamstall Ridware Rugeley WS15 3RS
☎ *088922 245*
Manufacture and sale of stoneware domestic pottery and hand-coiled decorative vases. Workshop and gallery selling directly to public and wholesale outlets. All work produced is hand-made and original. C/W/E. Credit cards.
Late March-Xmas open daily (except Mon) 10.30am-5.30pm. Xmas-late March only Sat-Sun and Bank Holidays 10.30am-5pm.
Directions: see under Craft Centres.

ARTIST
Studio Arts
(Geoff Selkirk)
Ridware Arts Centre, Hamstall Hall, Blithbury Road, Hamstall Ridware, Rugeley, WS15 3RS
C/W/E
Late March-Xmas open daily (except Mon) 12noon-5pm. Xmas-late March only Sat-Sun and Bank Holidays 12noon-5pm.
Directions: see under Craft Centres.

JEWELLERY
Allen G Brown
Ridware Arts Centre, Hamstall Hall, Blithbury Road, Hamstall Ridware Rugeley WS15 3RS
☎ *0889 22350*
Craftsman and assistant, designing and making contemporary precious jewellery. Award winner.
C/R&R/W/E. Credit cards.
Late March-Xmas open daily (except Mon) 10.30am-5.30pm. Xmas-late March only Sat-Sun and Bank Holidays 10.30am-5pm.
Directions: see under Craft Centres.

GLASSWORK
Sarah Richardson Glass Designs
Ridware Arts Centre, Hamstall Hall, Blithbury Road, Hamstall Ridware Rugeley WS15 3RS
☎ *0889 22297*
Workshop making all forms of studio glassware. Also gallery

selling products and complimentary glass by other designers.
C/R&R/W/E. Credit cards.
Late March-Xmas open daily (except Mon) 10.30am-5.30pm. Xmas-late March only Sat-Sun and Bank Holidays 10.30am-5pm.
Directions: see under Craft Centres.

Stafford

DOLLS' HOUSES & TOYS
Staffordshire Dolls Houses
(Mrs S C Seden-Fowler and Mr A J Fowler)
14 Brookside , Ranton ST18 9JA
☎ *0785 282726*
Craftswoman and husband, making dolls' houses and toys. Member of rural craft association and British Toymakers Guild. C/R&R/W/E.
Mon-Fri 9am-6pm, Sat-Sun 10am-6pm. Visit by appointment.

CABINET MAKING
David Hanlon
Amerton Working Farm
Stowe-by-Chartley ST18 0LA
☎ *0889 270294*
Workshop and small craft shop; cabinet making and woodturning by craftsman with part-time assistance. C/R&R. Credit cards.
Open daily 10am-5pm.

Stoke-on-Trent

RUSH & CANE WORK
S & M Cane & Rush Seating
(Stephen Massey)
143 Leek Road, Shelton ST4 2BW
☎ *0782 48852*
6 years' experience restoring cane and rush seating. C/R&R/W
Open at all reasonable times; please telephone first

Wolverhampton

GLASS ENGRAVING
The Glass Studio
(Mrs Catherine Downes)
Sunshine Farm Crafts, Hilton Lane
Essington WV11 2AU
☎ *0922 416948*
Hand-engraved and sandblasted decorated glassware. Mail order suppliers of glass blanks, especially glass paperweights with recess.
C/R&R/W/E. Credit cards.
Open daily (except Mon) 10.30am-5pm.
Directions: from M6 junc 11 take A462 Willenhall road, after 1 mile take 1st turn right. At T junc turn right; studio 30yd on left, then left behind wood.

CERAMICS
Paul Gooderham
Old Church Studios
The Old Church, Watling Street
Gailey, Nr Standeford ST19 5PR
☎ *0902 790078*
One-off and standard items in porcelain and stoneware, including large decorated panels. Workshop and small studio shop. C/W/E
Mon-Fri 9.30-3.30pm, Sat-Sun by appointment.
Directions: opposite Spread Eagle PH off A449, or through lychgate off A5.

NEEDLECRAFT
White Cottage Country Crafts
(Mrs J Taylor), White Cottage
24 Post Office Road, Seisdon WV5 7HA
☎ *0902 896917*
Workshop running weekend and day courses in patchwork and quilting, embroidery and stencilling. Needlecraft supplies including tapestry. C/W. Credit cards.
Wed-Sat 10am-5pm. (Closed Xmas-New Year)

SUFFOLK CRAFT CENTRE

Aldringham Craft Market
(Godfrey Huddle)
Aldringham, Nr Leiston IP16 4PY
☎ 0728 830397
Galleries exhibiting and selling East Anglian crafts including: pottery, baskets, corn dollies, drawings, etchings and paintings, prints, dolls and toys, glass, jewellery, leather, sculptured and turned wood. Special events. Coffee shop.
Mon-Sat 10am-5.30pm. Sun 10am-5.30pm, lunchtime closing; advisable to telephone first during winter. Admission to galleries, exhibitions and car parking free.

SUFFOLK CRAFT WORKSHOPS

Bungay

SHEEPSKIN CLOTHING
Nursey & Son Ltd
12 Upper Olland St, Bungay NR35 1BQ
☎ 0986 892821
Manufacturers of all lamb leather and sheepskin products. Pioneers of the sheepskin coat; hats, gloves, slippers, rugs, gifts. Repairs on own products. W/E. Credit cards.
Workshop by appointment only; factory shop Mon-Fri 9am-5pm, lunchtime closing.

Bury St Edmunds

LEATHERWORK
Kohl & Son
2 Finchley Avenue
Mildenhall Industrial Estate
Mildenhall IP28 7BG
☎ 0638 712069
Quality leather goods; handbags, purses, light luggage and gift items. C
Mon-Thur 10am-3pm, other times by appointment.

METALWORK
Brookes Forge Flempton
(Brian Brookes)
Flempton IP28 6EN
☎ 0284 728473
Decorative metalwork specialising in manufacture of chandeliers and rushlight/candle holders. Antique metalwork restoration. Fine sand castings in brass, bronze, bell metal, gun metal, nickel silver and iron, patinated to match original patterns. Very varied commissions undertaken. C/R&R/W/E. Credit cards.
Mon-Fri 2.30-6.30pm.

CERAMICS
Brackland Pottery
(David Alexander)
28A The Green
Barrow IP29 5AA
☎ 0284 811222
Pottery established 11 years ago producing ovenproof, slip decorated dishes and bowls, individual dishes and pots, specialising in traditional slip decoration and one-off items. C/W/E
Mon-Thur 9am-8pm, lunchtime closing. Also Fri & Sat Jan-Mar & Oct. (Closed Dec).
Directions: from A45 Cambridge-Bury St Edmunds road turn south to Barrow (2 miles); pottery is next to Spar shop on right.

Debenham

CERAMICS
Carters Ceramic Designs
(Tony & Anita Carter)
Low Road, Debenham IP14 6QU
☎ 0728 860475
Ceramic designers, internationally known for their unusual teapots. Nine people employed; work sold worldwide. C/W/E. Credit cards.
Mon-Fri 9am-5pm, Sat 10am-5pm, lunchtime closing. Advisable to telephone first.
Directions: approaching village from south, turn left at Cherry Tree Inn before bridge over river, at T junc turn right into Low Road, pottery on left.

Eye

CERAMICS
Robin Welch Ceramics
Stradbroke IP21 5JP
☎ 0379 384416
Ceramics artist producing one-off pieces, murals, garden pieces, paintings etc. 30 years' international reputation with many notable commissions and one-man exhibitions world-wide. C/W/E
Open daily 9am-6pm.

CERAMICS
Church Cottage Pottery
(Tom and Heather Baker)
Wilby IP21 5LE
☎ 0379 384253
Handmade domestic garden and decorative pots in terracotta and stoneware. Husband and wife team also providing home baking in tea rooms. C
Mon-Sat (closed Wed) 10am-6pm, Sun 2-6pm. (Closed during Jan).

Hadleigh

FURNITURE
Ben Gordon
The Workshop, Shelley Priory Farm Hadleigh IP7 5RQ
☎ 0206 37297
Furniture making, repairs and restoration work; general woodworking as required. Mainly working to commission. C/R&R
Mon-Fri 9am-5.30pm, Sat by arrangement. Please telephone first.
Directions: leave Hadleigh along Duke Street, then through Lower Layham following signs to Stoke. After nearly 4 miles from Hadleigh, go past pink farmhouse on left, past farm buildings and turn left into farm yard; workshop round the back.

Halesworth

CERAMICS
Dorothy Midson Ceramics
Blythburgh Pottery, Chapel Lane Blythburgh IP19 9LW
☎ 0502 70234
Individual hand-built ceramics. Member of regional craft and potters associations and Society of Designer Craftsmen. C/W
Mon-Sat 9am-5pm, lunchtime closing.
Directions: at rear of Blythburgh Post Office adjacent to A12.

CERAMICS
Chediston Pottery
(Mark Titchiner)
'The Duke'
Chediston Green IP19 0BB
☎ 0986 85242
Pottery specialising in hand-thrown terracotta garden pots using traditional techniques; local clays, large wood-fired kiln. Also some

glazed earthenware. C/W
Mon-Sat 9am-5pm, Sun by appointment.
Directions: from Halesworth on
B1123, turn off right through
Chediston; pottery on left.

Ipswich

CERAMICS
Kersey Pottery
(Fred Bramham and Dorothy Gorst)
The Street, Kersey IP7 6DY
☎ 0473 822092
Handmade tableware, vases, lamps,
bowls, plates and garden pots. The
work is high-fired and decorated
with various glazes and techniques.
C/E. Credit cards.
Tue-Sat 9.30am-5pm, Sun 11am-5pm.
Also Bank Holidays.
Directions: 2 miles from Hadleigh off
A1071.

STRAW CRAFT & DRIED FLOWERS
Corn-Craft
(Royston and Winifred Gage)
Bridge Farm, Monks Eleigh IP7 7AY
☎ 0449 740456
Making of traditional corn dollies
and dried flower products.
Appointment must be made for
demonstration. Also craft shop and
tea room. 70 acre site where flowers
can be picked for drying. C/W/E.
Credit cards acepted
Open daily 10am-5pm. Telephone for
demonstration appointment.
Directions: from Ipswich take A1071
to Sudbury, turn onto A1141 at
Hadleigh, Corn-Craft on right before
reaching Monks Eleigh.

FORGEWORK
Clover Forge
(Trevor Self)
Brimlin Cottage, Chattisham IP8 3QQ
☎ 0473 87441/87480

Traditional forge making individual
items to customers' requirements.
C/R&R/W
Mon-Fri 8am-4pm, Sat 9am-1pm.

Lowestoft

CERAMICS
Earth Pottery
(Giles and Carol Cattlin)
94 Norwich Road, Lowestoft NR32 7BD
☎ 0502 511396
Hand-thrown stoneware in several
different colour glazes and variety
of patterns. Plant pots, kitchenware,
lamps, candle holders, celebration
plates. Work also sold at Pleasure-
wood Hills Theme Park, Corton,
Lowestoft and Fritton Lake.
C/W/E. Credit cards.
Mon-Sat 9am-6pm.

BOAT BUILDING
Boatworld
(John Elliot)
Harbour Road, Oulton Broad NR32 3LZ
☎ 0502 569663
Boat building business on the
shores of Lake Lothing incorporating the world famous International
Boatbuilding Training College;
visitors welcome to watch students,
many from overseas, practising old
craft skills on a wide variety of
boats under the guidance of master
boat builders; 12-14 craft of various
sizes always under construction.
Tea rooms. Guided tours available
by appointment.
C/R&R/E. Credit cards..
May-Oct Mon-Fri 10am-4pm
Admission charge: adults £1.95, OAPs
and children £1.50 (free under 5 yrs).
Family: £7. Groups: special rates.
Directions: just off A1117 close to
centre of Oulton Broad.

Saxmundham

CERAMICS
Milestone House Pottery
(D Rose, B Rose and H Barclay)
High Street, Yoxford IP17 3EP
☎ *0728 77465*
Three potters, members of Suffolk
Craft Society, producing strong
reduction-fired stoneware in
natural colours; also porcelain.
Wide and unusual range of domestic and garden pottery; special
orders undertaken. Shop also
selling other crafts and gallery with
special exhibitions. C. Credit cards.
Mon-Sat (Wed half-day closing) 10am-
5pm. Jan-Easter closed Mon. Visitors
welcome by appointment only.
Directions: off A12 Lowestoft-Ipswich
road, take A1120 through Yoxford,
towards Peasenhall; on left after PH.

CERAMICS
Jonathan Keep Pottery
31 Leiston Road, Knodishal IP17 1UQ
☎ *0728 832901*
Handmade, hand decorated
domestic pottery; well crafted, well
designed domestic pots. C/W/E
Mon-Fri 9am-5.30pm, Sat 10am-5pm.

CABINET MAKING
John Barrett — Cabinet Maker
Cottons Yard, High Street
Yoxford IP17 3EY
☎ *0728 77652/77654*
Furniture designed and made to
order, both original and reproduc-
tion. Some notable commissions.
Many pieces in private collections.
Member of Suffolk Crafts Society. C
Mon-Fri 8.30am-6pm, Sat-Sun variable
times; advisable to telephone first.
Directions: next to Milestone House
Pottery (see above).

Stowmarket

CERAMICS
Pauline Bracegirdle
Lodge Cottage, Back Street
Gislingham IP23 8JG
☎ *0449 781470*
Trained artist and painter making
earthenware pottery; domestic
ware, decorative plates and ceramic
small earthenware sculpture. Good
designs and decoration. House
signs and commemorative plates. C
Open daily (except Fri mornings)
10am-5pm, lunchtime closing.

Sudbury

WOODWORK
Clarity 3D
(Michael Goffin and Maurice Fitzgerald)
Unit 8, Alexandra Road
Sudbury CO10 6XH
☎ *0787 76181*
Furniture and interior design in
wood; individual commissions and
historical and heritage projects
related to the leisure industry.
C/R&R/W/E. Credit cards.
Mon-Fri 9am-5pm, Sat 9am-2.30pm.

Woodbridge

FORGEWORK
Hector Moore AFCL
The Forge, Brandeston IP13 7AN
☎ *0728 685354*
Established family business offering
traditional and modern designs and
techniques in blacksmithing and
metalwork including architectural
and ecclesiastical. Sculpture.
Specialists in village signs. Many
designs available. C/R&R/W/E.
Mon-Sat 8am-5pm, advisable to
telephone first. Sat by appointment.

LEATHERWORK & SHEEPSKINS
Shasha Toptani
22-24 Fore Framlingham
Woodbridge IP13 9DF
☎ *0728 724206*
Made-to-measure leather, suede
and sheepskin; bags, belts etc. Other
crafts from local artists. C/R&R/E
Open daily 9am-5.30pm, some Sat
mornings; advisable to telephone first.

CERAMICS
Butley Pottery
(Honor Hussey)
Mill Lane, Butley IP12 3PA
☎ *0394 450785*
Pottery producing Majolica earthenware; hand-painted plant and
domestic pottery. Guided tours of
workshop by arrangement. Tuition:
five-day and weekend courses.
Other craftwork sometimes
demonstrated. Showrooms and tea
rooms. C/W/E. Credit cards.
Open daily 10.30am-5pm.

TOY MAKING
Ron Fuller
Willow Cottage, Laxfield IP13 8DX
☎ *0986 790317*
Wooden, tinplate and cardboard
toys made by sole craftsman; also
repairs of old toys. C/R&R/W
Mon-Sat 9am-5pm, lunchtime closing.

SURREY CRAFT WORKSHOP CENTRES

Manor Farm Craft Centre
Wood Lane, Seale
Nr Guildford GU10 1HR
☎ 02518 3661
Individual workshops with show-rooms, including woodturning, jewellery and stained glass, pottery, knitting, smocking, soft furnishings etc. Also restaurant with tea rooms.
Tues-Sat 12noon-5pm, Sunday 2-5pm.
Directions: from Guildford take A31 Hog's Back road to Farnham. After passing Hogs Back Hotel on right, take first turn left to Seale. Craft Centre is half a mile down Elstead road on left.

Smithbrook Kilns
Cranleigh GU6 8JJ
☎ 0483 276455
Art, craft and business centre where craftspeople work and sell their products (see entries below). Photograph library and David Shepherd Wildlife Charity Shop selling prints, books, videos etc to raise funds for conservation projects. Restaurant catering for functions, and home-made food shop. Picnic area and facitilities for public.
Individual business opening hours.
Directions: on A281 Guildford-Horsham road, near crossroad with Cranleigh-Dunsfold/Godalming road.

SURREY CRAFT WORKSHOPS

Cranleigh

SOFT FURNISHINGS & UPHOLSTERY
Fabric Fantasy
(Lesley Jacevicius), Unit 5
Smithbrook Kilns, Cranleigh GU6 8JJ
☎ 0483 275581
Made-to-measure curtains, loose covers, headboards and other upholstery, lampshades etc. Three employed in workshop. C/R&R
Mon-Fri 9.30am-5.30pm, Sat 9.30am-1pm.
Directions: see under Craft Centres.

DRESSMAKING
Buttons & Bows
(Helen Kolmar), Unit 7
Smithbrook Kilns, Cranleigh GU6 8JJ
☎ 0483 275688
Dressmaking specialising in bridal gowns and ball gowns. C/R&R
Tues-Fri 10am-5pm, Sat 10am-1pm
Directions: see under Craft Centres.

FURNITURE
Sean Fitzgerald
The Studio Workshop
Smithbrook Kilns, Cranleigh GU6 8JJ
☎ 0483 275900
Contemporary furniture in selected hardwoods and veneers. Work mainly to commission. C
Mon-Fri 10am-6pm, Sat 10am-5pm.
Directions: see under Craft Centres.

GOLD & SILVERSMITHING
Jewellery Studio
(K Rogers and S Parsons)
Unit 23-24, Smithbrook Kilns
Cranleigh GU6 8JJ
☎ 0483 277963/273111
Professional qualified partnership; exclusive handmade jewellery using precious metals and stones. C/R&R
Mon-Fri 9am-5pm, Sat 9.30am-5pm.
(Closed Xmas-New Year)
Directions: see under Craft Centres.

ARTIST
Annette Olney
The Smithbrook Art Studio
Unit 22, Smithbrook Kilns
Cranleigh GU6 8JJ
☎ *0483 271311*
Studio/gallery; pastel portraits of
animals and children. Line drawings
of houses etc. Freelance illustration
in colour and black-and-white. C
Tues-Sat 10.30am-6pm, Sat 11am-5.30pm.
Directions: see under Craft Centres.

ANTIQUE FURNITURE
RESTORATION
Smithbrook Restorations
(Stephen Woodger and Simon Wootton)
Unit 63a, Smithbrook Kilns,
Horsham Road, Cranleigh GU6 8JJ
☎ *0483 268376*
Furniture restoration service; French
polishing, etc. Cabinet making.
C/R&R/E.
Mon-Fri 8.15am-6pm.
Directions: see under Craft Centres.

DECORATIVE PAINT FINISHES
The Bermuda Collection
(Jamie and Betsy Sapsford)
Unit 63, Smithbrook Kilns
Cranleigh GU6 8JJ
☎ *0483 268418*
Decorative painting of furniture
and interiors; all aspects of broken-
colour work, stencilling and hand-
painted designs. In-house stencil
design service. Clients' furniture
decorated and new pieces.
C/R&R/W. Credit cards.
Open daily (except Mon) 10am-5pm.
Directions: see under Craft Centres.

FURNITURE MAKING
What Not Antiques
(J and M Wylie), Units 14 & 56
Smithbrook Kilns, Cranleigh GU6 8JJ

☎ *0483 271796*
Handmade and hand polished pine
furniture; small family business.
C/R&R/W/E. Credit cards.
Mon-Fri 9.30am-5.30pm, Sat-Sun
10.30am-5.30pm.
Directions: see under Craft Centres.

DRESSMAKING
Sew Simple
(Marianne Britton), Unit 83
Smithbrook Kilns, Cranleigh GU6 8JJ
☎ *0483 267917*
All types of dressmaking and
alterations; 8-lesson sewing course
working in knit fabrics. Large
selection of fabrics for sale. C/R&R
Mon-Fri 9.30am-5pm, lunchtime closing.
Directions: see under Craft Centres.

SOFT TOYS
Poppicats
(Mrs Margaret White), Unit 3A
Smithbrook Kilns, Cranleigh GU6 8JJ
☎ *0483 268021*
Workshop making soft toys; other
local crafts also for sale. C/R&R/W
Tues-Fri 10.30am-5pm, Sat-Sun
11.30am-3.30pm.
Directions: see under Craft Centres.

STONEMASONRY
Sussex Stone (Billingshurst) Ltd
(C G Kemp - Director), Unit 59-60
Smithbrook Kilns, Cranleigh GU6 8JJ
☎ *0483 277969*
Traditional, ecclesiastical and
contemporary stonework. Five
craftsmen employed in restoration
work, carving, lettering etc. Stone,
slate, marble supplies. C/R&R/E
Mon-Fri 8am-5pm. (Closed Xmas-
New Year)
Directions: see under Craft Centres.

UPHOLSTERY
Cygnet Designs
(David Picott), Unit 8
Smithbrook Kilns, Cranleigh GU6 8JJ
☎ *0483 276001*
Upholstery and restoration of new
and antique furniture. C/R&R/W/E
Mon-Sat 9am-5.30pm.
Directions: see under Craft Centres.

ENGRAVING
Tedstones
(Eric and Cindy George), Unit 49
Smithbrook Kilns, Cranleigh GU6 8JJ
☎ *0483 272010*
Engraving and manufacture of
sports trophies and specialist dart
shop. Engraving on glass; jewellery,
hot foiling etc. C/R&R/W/E
Mon-Fri 8am-8pm, by appointment.
Sat 10am-4pm.
Directions: see under Craft Centres.

FORGEWORK & LIGHT FITTINGS
Smithbrook Ltd
(R Cook)
Smithbrook, Cranleigh GU6 8LH
☎ *0483 272744*
Leading manufacturers of fine iron
light fittings with many different
finishes, some incorporating wood;
sizes range from domestic to over a
quarter-ton. Gates, railings, firetools
and accessories, garden items etc.
Showroom in mediaeval barn.
C/R&R/W/E. Credit cards.
Mon-Sat 8.30am-5.30pm.
Directions: see under Craft Centres.

SILVER JEWELLERY
Jewellery by Jon Dibben
Unit 40A, Smithbrook Kilns
Cranleigh GU6 8JJ
☎ *0483 278170*
Jewellery designed and made in
silver and precious metals with

semi- and precious stones. Evening
jewellery courses.
C/R&R/W/E. Credit cards.
Tues-Sat 10am-5.30pm.
Directions: see under Craft Centres.

CLOCKMAKING & REPAIRING
Peter Hopkins Clocks
Unit 45A, Smithbrook Kilns,
Horsham Road, Cranleigh GU6 8JJ
☎ *0483 278201*
Small traditional workshop
specialising in the repair, restoration and sale of quality and antique
clocks. Member of British Horo-
logical Institute and British Watch
and Clockmakers Guild. C/R&R
Wed-Sat 10am-4.30pm.
Directions: see under Craft Centres.

Dorking

CERAMICS
Joan Hepworth
Robin Cottage, Stones Lane
Westcott RH4 3QH
☎ *0306 880392*
Studio pottery making hand-built
stoneware and cast porcelain. C/W/E
Visitors welcome by appointment
only. Visiting charge: £5pp.

FORGEWORK
Newdigate Forge
(Kit Lambert)
Unit 2, Dean House Farm
Church Road, Newdigate RH5 5DL
☎ *0306 77736*
Traditional blacksmithing; orna-
mental ironwork, restoration,
sculptural work. Member of British
Artist Blacksmith Association.
C/R&R/W/E
Mon-Fri 8am-6pm, Sat-Sun 9am-1pm.

Farnham

GOLD AND SILVERSMITHING
Clare Street Jewellery & Engraving
*Little Orchard, Woodcut Road
Wrecclesham GU10 4QF*
☎ 0252 733232
Design and manufacture of precious metal jewellery. Also hand-engraving specialising in seal engraving of signet rings etc, (prizes have been won in Goldsmiths' Company annual competition). C
*Mon-Fri 9am-9pm, Sat 11am-1pm.
Please telephone for appointment at least 24hrs before visiting.*

KNITWEAR
Alison Ellen Knitwear
Jeffreys Cottage, Dockenfield GU10 4HS
☎ 025 125 2442
Individually designed and hand-knitted jumpers, jackets, waistcoats and hats in hand-dyed natural yarns. On Crafts Council selected index and member of regional craft guild. C/W/E. Credit cards.
Open daily by appointment.
Directions: from Farnham station take A287 to Frensham, turn right to Dockenfield. Jeffreys Cottage down turning on right just before turning left and church.

Godalming

FORGEWORK
M G Henke Wrought-ironwork
*(Marcus Henke BA)
Hambledon House, Van Lane
Hambledon GU8 4HW*
☎ 0428 684343
Individually designed, hand-forged wrought-ironwork made to commission. Gates, railings, balustrades, ornamental screens, fire-baskets and most items of decorative ironwork. C/R&R
Mon-Fri 9am-6.30pm, Sat 9am-1.30pm, by appointment only.
Directions: take A283 to Petworth, turn right to Hambledon.

FORGEWORK
Chiddingfold Forge
*(David Wright)
The Green, Chiddingfold GU8 4XR*
☎ 0428 684902
Wrought ironwork, restoration, repairs, fabrications and some non ferrous work. Metalworker with 30 years' experience making unusual pieces (wrought iron spinning wheel and threading hook, spinning and weaving box etc). C/R&R/W/E.
Mon-Sat 9am-5pm, lunchtime closing. Evenings best; advisable to telephone first.
Directions: between Milford on A3 and Petworth on A283, forge is on the Green between the Crown Inn PH and pond.

SPINNING & WEAVING
Thursley Textile Designs
*(Zoë O'Brien)
1 Moushill Lane, Milford GU8 5BH*
☎ 0483 424769
Three textile artists working with varied fibres to produce fabrics ranging from fine silk scarves to heavy floor rugs and wall hangings. Members of national and regional craft guilds. Textile shop. Groups by arrangement. C/W/E. Credit cards.
Mon-Sat 10am-5.30pm. (Closed Xmas-New Year).

Godstone

HARP MAKING & RESTORATION
Pilgrim Harps
*(John D Hoare and Michael Stevens)
Stansted House, Tilburstow Hill Road
Godstone RH9 8NA*

☎ 0342 893242

Harp makers and restorers producing a range of high quality instruments; Celtic, folk, chromatic pedal. All available with alternative finishes. Covers, cases and makers' materials supplied. C/R&R/E

Mon-Fri 8am-6pm and weekends by appointment.

Directions: 2 miles south of Godstone.

UPHOLSTERY RENOVATION
Robert Lines Quality Upholstery
Unit H5, Haysbridge Business Centre, Brickhouse Lane
South Godstone RH9 8JW
☎ 0342 844208

Period and contemporary soft furnishing restoration, also cars, carriages etc. Winner of several awards (including Master Upholsterers). C/R&R

Mon-Fri 8.30am-5.30pm.

Guildford

CERAMICS
Grayshott Pottery
(David Real and Philip Bates)
School Road
Grayshott, Hindhead GU26 6LR
☎ 0428 604404/606466

An extensive selection of plain and decorated stoneware made on the premises; porcelain clocks and oven-to-table ware. Shop with crafts and gifts. C/W/E. Credit cards.

Mon-Sat 9am-5pm (workshop: daily demonstrations until 4.30pm). Advisable to telephone first.

KNITWEAR
Scarab Knitwear
(Vivien Calleja)
14 Beechcroft Drive, Guildford GU2 5SA
☎ 0483 38653

Machine-knitted plain or Fair Isle coats, skirts, suits etc all from own designs. Hand-knits with crochet finish. Jackets in mohair a speciality, in unique design or colours. C

Visitors welcome by appointment only.

CERAMICS
Mary Wondrausch
The Pottery, Brickfields
Compton GU3 1HZ
☎ 0483 414097

Pottery workshop and shop specialising in slipware and scgraffito; commemorative pieces. Author of book and articles on slipware. Member of Craftsmen Potters Association. C/W/E

Mon-Fri 9am-5pm, Sat-Sun 2-5pm.

GOLD AND SILVERSMITHING
Lorraine van Papen
22 Halfpenny Close, Chilworth, St Martha's, Guildford GU4 8NH
☎ 0483 67886

Jewellery designed and made in gold, silver and precious stones. Also restoration, re-modelling and repairs undertaken. Award won for diamond jewellery design, member of Gemmological Associaton of GB and regional craft guild. C/R&R

Open daily 10am-5pm, advisable to telephone first. (Closed Xmas-New year).

PICTURE FRAMING
Country Pictures
(Tim Clarke)
Whipley Manor Farm, Horsham Road
Bramley GU5 0LL
☎ 0483 268310

Picture framing and restoration workshop with picture gallery. C/R&R/W/E. Credit cards.

Mon-Fri 10am-5pm, Sat 10am-1pm.

Directions: from Bramley on A281 Horsham road, at crossroad (Selhurst Common/Wintershall signed to right) turn left; Country Pictures on right.

RUSH & CANE SEATING
John Harriman
Bakers Cottage, Shere Road
Ewhurst GU6 7PQ
☎ *0483 272639*
Re-seating of cane and rush chairs, seagrass and Danish cord. Also antique furniture restoration. Maker of Bergére suites, screens, head-boards, grandfather clocks, tables etc. C/R&R
Visitors welcome by appointment only.

CERAMICS
Larchwood Pottery
(Mavis Yates)
Manor Farm Craft Centre, Wood Lane
Seale GU10 1HR
☎ *02518 3661*
Wide variety of handmade ceram-ics; vases, bowls, framed pictures, wall hangings.
C/W/E. Credit cards.
Tues-Fri 11.30am-4.30pm, Sat-Sun 2-5pm.
Directions: see under Craft Centres.

KNITWEAR
Seale Craft Shop
(Janice Midgley)
Manor Farm Craft Centre, Wood Lane
Seale GU10 1HR
☎ *02518 3661*
Machine-knitted dresses, suits, jumpers and jackets. Many original designs; orders welcomed. Winner of prizes in national competitions. C/R&R/E
Tues-Fri 11am-5pm, Sat 2-5pm.
Directions: see under Craft Centres.

WOODTURNING
Bob Cordell — Woodturner
Manor Farm Craft Centre, Wood Lane
Seale GU10 1HR
☎ *02518 3661/2103*
All types of woodturning; also other wooden items made.
C/R&R/W/E
Tues-Sat 10.30am-5pm. Some Suns 2-5pm.
Directions: see under Craft Centres.

SOFT FURNISHINGS
Sewing Box
(Miss S Morgan, Mrs S Carter and D Babayan)
Manor Farm Craft Centre, Wood Lane
Seale GU10 1HR
☎ *02518 2101*
All soft furnishings; curtains, loose covers, upholstery etc hand-finished. Also smaller items; cushions, peg bags, oven gloves etc. C/R&R/W
Tues-Fri 12noon-5pm, Sat-Sun 2-5pm.
Directions: see under Craft Centres.

DECORATIVE PAINT FINISHES
Angela Shaw Hand-Painted Furniture
Flexford House
Hog's Back GU3 2JP
☎ *0483 810223*
Workshop handling private commissions and courses in all types of decorative paint finishes including stencilling, antiquing, gilding, découpage etc.
C/R&R/W/E
Mon-Fri 9.30am-4.30pm.
Directions: from A3, fork off to A31 Farnham road. Take first slip-road to left signed Puttenham/Godalming. Turn right at T junc, taking you on A31 back towards Guildford. After 1 mile, sign to Flexford House on left.

Hindhead

ANTIQUE DOLL RESTORATION
Columbine Crafts & Curios
(Gillian Rawcliffe)
Crossways Road
Grayshott GU26 6HG
☎ *0428 605220*
Restoration of antique dolls and jewellery. Range of unusual gift items on sale. C/R&R/W
Mon-Sat (closed Wed) 9.30am-5pm, lunchtime closing.

Horley

CERAMICS WORKSHOP
Chris-Craft Hobby Ceramics
(Christine Skeggs)
The Workshop, Edolphs Farm,
Norwood Hill Road
Charlwood RH6 0EB
☎ *0293 863241*
Hobby ceramics classes run by qualified teacher. Firing service for classes and customers; materials and accessories available.
Tues-Fri 10am-4pm, Sat 10am-1pm.
Directions: Edolphs Farm is 1 mile from Charlwood village shops on road signed to Leigh.

Leatherhead

CERAMICS
The Pottery Studio
(Eileen Stevens)
The Birches, Park Close
Fetcham KT22 9BD
☎ *0372 375967*

Reduction stoneware pottery; domestic and individual pieces.
Open daily 2-5pm, please telephone first.

FORGEWORK
Richard Quinnell Ltd, Fire & Iron Gallery
(Richard Quinnell MBE)
Rowhurst Forge, Oxshott Road
Leatherhead KT22 0EN
☎ *0372 375148*
Expert consultant in design, construction and restoration of ornamental metalwork both in UK and overseas; Richard Quinnell received the MBE in 1989 for his central part in the revival of black-smithing. Nine craftsmen working on architectural metalwork commissions. Gallery exhibiting and selling work of leading artist-metalsmiths. C/R&R/E. Credit cards.
Mon-Fri 9am-5pm, lunchtime closing. Sat 9am-1pm.

Windlesham

CERAMICS
Elaine Coles Ceramics
Country Gardens, London Road
Windlesham GU20 6LL
☎ *0344 874181*
Pottery making a wide range of domestic ware, indoor plant pots and other unusual and interesting ceramics. Professional member of Craftsmen Potters Association and regional guild of craftsmen. C/E. Credit cards.
Wed-Sun 10am-5pm.

EAST SUSSEX CRAFT WORKSHOPS

Battle

CHARCOAL MAKING
Blackman, Pavie & Ladden Ltd
Marley Lane
Battle TN33 0RE
☎ 0424 870333
Lumpwood charcoal manufacturers, packers, distributors and importers, established in 1888.
Mon-Fri 8am-4.30pm by appointment.

CARRIAGE BUILDING &
WHEELWRIGHTING
R G Carey
Parish Farm, Hooe TN33 9HS
☎ 0424 892051
Carriage builder producing vehicles, many for use in driving trials. Wheelwrighting and carriage restorations. C/R&R/W/E
Mon-Fri 8.30am-5pm, lunchtime closing. Sat-Sun by appointment.

BRICK AND TILE MAKING
Aldershaw Tiles
(Richard Williams)
Aldershaw Farm
Kent Street (A21)
Sedlescombe TN33 0SD
☎ 0424 754192
Handmade buff and terracotta (Sussex Weald clay) mathematical tiles, bricks and tiles, floor paviors, decorative bricks, garden edging tiles. Nine employed, specialising in all types of handmade bricks and tiles in a variety of colours for conservation work. Winners of many small business awards and BBC Enterprise Award.
C/R&R/W/E
Mon-Fri 9am-5.30pm, Sat 9am-12noon.

CLAY PRODUCTS
Aldershaw Crafts
(Mrs Beverley Blunden)
Aldershaw Farm, Kent Street (A21)
Sedlescombe TN33 0SD
☎ 0424 754192
Handmade clay products, flowers, birds, houses etc. Naive paintings on wood. Miniature paintings. House signs to order. Hand-painted wooden candlesticks and frames, tuck boxes etc. Dried flowers and candles. C/W/E
Open daily 9am-5pm. (Closed Xmas-Easter)

Beckley

FORGEWORK
Great Knelle Forge
(Robert Booth)
Whitbread Lane
Beckley TN31 6UB
☎ 0797 260570
Forge situated on popular 'Childrens' Farm' as seen on TVS. Hot forge work; handmade items from smallest bracket up to heavy steel fabrication. Restoration and repair of all metals. Supply and fitting service. C/R&R/W/E
Mon-Fri 8.30am-5pm, Sat 9am-1pm.
Directions: on A268 between Beckley and Northiam. Childrens' farm is well signed.

Brighton

CERAMICS
Eileen Lewenstein
11 Western Esplanade
Portslade BN41 1WE
☎ 0273 418705
Eileen Lewenstein's ceramic bowls,

EAST SUSSEX · BATTLE · BECKLEY · BRIGHTON

219

vases, relief panels and sculptures have been widely exhibited and can be seen at the V & A Museum and other collections including Contemporary Ceramics, London. C/W
Visitors welcome by appointment only.

TRADITIONAL DOLL MAKING & RESTORATION
Recollect Studios
(Carol and Jeff Jackman)
The Old School, London Road
Sayers Common BN6 9HX
☎ *0273 833314*
Family business set up in an old village school, established over 20 years' ago, restoring antique dolls and making reproductions of traditional porcelain and wax dolls. Suppliers of dolls (in kit form or assembled) and parts, doll accessories and dolls' house miniatures. Regular teaching courses given in all aspects of doll making and restoration. Members of UK Institute for Conservation and the Doll Artisan Guild. Groups by appointment; demonstrations. C/R&R/W/E. Credit cards.
Tues-Sat 10am-5pm.
Directions: from Brighton on old A23 (now B2118) after B2116 Worthing/Lewes traffic lights, Recollect Studios car park on right.

Burwash

FORGEWORK
Wealden Ironcrafts
(D A Hedges and Mrs A J Watson)
The Forge, High Street
Burwash TN19 7EP
☎ *0435 883422*
Ornamental wrought ironwork; gates, balustrades, balcony rails etc. Two craftsmen employed in high

class blacksmithing of all types. C/R&R/W/E
Mon-Fri 7am-6pm, lunchtime closing. Sat 7am-1pm.

Ditchling

GOLD & SILVERSMITHING
Anton Pruden & Rebecca Smith — Silversmiths
1 Turner Dumbrell Workshops,
North End, Ditchling BN6 8TD
☎ *0273 846338*
Following a family tradition in silversmithing, work produced particularly for ecclesiastical use (commissions from all over UK and overseas for chalices, communion sets, cloak clasps etc.) Jewellery and silver and gold work for domestic use. C/R&R/W/E. Credit cards.
Tues-Sat 10am-6pm, lunchtime closing.

CERAMICS
Jill Pryke Pottery
8 High Street, Ditchling BN6 8TA
☎ *0273 845246*
Green and blue glazed earthenware pottery. Domestic items and named mugs, wedding and anniversary plates to order. C
Mon-Sat (closed Wed) 10am-5pm, lunchtime closing. (Closed Xmas-New Year)

SADDLERY
Dragonfly Saddlery
(Richard and Sue Paine)
Ditchling Crossroads
Ditchling
☎ *0273 844606*
Husband and wife team; Master Saddlers supplying and repairing saddlery for customers worldwide. C/R&R/E. Credit cards.
Open daily 9.15am-5.15pm.

PICTURE FRAMING
Roger Mills Picture Framer
7 High Street, Ditchling BN6 8SY
☎ 0273 844547
Bespoke framing by craftsman with
nearly 30 years' experience. C/R&R
Tues-Sat 9am-5pm, lunchtime closing.

Hailsham

ANTIQUE FURNITURE
RESTORATION
Richard Reading, Furniture
Maker and Restorer
Westwood, Trolilloes Lane
Cowbeech BN27 4QR
☎ 0435 830249
Fine cabinet making of all descrip-
tions and furniture restoration.
C/R&R/W/E
Mon-Fri 9am-5pm, Sat 9am-1pm.

THATCHING
Master Thatchers (South)
(John Hall)
Diplocks Farmhouse, Western Road
Hailsham BN27 3EJ
☎ 0323 841057
Thatching contractors; member of
National Society of Thatchers and
Guild of Master Craftsmen.

Hastings

GILDING & PICTURE FRAME
RESTORATION
Ian Pike
Little Hides Farm Cottage
Westfield TN35 4PH
☎ 0424 754104
Gilder specialising in restoration of
antique frames and all aspects of
decorative paint finishes (marbling
etc). C/R&R/W
Mon-Fri 9.30am-5.30pm.

FORGEWORK
Ironcrafts Ltd
(D A J Griffiths)
Wheel Lane, Westfield TN35 4SE
☎ 0424 751775
Four forgeworkers employed
producing wrought ironwork.
Commissions for many major
buildings in London. Suppliers of
ironwork to stores and galleries.
C/R&R/W/E
Mon-Fri 8.30am-5.30pm.

Heathfield

SPINNING & CARDING SERVICE
Diamond Fibres Ltd
(R E and P M Mobsby and A G and
D J Ferrige)
Diamonds Farm
Horam TN21 0HF
☎ 04353 2414
Production of finished yarn from a
wide variety of wool animals.
Woollen yarns, carded wool and
fleeces. Washing and carding of
other wools; spinning of wool and
wool blends. C/W
Open at any time, by appointment.

UPHOLSTERED FURNITURE &
RESTORATION
The Old Bakery Furnishing Co
(Alan and Ann Spencer)
Punnett's Town TN21 9DS
☎ 0435 830608
Hand-built elegant sofas and chairs
manufactured, custom-made to any
size or finish. Also upholstery
restoration work and soft furnish-
ing. Showroom selling wide range
of interior design supplies.
C/R&R/E. Credit cards.
Mon-Fri 9am-5pm, Sat 9am-1pm.
Directions: in village of Punnett's
Town on B2096 to Battle.

WHEELWRIGHTING
David Bysouth
The Hub, Three Cups
Heathfield TN21 9RD
☎ 0435 830776
Two wheelwrights employed making wooden wheels and shafts for veteran cars, carts, waggons and horse-drawn vehicles. C/R&R/E
Mon-Fri 8.30am-5.30pm.

FORGEWORK
Ragged Dog Forge
(David Pettitt)
Ragged Dog Lane, Waldron TN21 0NJ
☎ 04353 2015
Forgeworker trained in all aspects of blacksmithing and design. Won Worshipful Company of Blacksmiths' certificate and winner of competition for first International Festival of Iron. C/R&R/W/E
Mon-Fri 8am-5.30pm, Sat 8am-12.30pm.

Herstmonceux

TRADITIONAL CHAIR MAKING
Sussex Windsors
(Barry and Mary Murphy)
Dormer's Farmhouse, Windmill Hill
Hailsham BN27 4RY
☎ 0323 832388
Makers of traditional-style Windsor chairs, stools and high-chairs in ash and elm; also country style pieces in English hardwoods. C
Mon-Fri 10am-5pm, lunchtime closing. Sat-Sun 10am-1pm.
Directions: 1 mile east of Herstmonceux on A271.

BASKETWORK
Thomas Smith (Herstmonceux)
(South Down Trugs & Crafts Ltd)
The Trug Shop, Hailsham Road
Herstmonceux BN27 4LH
☎ 0323 832137/833801
The original makers of traditional Sussex trug baskets made from sweet chestnut and willow. Varied sizes and shapes, plain and decorated. Can be dated and named for special occasions. Winners of Export Award. Work sold by department stores, garden centres, crafts shops and mail order companies. R&R/W/E. Credit cards.
Mon-Fri 9am-5pm (Sat, shop only: 10am-5pm). (Closed Xmas-New Year).
Entry fee between £1 and 50p.

BASKETWORK
The Truggery
(Mrs S Sherwood)
Coopers Croft
Herstmonceux BN27 1QL
☎ 0323 832314
Business established over 100 years ago, making Sussex trug baskets using traditional methods. C/R&R/W. Credit cards.
Open daily 10am-5.30pm. (Closed Xmas-New Year).
Directions: on A271 on outskirts of Herstmonceux.

Lewes

SMOCKING & CHILDRENS' CLOTHE
Specially for You
(Mrs Julian Akers-Douglas)
Warnham Cottage
East Hoathly BN8 6DP
☎ 0825 840397
Unique contemporary and traditional designs for adults and childrens' clothes with hand-smocking and embroidery. All custom made in busy workshop. C/W/E. Credit cards.
Thur-Sat 10am-4pm. Other times by appointment.

FURNITURE
Wales & Wales
(Rod and Alison Wales)
Longbarn Workshop, Muddles Green
Chiddingly BN8 6HW
☎ 0825 872764
Commissioned furniture for public,
commercial and private clients.
Many projects in collaboration with
architects and designers (including
Canary Wharf). C/W/E
Mon-Fri 8.30am-6pm.

FURNITURE
John Wyndham Furniture Designs
Westgates, Muddles Green
Chiddingly BN8 6HW
☎ 0825 872036
Design and manufacture of furni-
ture to commission, mostly in
European hardwoods. Winner of
South East Arts Award 1982. C/E
Open daily (except Sat) 9am-1pm.
Visitors welcome by appointment
only. Visiting charge £10.

BATIK & COLLAGE WORK
Mary Potter Studio
Hunters Wood, Laughton BN8 6DE
☎ 0825 840438
Batik and collage pictures and
panels; screen printed silk scarves
and cards. Workshop/studio with
gallery, work widely exhibited in
UK and overseas. C/W/E
Open during normal working hours,
lunchtime closing. (Closed during May).
Directions: from Ringmer on Lewes-
Eastbourne road at Laughton turn left at
Roebuck Inn. Studio on right up lane.

MUSICAL INSTRUMENT MAKING
Malcolm Rose and Karin Richter
The Workshop, English Passage
Lewes BN7 2AP
☎ 0273 481010

Makers of harpsichords and clavichords. Conservation of early keyboard instruments. C/R&R/W/E
Visitors welcome by appointment.

MOHAIR PRODUCTION
Brickfield Angoras — The Mohair Centre
(Martin and Jacky Webb)
Brickfield Farm, Laughton Road
Chiddingly BN8 6JG
☎ 0825 872457
Goat rearing, spinning, weaving,
dyeing and knitting of mohair
garments. Visitors welcome to see
fleeces being processed. Groups
welcome; childrens' parties by
arrangement. Childrens' holiday
workshops. Spinning and dyeing
weekend tuition. National award
for fleeces. C/W. Credit cards.
Open daily 10am-5pm.
Directions: from A22 Uckfield-
Eastbourne road turn onto B2124
Ringmer road and take first turn left
to Brickfield Farm.

WOODTURNING & FURNITURE
E & I Furniture Makers
(I H Setterfield and E Harrison)
Ailies Buildings, Whitesmith Lane
East Hoathly BN8 6QP
☎ 0825 872527
Furniture designed and built to
commission. All types of hand
turning including balasters. Recent
work includes various four-poster
beds, 10ft mahogany extending
table and 12 chairs, panelled doors,
oak panelling, dressers, desks,
music units, garden furniture. Also
repairs and restoration work on
antique and modern furniture.
C/R&R/W/E. Credit cards.
Mon-Fri 8.30am-5.30pm, Sat 10.30am-
4pm.

Directions: from A22 at East Hoathly turn into village centre, then 1st right; continue for 1 mile until sign on right.

ANTIQUE FURNITURE
RESTORATION
Maxwell Black
(Maxwell and Hilary Black)
Brookhouse Studios, Novington Lane
East Chiltington BN7 3AX
☎ *0273 890175*
Any furniture restoration and conservation work undertaken; also some unusual work on metal objects. C/R&R/E
Mon-Fri 8am-6pm. (Closed Aug).

FORGEWORK
G W Day & Co — Blacksmiths
(David William Cox & Brian Pettitt)
East Chiltington Forge, Highbridge Lane
South Chailey BN7 3QY
☎ *0273 890398*
Business employing seven forge-workers. Projects ranging from new developments to restoration of stately homes, including staircase at Saint Hill Castle, general ironwork at Battle Abbey, Arundel Castle and Bodiam Castle. C/R&R/W/E
Mon-Fri 9am-5pm, lunchtime closing.
Sat 9am-12noon. (Closed Xmas-New Year).
Directions: from Chailey on A275 heading towards Lewes, take 1st right turn. At T junc turn left, then take 1st left turn. Forge on left.

WOODTURNING & CARVING
The Beautiful Bowl & Carving Co
(Rosemarie Yeh)
Unit 14, Star Brewery Workshops,
Castle Ditch Lane, Lewes BN7 1YJ
☎ *0273 486329*
Turned decorative bowls in English and foreign hardwoods (not rain

forest timbers). Also hand-carved furniture. C
Mon-Sat 10am-4pm, advisable to telephone first.

CERAMICS
Mo Hamid BA(Hons) Ceramics
Unit 7, Star Brewery Workshops,
Castle Ditch Lane, Lewes BN7 1YJ
☎ *0273 483295*
Pottery workshop with small showroom set up with Crafts Council grant, producing hand-thrown decorated functional stoneware pottery. Work also sold through craft shops and galleries. C/W. Credit cards.
Mon, Thur-Sat 11am-5.30pm, advisable to telephone first.

Mayfield

WOOD CARVING
Greta E Chatterley
Timewell Cottage, South Street
Mayfield TN20 6BY
☎ *0435 872435*
All types of wood carving, sculpting and letter-cutting undertaken in wood; also repairs. Human and animal models, ecclesiastical figures, crosses, house signs, memorial plaques, club shields, coats of arms and collectors' caddy spoons. C/R&R/E. Credit cards.
Open at any reasonable time, advisable to telephone first.
Directions: Mayfield is 9 miles south of Tunbridge Wells on A267.

Pevensey

BRASS, COPPER AND METALWORK
Glynleigh Studio
(Sam Fanaroff)
Peelings Lane, Westham BN24 5HE

☎ *0323 763456*
A variety of work in non-ferrous
metals such as brass and copper.
Plates, dishes, jugs, lamps, candlesticks, bowls etc. Also church work
of any description. Antique
restoration involving metal. C/R&R
Mon-Fri 9am-5pm, lunchtime closing.
Visit by appointment.

Rye

CERAMICS
The John Solly Pottery
Goldspur Cottage, Flackley Ash
Peasemarsh TN31 6YH
☎ *079 721 276*
High fired earthenware and
slipware produced by potter with
nearly 40 years' experience. Tuition
courses are internationally known.
C
Mon/Wed/Fri & Sat 10am-6pm.
Directions: on A268 Rye-Hawkhurst
road at western end of village opposite
Flackley Ash Hotel (parking).

CERAMICS
David Sharp Ceramics
(D T and D H Sharp)
55 High Street, Rye TN31 7EN
☎ *0797 222620*
Handmade and hand-painted
pottery; house plaques and cats a
speciality. C/E. Credit cards.
Mon-Fri 9am-5pm, Sat-Sun 10.30am-
5pm.

FURNITURE
Rustic Craft Workshop
(Barry Jones)
Dixley Lane, Beckley TN31 6TH
☎ *0797 260522*
Country furniture making using
solid home-grown timber to
traditional and functional rural

designs for homes/gardens.
Wheelbarrows, garden saw-horses,
flower planters, tables, coffers,
stools. Roundwood furniture of
unique design. Sawmilling and
local wood machining service to
fellow craftsmen. C/R&R
Open daily 10am-6pm; advisable to
telephone first.

HAND-PAINTED CERAMICS
Rye Pottery Ltd
(Tarquin and Elizabeth Cole)
77 Ferry Road, Rye TN31 7DJ
☎ *0797 223038*
Fine ceramic figures hand-painted
and sold in UK and overseas;
Chaucer figures, lovers, American
folk heroes, pastoral primitives,
animals etc. Pottery employs eight
craftspeople and has won many
awards from 1959 onwards.
C/W/E. Credit cards.
Mon-Fri 9am-5pm, lunchtime closing.
Sat 9.30am-5pm, lunchtime closing.

TILE MAKING
Rye Tiles Ltd
(Tarquin and Elizabeth Cole)
Wishward, Rye TN31 7DH
☎ *0797 223038*
Wide range of floor, wall and
fireplace tiles screen printed and
hand-painted. Six people employed;
twice winners of Design Award.
C/W/E. Credit cards.
Mon-Fri 9am-5pm, lunchtime closing.
Sat 10.30am-4.30pm, lunchtime closing.

Tunbridge Wells

FURNITURE
David Haugh Furniture
Unit 7, Noblesgate Yard
Bells Yew Green TN3 9AT
☎ *0892 750310*

Furniture maker with a standard range of furniture also working to commission; design service available; drawings to customers' requirements.
C/W/E. Credit cards.
Mon-Fri 8.30am-5.30pm, Sat 8.45am-5pm.

Uckfield

STAINED GLASS
Glass Graphics
(Richard Hobley)
Gate House Cottage, Chapel Wood
Nutley TN22 3HE
☎ 0825 740503
Contemporary stained glass window manufacturers. Three employed; supplier to major local window manufacturers and installers. Private enquiries welcome. Member of Guild of Master Craftsmen. C/R&R/W/E
Open daily 9am-5pm.

CERAMICS
Living Ceramics
(Clare McFarlane)
Bird-in-Eye Farm, Framfield Road
Uckfield TN22 5HA
☎ 0825 890163
Hand-painted semi-porcelain realistic animal figures and sculpture, specialising in cats, pigs, frogs, sheep, chicken and duck. Some pieces exhibited in Society of Wildlife exhibition in The Mall, London. W/E.
Mon-Fri 9.30am-5pm, Sat 10am-12noon. (Closed Xmas-New Year).
Directions: from Uckfield take Framfield Road. After about quarter mile go up hill and look out for Craft Workshops sign on right.

WOODLAND MANAGEMENT & FURNITURE
Wilderness Wood and Adrian Chaffey Furniture
(Chris and Anne Yarrow)
Hadlow Down TN22 4HJ
☎ 0825 830509
60 acre working wood open to the public for recreational and educational purposes, where wood is grown in traditional coppices and plantations. The harvested wood is used to produce garden items; rose arches, garden furniture etc. Workshop producing indoor furniture and other wooden items to original designs. C/R&R/W/E
Open daily 10am-dusk. Advisable to telephone for appointment if visiting the workshop. Visiting charge: £1.30 adults, £1 OAPs, 70p children.
Directions: 12 miles south of Tunbridge Wells, in Hadlow Down village on south side of A272 road.

Wadhurst

HEDGE LAYING
G W Streete
Buckhurst Bungalow
Buckhurst Lane, Wadhurst TN5 6JY
☎ 0892 882553
Hedge laying and advisory service. Southern England Champion 1989-91.

Wivelsfield Green

BATIK WORK
Rosi Robinson
High Pines, Hundred Acre Lane
Wivelsfield Green RH17 7RS
☎ 0444 84584
Batik pictures, cards and scarves. Work exhibited overseas; sold through exhibitions and demonstrations. Tuition given. C
Visitors welcome by appointment.

WEST SUSSEX CRAFT WORKSHOP CENTRE

Amberley Chalk Pits Museum
Amberley, Arundel BN18 9LT
☎ *0798 831370*
Working museum set in a former chalk quarry in the South Downs. Traditional craft workshops include blacksmithing, potting, printing, boat building, woodturning, engineering. Also traditional ironmonger's shop, vintage wireless exhibition, steam road vehicles, brickmaking display, concrete exhibition, lime kilns and grinding mill. Programme of special events available. Restaurant and picnic area; nature trail.
1Apr-1Nov Wed-Sun 10am-6pm, daily during Summer holidays (24 Jun-13 Sept) 10am-6pm.
Charges: Adults £3.90, OAPs & students £3, Children (5-16 yrs) £1.80. Family and party rates available.
Directions: on B2139 between Arundel and Storrington, signposted 'Industrial Museum' from A24 and A29. Adjacent to railway station (London/Victoria and south coast) and travel by launch on River Arun from Littlehampton.

WEST SUSSEX CRAFT WORKSHOPS

Arundel

FORGEWORK
Chalk Pits Forge
(A J Breese)
Chalk Pits Museum
Houghton Bridge, Amberley BN18 9LT
☎ *0798 831370*
High quality forged ironwork. Workshop within grounds of industrial museum (entry fee £3.90, but free entry for prior appointments at forge). C/R&R
Mon-Fri 8am-5pm, Sat-Sun by appointment.

CLOCK RESTORATION
Arundel Dials Service
(Rob Gillies)
48 Maltravers Street
Arundel BH18 9BU
☎ *0903 882574*
Clocks re-painted, re-silvered and antique dials re-written. Also new dials designed and large coloured outdoor clocks made with gold leaf numbers. C/R&R/E
Open daily 9am-5pm, telephone for appointment.

GLASS ENGRAVING
Arundel Fine Glass Studio
(Jacques Ruijterman)
Tarrant Square
Arundel BN18 9DE
☎ *0903 883597*
Individually designed engraving on hand-blown lead crystal. Old and new shapes, studio glass and replica 18th century British glass. Fellow of Guild of Glass Engravers. C/W/E. Credit cards
Tues-Fri 9.15am-5pm, Sat 9.15am-4pm, lunchtime closing.

Billingshurst

FURNITURE
Roger E Smith
Barflies, Broadford Bridge
Billingshurst RH14 9EB
☎ *0798 813695*

Traditional and contemporary furniture made to customers' individual requirements. Particularly work in English oak. Commissions include church, commercial and domestic furniture. Small production runs undertaken. C/E
Open daily; advisable to telephone first.
Directions: from Billinghurst take A29 to Pulborough. At A2133 crossroads take left turn. Turn right (south) at next crossroad; Barflies 100yd on left.

Burgess Hill

GOLD & SILVERSMITHING
Jack Trowbridge
75 Royal George Road
Burgess Hill RH15 9SG
☎ 0444 232208
Design and manufacture of contemporary gold and silver jewellery; rings, bracelets, chains etc. Stock items, but will accept commissions. Freeman of the Goldsmith's Company. C/W/E. Credit cards
Open daily 9am-5pm. Visitors welcome by appointment.
Please telephone for directions.

Chichester

CERAMICS
Donald & Jacqueline Mills
Paddock Cottage, Itchenor Road
West Itchenor PO20 7DH
☎ 0243 512025
Decorated porcelain and some wax resist stoneware produced by experienced potter and decorated by trained artist. C/W
Visitors welcome by appointment.

MUSICAL INSTRUMENT MAKING
John Storrs Workshop
North Mundham PO20 6NR
☎ 0243 776263

Makers of early keyboard instruments and kits; harpsichords, spinets and clavichords. Well equipped workshop which visitors are welcome to visit to see how instruments are designed and made with the very latest computer aids. Business established 20 years ago, exporting about half of the 1,500 instruments and kits made. Also makers of music stands and cabinets. C/R&R/W/E
Mon-Fri & most Sats 9am-6pm.
Visitors welcome by appointment.

Hassocks

CABINET MAKING & FURNITURE RESTORATION
The Cabinet Shop
(Chris Batchelor)
Unit 4, Station Industrial Estate, Keymer Road, Hassocks BN6 8JA
☎ 0273 846388
Two craftsmen employed in workshop; restoration work, French polishing etc. Featured in many local and national publications. C/R&R/W/E
Mon-Fri 8am-5.30pm, Sat 8.30am-1pm.

GOLD & SILVERSMITHING
Guen Palmer Designs
11 The Twitten, Ditchling
Hassocks BN6 8UJ
☎ 0273 845086
Contemporary jewellery produced by graduate from Central School of Art & Design with several years' experience working in the jewellery trade in Africa and UK. Gold, silver, exotic materials, gemstones and pearls. Range of work is available at workshop and various other outlets C/W/E. Credit cards
Visitors welcome by appointment.

Haywards Heath

FORGEWORK
Wyvern Smithy
(Stephen Darby)
The Old Forge, Quick's Yard
High Street, Handcross RH17 6BJ
☎ *0444 400491*
Traditional village forge still using
hand-pumped bellows, specialising
in hand-crafted gates, railings,
weather vanes, fire irons etc. Small
gift items available. C/R&R/E
Thur-Sat 9am-5.30pm.

FORGEWORK
John Franks Blacksmith
Warninglid Forge, Cuckfield Lane
Warninglid RH17 5SN
☎ *0444 85359*
Five craftsmen designing, making
and restoring wrought ironwork.
Six times winner of Worshipful
Company of Blacksmiths' silver
cup. C/R&R/W. Credit cards
Mon-Fri 8am-5pm, Sat 8.30am-1pm.

FURNITURE
Tom Taylor of Lindfield
37 Sunte Avenue
Lindfield RH16 2AB
☎ *0444 455546*
Hand-crafted furniture for bedrooms, bathrooms and kitchens
designed and made in many
different timbers and finishes.
Interior design service . C/E
Mon-Fri 8.30am-6pm.

Littlehampton

FURNITURE
Wood Design
(Brendan J Devitt-Spooner)
The Acre, Dappers Lane
Angmering BN16 4EN

☎ *0903 776010*
Fine contemporary furniture
designed and made, much of it
from 1987 hurricane wood. Featured on TV and national publications. A woodturner shares part of
workshop. C/E
*Mon-Fri 8.30am-5.30pm, lunchtime
closing. Sat by appointment.*
Directions: north of Angmering, between
A259 Littlehampton-Worthing road and
A27 Arundel to Brighton road.

Midhurst

CABINET MAKING
Sussex Furniture Design Ltd
(Trevor Baker, Hugh and Anne Mitchell)
The Wharf, Midhurst GU29 9PX
☎ *0730 815816*
Three craftsmen including an
apprentice, making furniture to
contemporary designs utilising
traditional methods. Young
Craftsmen of the Year 1987.
C/W/E. Credit cards
*Mon-Fri 7.30am-6pm, lunchtime
closing. Sat-Sun 8am-1pm.*

HAND-PAINTED FURNITURE
Stanwater Designs Ltd
(J E and D J Garstin)
June Lane Barn, June Lane
Midhurst.
☎ *0730 812578/814900*
Hand-painted beds and custom
built furniture supplying many
notable people. Rural business
employing four craftspeople,
established 14 years ago. A wide
range of designs and finishes
available; decorated four-poster
beds and half-testers a speciality.
Unique painting is by Janette
Garstin, a textile designer who
decorates individual pieces with

flowers and country scenes. Also suppliers of handmade mattresses and bed hangings. C/W/E

Mon-Fri 9am-5pm, lunchtime closing. Sat-Sun by appointment.

Directions: from Midhurst take Petworth road and turn left into June Lane; workshop on right.

Petworth

WEAVING
Graffham Weavers Ltd
(Gwen and Barbara Mullins)
Shuttles, Graffham GU28 0PU
☎ 07986 348

Variety of hand-woven products; floor rugs (flat and pile), saddle blankets, cushions, bags, stoles and scraves made on the premises. C/E
Visitors welcome by appointment.

Pulborough

FORGEWORK
Charlton Ironwork
(M V Jones)
North Heath Farm, Gay Street Lane Pulborough RH20 2HW
☎ 0798 875474

Forge employing two craftsmen manufacturing ornamental hand-forged wrought ironwork. Cast iron benches, fire screens, baskets and furniture, firebacks, mangers, flower stands, barbeques etc. Iron restoration, shot blasting and general repairs. C/R&R/W/E
Mon-Sat 8am-5.30pm.

Steyning

GOLD & SILVERSMITHING
Gordon Lawrie
Hammes Farm, Washington Road Wiston BN44 3DA

☎ 0903 814056

Interesting, figurative jewellery made in gold, silver and titanium. Designs inspired by the landscape, myths and magic. Work widely exhibited in the UK and abroad. C/E. Credit cards

Mon-Sat 10am-5pm; advisable to telephone first. Other times by appointment.

Directions: on A283 1 mile out of Steyning opposite Wiston Pond.

Storrington

ANTIQUE FURNITURE RESTORATION
Thakeham Furniture Restorers
(Tim Chavasse)
Rock Road, Storrington RH20 3AE
☎ 0903 745464

A team of skilled craftsmen restoring English 18th and 19th century furniture. Some items for sale. C/R&R/W/E

Mon-Fri 8.30am-5pm.

Directions: 2 miles north of Storrington off B2139.

Worthing

FURNITURE MAKING
Country Pine Furniture
(Colin Holmes and Wayne Clear)
The Woodyard, France Lane Patching BN13 3UP
☎ 090 674 202

Designers and makers of hardwood and softwood furniture to customers' requirements. C/R&R/E

Mon-Sat 9am-5.30pm.

Directions: from A27 Arundel-Worthing road at A280 crossroads, turn north beside Horse & Groom PH workshop on right.

WARWICKSHIRE CRAFT WORKSHOP CENTRES

Middleton Hall Craft Centre
Middleton, Nr Tamworth B78 2AE
Opened in 1990, the centre comprises individual workshops in converted stables (see entries below). The project was undertaken in conjunction with the Rural Development Commission, and work continues on restoring the hall and grounds, which are open to the public; walled gardens, lake encircled by nature trails, restored smithy, shop and tea rooms.
Craft Centre: Open all year, Wed-Sat 11am-5pm, Sun 1-5.30pm. Hall & grounds: Sun & Bank Holidays 2-5.30pm 1 Apr-30 Oct.
Directions: on A4091 4 miles south of Tamworth, between Belfry and Drayton Manor Park.

Manor Farm
(T Osborne & Son)
Wood Lane, Earlswood B94 5JH
☎ 056 46 2729
With support from the Rural Development Commission, a dairy farm has developed craft workshop units (see entries below). The farm is also well-known for its ice cream, which is sold through the farm shop and many other outlets, and is a member of Warwickshire Fine Foods Ltd.
Craft Workshops are open at weekends throughout the year and Bank Holidays. See entries for further opening times or telephone for information.
Directions: 2 miles from M42 junc 3; take Foreshaw Heath exit. Follow to T-junc, turn left. After 2 miles take 1st turn right into Wood Lane.

Hatton Country World
Dark Lane, Hatton
Nr Warwick CV35 8XA
☎ 0926 842436
Craft village with 35 workshops (see some entries below) in converted Victorian farm buildings. Also rare breeds and childrens' farm, adventure playground, nature trail to canal locks and café.
Open daily 10am-5pm. Free entry to craft village.
Directions: just off A4177 Solihull-Warwick road (signposted).

WARWICKSHIRE CRAFT WORKSHOPS

Alcester

FORGEWORK
Harry Green
The Smithy, Dunnington B49 5NN
☎ 0789 490125
All types of ironwork undertaken; ire baskets and tools, wrought iron gates, weather vanes etc. Kango and steel sharpening. C/R&R
Mon-Fri 9am-5.30pm, Sat 9am-2pm.

Atherstone

CERAMICS
Carter Pottery
(C J and J Carter)
Highfields Farm, Grendon CV9 3QL
☎ 0827 713307/715963
High quality hand-thrown stoneware. Individual pieces with strong, lively character and use of traditional decoration. C/W/E

Visit by appointment.
Directions: from Atherstone take
B4116, turn onto B5000.

CERAMICS
Jo Connell
Witherley Lodge, 12 Watling Street
Witherley CV9 1RD
☎ *0827 712128*
Hand-built decorative ceramics in
colours ranging from pastels to
black, unglazed; vases, jugs, bowls,
lamps, tiles for fireplaces etc. Work
on display in workshop, also sold
through shops and galleries. Widely
exhibited. Member of Craft Potters
Association and regional potters'
association. C/W/E
Visitors welcome by appointment.
Directions: on A5 between M42 junc
10 and M69 junc 1. Witherley Lodge at
end of small stretch of dual carriage-
way 1 mile east of Atherstone.

Coventry

CARRIAGE RESTORATION &
DRIVING EQUIPMENT
Withybrook Carriages &
Ousbey's Harness Room
(Rod and Gill Ousbey)
Stable Cottage, Featherbed Lane
Withybrook CV7 9LY
☎ *0455 220127*
Restoration of coaches and carriages; also vintage cars. 25 years'
experience in brush finish coach
painting and coach trimming;
experience of restoration on
virtually every type of carriage.
Shafts and poles made from timber
self-cut and seasoned to ensure
quality. Work for museum and
private collectors. Accident repairs
and insurance work undertaken.
Vehicles for sale and a wide range

of driving accoutrements. Also
fibreglass display horses.
C/R&R/W/E. Credit cards.
Mon-Fri 10.30am-5.15pm. Some Sats-
Suns; advisable to telephone first.

WOODTURNING
Exclusive Wood Products
(Darren Guy)
27 Lammas Court, Wolston CV8 3LP
☎ *0203 544934*
Handmade beds, headboards, stair
spindles, childrens' playhouses,
household accessories. Copy
turning undertaken. Prizewinner;
Livewire 1990 county finals. C/W
Mon-Sat 9am-5pm, lunchtime closing.

CERAMICS
Cheryl Gibbons Ceramics
Rose Cottage, The Row
Ansty, Coventry CV7 9JA
☎ *0203 619505*
Pottery and shop in a converted old
building next to the Oxford Canal.
Unusual hand-built ceramics and
thrown pottery decorated with
irridescent lustre colours and
under-glazes. Shop selling other
crafts and work by local artists. C
Mon-Fri 10am-3.30pm, telephone
appointments welcome. (Closed Jan-
Feb).

CAR RESTORATION
Available Austins
(Bob and Eileen Glenister)
Old Arley CV7 8FG
☎ *0676 41276*
Workshop re-building traditional
'20s and '30s vehicles; skills
including paint, trim, mechanical,
metal and woodworking. Large
range of spares for pre-war Austins.
C/R&R/W/E. Credit cards.
Mon-Sat (closed Tues) 9am-5pm.

WOODEN TOYS & DOLLS' HOUSES
Beecham Toys
(M H Beecham)
D4 Wolston Business Park, Main St
Wolston CV8 3FU
☎ *0203 543974*
Workshop employing four crafts-
men making a wide range of wooden
toys; traditional style dolls' houses,
farms, forts, castles, garages. C/E
Mon-Fri 9am-5pm, Sat 8am-12noon.

Henley-in-Arden

ETCHING
**Capelli Printmaking Studio/
Gallery**
(Peter Burridge and Carl March)
3B Arden, Little Alne
Nr Henley-in-Arden B95 6HW
☎ *0789 488437*
Artists' etching workshop and
studio. Established artists repre-
sented in many public and private
collections. C/W/E
Open daily 10am-6pm.

CERAMICS
Torquil Pottery
(Reg Moon)
81 High Street,
Henley-in-Arden B95 5AT
☎ *0564 792174*
Handmade porcelain and stone-
ware. Workshop, retail shop and
gallery exhibiting fine art and crafts.
W/E. Credit cards.
*Mon-Sat 9am-6pm. (Closed Bank
Holidays).*

Leamington Spa

SMOCKING AND PATCHWORK
Traditional Cottage Crafts
(Thelma Watts)
Sinclair, Avon Dassett CV33 0AL

☎ *0295 89362*
Small business specialising in quilts,
cushions and wall hangings.
Sampler/needlepoint kits available.
Demonstrations given in local
tourist centres. C/R&R
Visit by appointment only.

ECCLESIASTICAL EMBROIDERY
Revelations
(Mrs Felicity J Howatson)
Colecraft Building, Southam Road
Long Itchington CV23 8QL
☎ *0926 815791/612773 (evenings)*
Vestment manufacture, banners,
alter linen and frontals etc. Ten
years' experience in producing
ecclesiastical work for UK and
overseas using a wide range of
fabrics at prices affordable to small
parishes. Lecturer and demonstrator, Member of Embroiderers Guild.
C/R&R
*Mon-Fri 9.30am-3pm; visit by
appointment.*

Nuneaton

CLOCK CASES
Middleway
(David and Carole Herbert)
Old Village Hall, Shenton CV13 6DA
☎ *0455 212372*
Clock cases individually designed
and handmade from the finest
materials. Commissions welcome.
Also some turned furniture and
restoration of cane and rush seating.
C/R&R/E. Credit cards.
Open Sat-Wed.

SCULPTURE
John Letts & Keith Lee Sculptures
The Old School, Church Lane
Astley CV7 8EW
☎ *0676 42073*

John Lett's sculptures are fluid and romantic; Keith Lee concentrates on character studies and men at work. The pieces are modelled in clay and cold-cast in bronze; limited editions and originals from £5 to £400. Special commissions undertaken. C/W/E. Credit cards.
Mon-Sat 10am-6pm, Sun by appointment.

Rugby

CERAMICS
Pailton Pottery
(Sylvia Edon Langham)
12 Lutterworth Road
Pailton CV23 0QE
☎ *0788 832064*
Large selection of kitchen and tableware suitable for oven, microwave and dishwasher. Many different ornamental animals and a range of decorative vases, wine sets etc. C
Mon-Sat 9am-6pm, visit by appointment only.

ANTIQUE FURNITURE RESTORATION
Wood Restorations
(P N and S Wood)
Eastfield Farm, Crick Road
Hillmorton CV23 0AB
☎ *0788 822253*
Established workshop with 25 years' experience in restoration work, upholstery and cabinet making. Five craftsmen working for regular clientele. C/R&R/E
Mon-Fri 9am-6pm, Sat 9am-12noon.

CERAMIC SCULPTURE
Paul Gandy Ceramics
Marton Farm House
Marton CV23 9RH
☎ *0926 632923*

Architectural and landscape studies in matt glazed, high-fired stoneware. Work widely exhibited in UK and overseas; many notable commissions including V&A Museum Trustees' retirement presentation and 'Young Linguist of the Year' trophy. C/W/E
Mon-Fri 9am-5pm, Sat-Sun by appointment.
Directions: workshop in former Georgian farmhouse on A423 Oxford-Coventry road in Marton.

THATCHING
Max Grindlay — Thatcher
Riverside Cottage, Church Road
Long Itchington CV23 8PR
☎ *0926 812659*
Thatching in combed wheat and water reed.

CERAMICS
Brinklow Pottery
(George and Diane Lindsay)
11 The Crescent, Brinklow CV23 0LG
☎ *0788 832210*
Studio pottery and gallery shop, two resident potters making functional and decorative stoneware. Table and ovenware, house name plaques, castles, dragons etc. Commemorative ware and commissions. Makers of the baptismal ewer in Coventry Cathedral. C
Wed-Sun 10am-6pm. Visitors to workshop by appointment.

TRADITIONAL CHAIR MAKING
Neville Neal
(Neville and Lawrence Neal)
22 High Street, Stockton CV23 8JZ
☎ *0926 813702*
Ladder and spindle back chairs with rush seats. A village craft, father (who was a pupil of Ernest

Gimson's partner) and son continuing the making of Gimson chairs from local grown ash and rushes. C/E
Mon-Fri 8.30am-5.30pm, lunchtime closing 12.30-1.30pm. Sat 9am-12noon.
Directions: 2 miles from Southam, just off A426 Southam to Rugby road.

FORGEWORK
Mitchell Metal Craft
(Stuart Mitchell)
Manor Farm, School Lane
Stretton-on-Dunsmore CV23 9NB
☎ 0203 544938
General fabrication, welding and ornate metalwork. Metal and wood garden benches, weathervanes and heavy duty barbecues. C/R&R
Mon-Fri 8am-5.30pm, Sat 8am-1pm.

Shipston-on-Stour

FORGEWORK
Halford Forge
(Charles and Toby Harness)
Queen Street, Halford CV36 5BT
☎ 0789 740026
Family business operating in forge known to have existed for over 200 yrs. Traditional and modern decorative and functional metalwork for interiors and outdoors.
C/R&R/W/E. Credit cards.
Mon-Fri 8am-6pm, lunchtime closing. Sat 8am-12.30pm.

CERAMICS
Whichford Pottery
(Jim Keeling)
Whichford CV36 5PG
☎ 0608 84416
Handmade terracotta flowerpots in a wide variety of shapes and sizes and floor tiles (in/outdoors). Also traditional English hand-painted 'Delftware' tiles. Winner of Small

Rural Business award. C/W/E
Mon-Fri 9am-5pm, lunchtime closing. Sat 10am-5pm. (Closed Xmas-New Year).

HAND-PAINTED SILK
Dunford Wood Designs Ltd
(Andrew Hardwick and Hugh Dunford Wood)
Nineveh Farm, Whatcote CV36 5EH
☎ 060 885 787
Hand-painted silk ties and accessories for men; brightly coloured ties, bow ties, waistcoats, braces and cummerbunds. Featured on BBC TV *The Clothes Show.*
C/W/E. Credit cards.
Mon-Sat 10am-4pm.
Directions: from M40 junc 12, follow road through Gaydon to Kineton. Turn right at Swan PH and immediately left signed Halford. Turn left signed Tysoe/Oxhill/Compton Wynyates. Turn right across A422, continue through Oxhill. In Whatcote, turn left at Royal Oak PH. After 1 mile turn right up drive marked Nineveh.

HURDLE MAKING
Michael Vincent — Hurdle Maker
Ashbourne Cottage, Ilmington CV36 4LJ
☎ 060 882 232
Manufacture of traditional flake (six-bar gate type) sheep hurdles. C/W
Mon-Fri 8am-5pm, lunchtime closing. Sat 8am-12noon.

FURNITURE
DSE Studios
(Roland Day, Andrew Shenton & Kevin Englefield)
Great Wolford CV36 5NQ
☎ 0608 74312
Work for corporate clients, private clients, ecclesiastical work and contract work. C
Mon-Fri 7.30am-6.30pm. Most Sats.

Solihull

FORGEWORK
Forgecraft
(B Tomkinson)
44 Milverton Road, Knowle B93 0HY
☎ 0564 773352
Wrought ironwork in traditional
and modern styles. General
forgework and allied crafts. C
Mon-Fri 9am-4pm, by appointment.

HEDGE LAYING
Midland Hedging
(D A Blissett)
90 Yew Tree Lane, Solihull B91 2RA
☎ 021 705 8071
Hedge laying contractors. Prize-
winner in national competitions.

STAINED GLASS & FIREPLACE
INSTALLATION
The Art of Glass
(James Scanlon)
Manor Farm Craft Centre
Wood Lane, Earlswood B94 5JH
☎ 05646 3992
Stained glass and antique fireplace
installations; period property
restoration work. C/R&R/E.
Mon-Fri 9.30am-5.30pm, Sat-Sun
12noon-6pm.
Directions: see under Craft Centres.

DRIED & FABRIC FLOWERS
Woodland Flowers
(Hazel McGann)
Unit 8, Manor Farm Craft Centre,
Wood Lane, Earlswood B94 5JH
☎ 05646 3134
Dried and silk flower arrangements;
wedding work, funeral work etc. C
Thur 2-7.30pm, Fri 2-6pm, Sat 12noon-
6pm, Sun 2-6pm.
Directions: see under Craft Centres.

NEEDLECRAFT
The Doll's House
(Mrs Stephanie Johnson)
Unit 7, Manor Farm Craft Centre,
Wood Lane, Earlswood B94 5JH
Dressed porcelain dolls, dressed
cuddly toys, satin cushions, pot
pourri, lacy tissue boxes. C
Wed-Fri 2.30-5.30pm, Sat-Sun 2-5.30pm.
Directions: see under Craft Centres.

NEEDLECRAFT & DECORATED EGGS
Petite Designs
(Linda Hargreaves)
Unit 1, Manor Farm Craft Centre,
Wood Lane, Earlswood B94 5JH
☎ 05646 3955
Childrens' dresses. Also decorated
eggs. C
Tues/Thur 1-7.30pm, Wed/Fri/Sat/Sun
1-5.30pm
Directions: see under Craft Centres.

UPHOLSTERY
Manor Farm Upholstery
(G Treagus)
Manor Farm Craft Centre
Wood Lane, Earlswood B94 5JH
☎ 05646 3982
Complete upholstery service on
modern and antique furniture. R&R
Mon-Fri 9am-5pm, Sat am and Sun pm.
Directions: see under Craft Centres.

CERAMICS & PAINTING WORKSHOP
Barn Studio Ceramics
(Mrs P Pittaway and Mrs G Savage)
Manor Farm Craft Centre
Wood Lane, Earlswood B94 5JH
☎ 05646 3965
All types of ceramics made; figures,
pots etc. Studio workshop for
people to hand-paint ceramics; no
charge for studio use. C
Tues 11am-9pm, Wed 1-5pm, Thur 1-

9pm, Sat 10am-1pm, Sun 2.30-5.30pm.
Directions: see under Craft Centres.

Stratford-upon-Avon

FURNITURE MAKING &
RESTORATION
Nicholas Wood Furniture Makers
(Graham Wood)
The Brickyard
Preston-on-Stour CV37 8BN
☎ *0789 450883*
Furniture handmade from English
hardwoods, combining the skills of
a boatbuilder and French polisher
with a trained furniture restorer.
Traditional tables, chairs, refectory
tables, dressers etc. Members of
Guild of Master Craftsmen. C/R&R/E
Mon-Fri 8am-5pm, lunchtime closing.
Sat 9am-4pm.
Directions: leave Stratford-upon-Avon
on A34 heading south to Oxford. Take
4th turn on right to workshop.

WOODCARVING
Carol & Peter Moss (Woodcarvers)
The Knapp, Armscote CV37 8DH
☎ *060 882 247*
Carving ranging from small gift
items, figures and plaques to carved
furniture; four-poster beds, dressers, tables etc. C/R&R/E
Mon-Fri 9am-5pm, Sat-Sun 9am-4pm.
Advisable to telephone first.
Directions: Armscote is 7 miles south
of Stratford-upon-Avon, a right turn
off A3400 at Newbold-on-Stour.

CABINET MAKING
Dennis Tomlinson Ltd
(D Tomlinson, P Handy & R D Tuplin)
The New Carpenters Shop
Beechams Buildings
Atherstone Industrial Estate
Stratford-upon-Avon CV37 8DX
☎ *0789 450588*

Workshop employing ten craftsmen; wood machinists, cabinet
makers, designer, polishers etc
making free-standing furniture.
Tables, desks, wardrobes, bathroom
cabinets etc. C/R&R/W
Mon-Fri 9am-5pm, Sat-Sun 9.30am-4pm.

QUILLING
Comfort Farm Crafts
(Mrs Brenda Morley)
Comfort Farm
Clifford Chambers CV37 8LW
☎ *0789 292163*
Rolled paper filigree work; mainly
quilling tuition. Also small quilled
items; cards, pictures. Mail order. C
Workshop dates and by appointment.

CERAMICS
Louise Darby BA(Hons)
Redhill Farmyard
Redhill B49 6NQ
☎ *0789 765214*
Hand-thrown ceramics; porcelain
and stoneware. Small workshop
with electric wheel and ceramic
fibre gas kiln. Also showroom.
Member of Craftsmen Potters
Association. C/W/E
Variable working hours due to
attendance at exhibitions, galleries
etc; advisable to telephone first.
Directions: 4 miles west of Stratford
on A46 to Worcester. Turn left (south)
at Stags Head PH (before Redhill
descent). Turn right into farmyard
next to PH; continue to workshop.

Sutton Coldfield

GLASS BLOWING
D Foote Glass Sculptures
Unit 1, Middleton Hall Craft Centre
Middleton B78 2AE
☎ *0827 260654*

Individually handcrafted glassware made by award-winning glass artist. C/R&R/W. Credit cards. *Wed-Fri 11am-4.30pm, Sat 11am-5pm, Sun 2pm-5pm.*
Directions: see under Craft Centres.

DRIED FLOWERS
Hollyfield Dried Flowers
(Jane Ridley)
Unit 4, Middleton Hall Craft Centre
Middleton B78 2AE
☎ 0213 843080
Arrangements for all occasions and made to order. Baskets and accessories and pot pourri. C
Thur-Sun 11am-5pm.
Directions: see under Craft Centres.

LACE BOBBINS
Malcolm and Margaret Thorpe
Unit 9, Middleton Hall Craft Centre
Middleton B78 2AE
Leading suppliers of hand-turned lace bobbins in plain woods and laminated styles; unique burr inserts and cat range.
C/R&R/W/E
Fri&Sun 2-5pm.
Directions: see under Craft Centres.

DECORATED GIFTWARE
Petalcraft
(Mrs Lyn Wright)
Unit 12, Middleton Hall Craft Centre
Middleton B78 2AE
☎ 0827 282857
Basketware and caneware decorated with bows and 'porcelained' silk flowers; also mirrors, wastebins, photoframes, caskets, lidded boxes etc. C/W
Wed-Sat 11am-5pm, Sun 1.30-5.30pm.
Directions: see under Craft Centres.

SPINNING & WEAVING
Lazy Kate's
(Carol Needham and Sylvia Sexton)
Unit 8, Middleton Hall Craft Centre
Middleton B78 2AE
☎ 0675 81528/81215
Luxury hand-spun and hand-woven goods; sweaters, jackets, cushions, shawls etc. Spinning and weaving supplies including fleeces, fibres and yarns. Tuition by appointment. C/R&R
Wed-Fri 11am-4pm, Sat 11am-5pm, Sun 11am-5.30pm.
Directions: see under Craft Centres.

KNITWEAR
Janet's Designer Mohair
(Janet Showell)
Unit 5, Middleton Hall Craft Centre
Middleton B78 2AE
☎ 0213 293931
Ready-made sweaters and jackets in mohair. Men's Aran sweaters and childrens' original knitwear.
C. Credit cards.
Sat-Sun 1-5pm.
Directions: see under Craft Centres.

GOLD & SILVERSMITHING
K Hamilton Gabb
Unit 11, Middleton Hall Craft Centre
Middleton B78 2AE
☎ 0827 260041
Hallmarked jewellery and silverware; gifts and ecclesiastical items. Exclusive designs for christenings, weddings and anniversaries undertaken. Also scale models in silver and gold. Repair work undertaken. Freeman of Worshipful Company of Goldsmiths.
C/R&R/W/E. Credit cards.
Wed-Sat 11am-5pm, Sun 1-5.30pm.
Directions: see under Craft Centres.

ARTIST
W M Brown ARBSA, NDD, ATC
Unit 2, Middleton Hall Craft Centre
Middleton B78 2AE
☎ *0675 463865*
Paintings in oil and watercolour;
quick portraits in pastel (b&w or
colour). Wood engravings, etchings
and calligraphy. C
Thur-Sat 11am-5pm, Sun 2-5pm.
Directions: see under Craft Centres.

CERAMIC SCULPTURE
Merlyn Lee
Unit 3, Middleton Hall Craft Centre
Middleton B78 2AE
☎ *0213 783932*
Ceramic sculpture based on bark
and fungi, slab pots, animals.
Exhibition work. C
Wed-Sat 11am-5pm, Sun 2-5pm.
(Closed Xmas-New Year)
Directions: see under Craft Centres.

UPHOLSTERY
Cyril Beard — Upholstery
  Restoration Specialist
Unit 10, Middleton Hall Craft Centre
Middleton B78 2AE
☎ *0827 261448*
Restoration of all kinds of antique
upholstered furniture including
repairs to woodwork, woodturning,
deep buttoning and respringing.
Complete range of antique velvet
patterns. R&R. Credit cards.
Wed-Sat 11am-5pm, Sun 1-5.30pm.
Directions: see under Craft Centres.

SOFT FURNISHING
Tudor Barn Interiors
(Shane Young)
Unit 6, Middleton Hall Craft Centre
Middleton B78 2AE
☎ *0213 843226*

Exclusive curtains, drapes, table
and bed linen designed and made
on the premises. Consultant and
adviser on interiors, furnishings
and lighting. Also many small gift
items. C/W
Wed-Sun 12noon-5.30pm. (Jan-Feb
Fri-Sun 1-5pm).
Directions: see under Craft Centres.

Warwick

CERAMICS
Jan Bunyan
4 Bridge Road
Butlers Marston CV35 0NE
☎ *0926 641560*
Hand-thrown decorated earthen-
ware pottery. Work exhibited with
Midlands Potters Association. C/W
Mon-Fri 10am-5pm, Sat-Sun by
appointment.

SADDLERY
A D R May (Saddler)
(David May)
Hatton Country World
Hatton CV35 8XA
☎ *0926 842760*
Master Saddler making and
repairing saddlery. Also boxwork,
cases, belts etc. C/R&R
Mon-Fri 9am-5pm, Sat 9am-12.30pm.
Directions: see under Craft Centres.

GLASS ENGRAVING
Crystal Maze
(David and Rosemary Thornhill)
Unit 18c, Hatton Country World
Hatton CV35 8XA
☎ *0203 335732*
Engraving of crystal, mirrors,
paperweights etc. C/R&R/E
Wed-Fri 9.30am-4.30pm, Sat-Sun
10am-4.30pm.
Directions: see under Craft Centres.

ANTIQUE CLOCK/BAROMETER
RESTORATION
Summersons
(Peter Lightfoot)
15 Carthouse Walk
Hatton Country World
Hatton CV35 8XA
☎ *0926 843443*
Second generation horology-trained
clockmaker undertaking all aspects
of restoration work on antique
clocks and barometers. Dial
restoration, wheel and hand cutting,
cabinet work and fretwork, polish-
ing, silvering etc. Parts mail order.
C/R&R/W/E. Credit cards.
Mon-Fri 9am-5pm, Sat 11am-4pm.
Directions: see under Craft Centres.

SIGN MAKING
Southam Signs
(S P & C Lambert)
Unit 20B, Hatton Country World
Hatton CV35 8XA
☎ *0926 842288*
Cast metal house name and number
signs. C/R&R/E
Open daily 9am-5pm.
Directions: see under Craft Centres.

BASKETWORK & DOLLS' HOUSES/
MINIATURES
Basketry World
(D J Baker and A T Bachelor)
Unit 19C, Hatton Country World
Hatton CV35 8XA
Basketry business employing four;
demonstrators for National Trust.
Baskets, rush, cane and straw
seating. Dolls' houses and mini-
atures. Members of Basketmakers'
Association and British Toy Makers
Guild. C/R&R/W/E. Credit cards.
Tues-Sun 10.30am-5.30pm.
Directions: see under Craft Centres.

WOODWORK
Line Nine Designs
(Tony Collins)
Unit 23, Hatton Country World
Hatton CV35 8XA
☎ *0926 842789*
Manufacturers of traditional
transport related models, bi-planes
and shelves made in wood and
metal. Six craftsmen employed in
making unique products (Lawrence
bi-plane video cassette shelf and
compact disc shelf, small train
cassette shelf, petrol pump book-
shelf etc) sold in large stores and
exported. Retail and mail order
service. C/W/E. Credit cards.
Open daily 9am-6pm.
Directions: see under Craft Centres.

STONE CASTING
Neil's Crafts
(Neil Grimmett)
Unit 24A/B Hatton Country World
Hatton CV35 8XA
☎ *0926 842913*
Concrete and stonecast ornaments.
Also plaster casting materials,
moulds and paints. R&R
*Open daily 9.15am-5.30pm, lunchtime
closing.*
Directions: see under Craft Centres.

KNITWEAR
Helen Reid Jackson Designs
Studio 3, Hatton Country World
Hatton CV35 8XA
☎ *0926 842496*
Made-to-measure knitwear. Hats
for all occasions. Various lace items.
Hand-painted T shirts.
C/R&R. Credit cards.
Tues-Sun 10am-5pm.
Directions: see under Craft Centres.

WOODTURNING
Hawkwoods
(Gavin Jones)
Unit 12A, Hatton Country World
Hatton CV35 8XA
☎ *0926 843301*
Fine woodturning; decorative,
functional and constructional.
C/R&R/W. Credit cards.
*Mon-Fri 10am-5.30pm, Sat-Sun
10.30am-6pm.*
Directions: see under Craft Centres.

PICTURE FRAMING & ENGRAVING
Mesdames Collections
(Roger and Diane Milnes)
Studio 11, Hatton Country World
Hatton CV35 8XA
☎ *0926 842021*
Picture mounting and framing. Also
engraving on flasks, tankards and
plaques. C/R&R/W. Credit cards.
Open daily 10am-5pm.
Directions: see under Craft Centres.

DRIED & FABRIC FLOWERS
Jintz Flowers
(Janet Arrowsmith)
Studio 7, Hatton Country World
Hatton CV35 8XA
☎ *05643 3281 (evenings)*
Dried flower arrangements and
bunches. Pressed flower work, silk
flowers etc. C
Tues-Sun 10.30-4.30pm.
Directions: see under Craft Centres.

FORGEWORK
The Wrought Iron Shop
(Roger Keith Smith)
Units 17B/C, Hatton Country World
Hatton CV35 8XA
☎ *0926 842278*
General blacksmithing and orna-
mental ironwork. C/R&R/W
Open daily 9am-6pm.
Directions: see under Craft Centres.

RESIN WORK
Field & Forest
(A L MacLeod)
Unit 17A, Hatton Country World
Hatton CV35 8XA
☎ *0789 267952 (evenings)*
Award winning hand-painted resin
figurines; birds, animals and
figures. C/R&R/W/E. Credit cards.
*Mon-Fri 11am-4pm, Sat-Sun 11am-
5.30pm. (Closed Jan-Feb).*
Directions: see under Craft Centres.

GOLD & SILVERSMITHING
Charles Goodman Jewellery
Hatton Country World
Hatton CV35 8XA
☎ *0926 842506*
Hand-crafted gold, silver and
enamel jewellery; repairs under-
taken. C/R&R/W/E. Credit cards.
Wed-Sun 10am-5pm.
Directions: see under Craft Centres.

WEST MIDLANDS CRAFT WORKSHOPS

Birmingham

WOODEN TOYS & NURSERY
EQUIPMENT
Paintwood Toys
(Louisa C Hills)
147 Prince of Wales Lane
Warstock B14 4LR
☎ *021 436 6289*
Wooden toys, nursery equipment
and hand-painted childrens'
furniture bearing the CE (European
Safety Standards) label. Discount
given to nurseries, schools etc for
large orders. C/R&R
Mon-Sat 9am-5.30pm.
Directions: from M42 junc 3 cross over
2 roundabouts, take 3rd turn (May-
pole Lane) from next roundabout,
straight over next roundabout;
opposite Prince of Wales PH on right
turn left into Prince of Wales Lane.
Workshops on left.

Dudley

WOODCARVING
**Deborah Fownes Rocking Horse
  Maker**
*The Rocking Horse Stable, Cotwall
End Nature Reserve, Catholic Lane
Sedgley DY3 3YE*

☎ *0902 887657*
Fine, hand-carved rocking horses
with leather saddles and harness,
hair manes and tails. Also carousel
animals, figureheads for ships,
anatomical studies etc. All work to
commission.
C/R&R/W/E. Credit cards.
*Mon-Fri (closed Thur) 9am-4pm, Sun
11am-4pm.*

Walsall

LEATHERWORK
Walsall Leather Centre Museum
54-57 Wisemore
Walsall WS2 8EQ
☎ *0922 721153*
Award-winning museum demonstrating leatherwork; saddlery and
harness making, small leather
goods, etc. Training and educational programmes, exhibitions and
publications, library, shop selling
leather work (including 'seconds').
Coffee shop. C
*Tues-Sat 10am-5pm, Sun 12noon-5pm
(4pm closing Nov-Mar) .*

WILTSHIRE CRAFT WORKSHOPS

Calne

UPHOLSTERY & SOFT FURNISHINGS
Shelburne Furnishings
(Ian and Lesley Hammond)
Unit 9 Harris Road
Porte Marsh Industrial Estate
Calne SN11 9PT
☎ *0249 814451*
Traditional upholstery and new commissions including office seating. Loose covers and curtains, four-poster bed 'dressing', pelmets and blinds. C/R&R/W
Mon-Fri 8.30am-6pm, Sat 10am-6pm. (Closed Xmas-New Year).

Chippenham

FORGEWORK
Hector Cole Ironwork
The Mead
Great Somerford SN15 5JB
☎ *0249 720485*
Ironwork, modern and traditional: work undertaken on commission basis. Specialist in reproduction mediaeval ironwork and blade smithing. Association of The Worshipful Company of Blacksmiths and official judge, National Blacksmiths' Competition committee. C/R&R/E
Open to visitors all evenings.

SCULPTURE
Andrea Garrihy - Sculptor
1 Old Pottery Studio
Corsham Court Gatehouse
Corsham SN13 0HF
☎ *0249 714030*
Stone carving, clay modelling, plaster casting, wax modelling for bronze work. Also drawings and design work. C/R&R/E
Tues-Fri 10am-4pm, lunchtime closing 12.30-2pm. Sat-Sun 2-4pm. Advisable to telephone first.
Directions: from A4 Bath-Chippenham road follow signs to Corsham Court.

CERAMICS
Lacock Pottery
(David and Simone McDowell)
1 The Tanyard, Church Street
Lacock SN15 2LB
☎ *0249 730266*
Ceramics, mainly stoneware using reduction glazes. Two well-equipped workshops are used for Pottery Summer School offering courses for all levels of ability. Residential and non-residential courses include meals etc and accommodation is provided in the mediaeval village surrounded by water meadows. C/W/E
Open daily 10am-5.30pm. (Closed Jan).
Directions: from M4 junc 17 take Chippenham turn-off then follow A350 to Lacock. Pottery is down lane beside village church.

RUG RESTORATION
Castle Combe Oriental Rugs Ltd
(Kathryn Jordan)
The Barn, Estate Yard
Castle Combe SN14 7HU
☎ *0249 782142*
15 years' experience in oriental rug restoration and conservation; two people employed.
R&R. Credit cards.
Mon-Fri 9am-4pm; advisable to telephone for appointment. (Closed Aug & part school holidays)

Devizes

CERAMICS
Cuckoo Craft
(Mrs Clare Milanes)
Breach House, Cuckoo Corner
Urchfont SN10 4RA
☎ 0380 840402
Pottery housed in a restored Wiltshire barn; ceramic sculpture, hand-built pots and thrown pots. Also pot-throwing and sculpting workshops for adults and children. C
Variable working hours; advisable to telephone first.

THATCHING
Higham Thatching
(Matthew A Higham)
The Maltings, Oak Lane
Easterton SN10 4PD
☎ 0380 818419
Traditional and ornamental thatching of old and new roofs using combed wheat reed and water reed. Member of regional Master Thatchers' Association.

FURNITURE MAKING & RESTORATION
Neil McGregor
The Old Chapel,Horton SN10 3LX
☎ 0380 723555
Furniture made to individual requirements, fitted and free-standing. Antique and modern furniture repaired and restored. C/R&R
Visitors welcome by appointment.

TEXTILES
Julieann Worrall Hood
29 High Street
Market Lavington SN10 4AG
☎ 0380 812861
Woven tapestry; miniatures to wall hangings. Crafts Council Setting-Up award in 1989, work exhibited at the V&A Museum. Many notable commissions undertaken. C/W/E
Only Sat-Sun by appointment until New Year, due to working in situ on various projects.
Directions: 6 miles south of Devizes on B3098, opposite Green Dragon PH.

Malmesbury

ANTIQUE FURNITURE & CLOCK RESTORATION
Restorations Unlimited
(R J Pinchis)
Pinkney Park
Malmesbury SN16 0NX
☎ 0666 840888
Restoration of all periods of antique furniture, long case and bracket clocks, porcelain and ceramics. Also specialists in reproducing items of furniture of all periods to match existing pieces; chairs, tables, cupboards etc. C/R&R/E
Mon-Fri 9am-5pm, lunchtime closing 12.30-1.30pm. Sat by appointment.
Directions: from Malmesbury take B4040 westwards (Old Bristol Road). After 6 miles pass Eagle PH on right, 200yd beyond turn left into drive to workshops.

Marlborough

DOLLS' HOUSE MINIATURES
Headley Holgate Impi
(Headley Holgate)
North Wing, Wolfhall Manor
Burbage SN8 3DP
☎ 0672 810519
Dolls' house furniture for collectors in waxed pine and French polished mahogany. W/E
Open daily 9am-5pm, lunchtime closing.

THATCHING
Chisbury Thatchers
(W B and R D Bacon)
The Castle , Great Bedwyn SN8 3LU
☎ *0672 870225*
Thatching contractors — visitors
should telephone for site in
operation where work can be
viewed.

UPHOLSTERY
Jan Turner Upholstery
The Old School House
Woodborough, Pewsey SN9 5PL
☎ *0672 851417*
Two employed upholstering
antique and modern furniture and
making curtains, loose covers etc.
C/R&R/W
Mon-Fri 9am-5pm, lunchtime closing
12.30-1.30pm. Sat 9am-1pm.

NEEDLECRAFT
The Painted Thimble
(Angela Sibthorp and Maureen
Preston)
The Square
Aldbourne SN8 2DU
☎ *0672 41045*
Hand-painted and appliqued
clothes for all sizes from babies to
size 22+ ladies; a wide range to suit
all tastes. Happy to discuss indi-
vidual requirements. Also giftware.
C/W/E. Credit cards.
Mon-Sat 10am-5pm.

THATCHING
J C Foster & Sons
(Colin and Christopher Foster)
Spring Cottage
Lyfield SN8 1TY
☎ *0672 86688*
Master Thatchers; thatching
contractors.

Melksham

FORGEWORK
A M Engineering
(Andrew Missen)
Old Forge, 206 Woodrow Road
Lower Forest SN12 7RD
☎ *0225 704230*
Traditional ornamental ironwork of
all kinds including gates, fire-
baskets, lamps, canopies, security
grills, room dividers and balus-
trades. Special commissions
undertaken. C/R&R
Open daily 8am-5pm.
Directions: Lower Forest lies between
Melksham and Lacock.

Salisbury

BRIAR PIPES
Tilshead Pipe Co Ltd
(B J Jones)
19 Candown Road
Tilshead SP3 4SJ
☎ *0980 620679*
Handmade briar smoking pipes
which have been individually
turned from solid blocks of natu-
rally dried Greek plateau briar.
Several styles, all based on tradi-
tional English shapes. C/R&R/W/E
Mon-Fri 8.30am-5.30pm.
Directions: 14 miles north west of
Salisbury on A360.

CERAMICS
Braybrooke Pottery
(Sally Lewis)
17 Andover Road, Upavon SN9 6AB
☎ *0980 630466*
Potter at work making commemorative, decorative and domestic
earthenware and stoneware thrown
pots, from egg cups to fountains. C
Open daily; telephone before visiting.

ANTIQUE FURNITURE
RESTORATION
Paul Winstanley
213 Devizes Road, Salisbury SP2 9LT
☎ *0722 334998*
Restoration work including turning,
veneering, French polishing and
traditional joinery. All periods of
English furniture repaired; tutor of
restoration. Reproduction furniture
sold retail and wholesale. C/R&R/W/E
*Mon-Fri 8.30am-5.30pm, lunchtime
closing.*

BOOKBINDING
Salisbury Bookbinders
(Alan Winstanley)
213 Devizes Road, Salisbury SP2 9LT
☎ *0722 334998*
Bookbinding and restoration of
antiquarian books; fine leather
binding, gold leaf tooling, manu-
script and artwork. Wide range of
work from specialist reproduction
binding of first editions to novelties,
presentation gifts etc. Notable
commissions includes Mountbatten
Memorial Donors Book. C/R&R/E
*Mon-Fri 8.30am-5.30pm, lunchtime
closing.*

THATCHING
Jim Goodland
6 Newton Close, Whiteparish SP5 2SP
☎ *0794 884636*
Business established for 12 years
employing three Master Thatchers;
altogether six people employed
thatching with combed wheat
straw, water reed and long straw.

TOY MAKING
Nippy Company
(Mr and Mrs A Holder)
43 Russell Street , Wilton SP2 0BG
☎ *0722 744143*

Manufacture of handmade wooden
toys; solid wood buses and vans
based on traditional vehicles
including post vans, double-decker
buses, ice-cream vans etc. Robust,
hand-painted and meeting safety
requirements. W/E
Mon-Fri 9am-4pm.

WOODWORK
Rackmaster
(Mr and Mrs A Holder)
43 Russell Street, Wilton SP2 0BG
☎ *0722 744143*
Specialist traditional beechwood
plate racks; the design, tried and
tested over eight years, ensures that
the widest plates and bowls are
accommodated and wine glasses
stand upright. Largest size takes 60
plates; wall fixings or draining
board feet. W/E
Mon-Fri 9am-4pm.

FURNITURE MAKING & UPHOLSTERY
Wilton Traditional Furniture
(Stephen Whittock and David White)
*Units 32-33 Barnack Trading Estate
Wilton SP2 0AW*
☎ *0722 744329*
Traditional furniture frames and
upholstery to suit customers' needs;
sofas, armchairs, dining chairs,
reclining chairs etc. Workshop and
showroom. C/W
Mon-Fri 8am-7pm, Sat-Sun 9am-6pm.
Directions: on A36 Wilton-
Warminster road, go under 1st
railway bridge, immediately turn
right, then left into industrial site.

PAPER MARBLING
Compton Marbling
(Solveig Stone)
*Lower Lawn Barns
Tisbury, Nr Shaftesbury SP3 6SG*

☎ 0747 871147
Manufacturers of hand-marbled paper in a variety of patterns and colours. Demonstrations given by appointment. Products include lighting, stationery boxes, books, albums, frames and wrapping paper. Catalogue and sample book available. C/W/E. Credit cards.
Mon-Fri 9am-5pm.
Directions: 16 miles west of Salisbury.

FURNITURE MAKING
Nicholas Turner
Old Well Studio
Ludwell, Nr Shaftesbury SP7 9ND
☎ 0747 828178
Design, manufacture and installation of one-off fitted interiors and free-standing furniture. A wide range of timbers and a variety of different finishes used. Work undertaken for private customers, interior designers, architects and builders. Also small batches of free-standing furniture. C/W/E
Mon-Sat 9am-5pm, by appointment.
Directions to workshop: from Shaftesbury on A30 to Salisbury turn right onto B3081 to Ringwood. After 150yd turn right into Mayo Farm; workshop at back of yard.

GLASS SANDBLASTING
Ruth Dresman Glass
54 East Hatch
Tisbury, Nr Shaftesbury SP3 6PJ
☎ 0747 871429
Cutting, masking and sandblasting crystal glass bowls and dishes with designs inspired by the countryside. Also decorated glass panels for windows and doors and special commissions. Work widely exhibited. C/W/E
Visitors welcome by appointment.

Swindon

UPHOLSTERY
Chiaroscuro
(Mrs Sandra Middleton)
Wootton Fields Farm
Marlborough Road
Wootton Bassett SN4 7SA
☎ 0793 848085
Small upholstery business (including some rush and cane work) and soft furnishing. Interior design service available. C/R&R/E
Mon-Fri 9am-5pm, lunchtime closing. Sat-Sun by appointment.

CERAMICS
Courtyard Pottery
(John Huggins)
Groundwell Farm, Cricklade Road
Swindon SN2 5AU
☎ 0793 727466
Pottery producing a wide range of quality terracotta plant pots decorated with various motifs. Pots are hand-thrown or hand-pressed, carrying 10 year frost-proof guarantee. Pots by post service; catalogues available. C/W/E
Mon-Fri 9am-5.30pm, Sat 9am-5pm.

ROCKING HORSES
Robert Mullis
55 Berkeley Road
Wroughton SN4 9BN
☎ 0793 813583
Hand-carved rocking horses in wood; large horses and miniatures on bow rockers or stands. Real hair used and stirrups; saddles made by Master Saddler. Also restoration work. C/R&R/W/E. Credit cards.
Very small workshop; visitors welcome by appointment only.

Warminster

FURNITURE
Horner's Furniture Makers
(Richard Horner)
The Old Stores, The Street
Kilmington BA12 6RG
☎ 0985 844218
Fitted and freestanding furniture
designed and custom made to blend
with period properties. Work to
commission only. C/E
Mon-Fri 9am-5pm, lunchtime closing.
Sat 9am-1pm.
Directions: turn off B3092 between
Frome and Mere, 300yd north of Red
Lion PH.

CERAMICS
Krukker Ceramics
(Vicky and Rob Whelpton)
1 Bull Mill, Crockerton BA12 8AY
☎ 0985 219577
Raku fired pottery and domestic
stoneware. Work sold in galleries
throughout UK. Members of
regional craft guild. C/W/E
Visitors welcome by appointment.
Directions: from A350 Warminster-
Blandford road heading south, at
Crockerton turn left opposite bus
shelter and follow to pottery on right.

FURNITURE
Matthew Burt Splinter Group
(Matthew Burt)
Albany Workshops
Sherrington BA12 0SP
☎ 0985 50996
Design and manufacture of one-off
fine contemporary furniture to
commission. Portfolio work (small
batch repeats). Also 'fantasy
pavilions' — imaginative houses for
children to play in and on. Work
exhibited at The Barbican, Edin-

burgh Festival etc. C
Mon-Fri 9am-6pm, by appointment.

ANTIQUE FURNITURE
RESTORATION
The Workshop
(Peter Heaton-Ellis)
c/o Cooper & Tanner
13a Market Place, Warminster
☎ 0985 50494 *(evenings)*
Antique furniture restoration
service including valuation work
and buying/selling (shop under
workshop open Sat). C/R&R
Workshop: variable times; visit by
appointment. Shop: Sat 10am-5.30pm.

GLASS ENGRAVING
Frank Grenier
Keyford
Upton Scudamore BA12 0AQ
☎ 0985 215401
Artist with 30 years' experience
painting and sketching, hand-
engraving in line and stipple on all
sizes and shapes of glasswork;
small presentation pieces to church
windows. Illustrations copied from
drawings or photographs. C/W/E
Open at any reasonable time.

Westbury

CERAMICS
The White Horse Pottery
(S and A Humm)
Newtown
Westbury BA13 3EE
☎ 0373 864772
Husband and wife team working in
a converted Victorian church school
building, making terracotta
gardenware and glazed domestic
ware. C/W
Tues-Sat 10am-5pm.

Bedale

SADDLERY
North of England Saddlery Co
(Lee Broadway)
27a North End, Bedale DL8 1AF
☎ 0677 422213
English leather saddles made to commission and all tack repaired.
Mon-Fri 9am-5.30pm, lunchtime closing.

CERAMIC TILES
Stephen Cocker
Beckside Cottage
The Bridge, Bedale DL8 1AN
☎ 0677 424658
Husband and wife team handmaking 'Delft' tiles. W/E
Mon-Sat 9am-4.30pm, lunchtime closing.

Easingwold

BRICK AND TILE MAKING
The York Handmade Brick Co Ltd
(G D H Armitage and T Bristow)
Forest Lane, Alne YO6 2LU
☎ 03473 8881
Handmade bricks, pavers, specials and floor tiles. 20 people employed manufacturing up to 80,000 bricks per week from the company's quarried clay. C/W/E
Mon-Fri 8am-4.30pm, Sat 9am-12noon.

FURNITURE
Acorn Industries
(A W Grainger)
Brandsby YO6 4RG
☎ 0347 5217
Small family business employing five furniture makers. Working mainly to order, all types of furniture designed and made in a variety of woods, each piece carved

with an acorn. C/R&R/E
Mon-Fri 8.15am-5.15pm, Sat 9am-1pm.
Directions: on B1363 York-Helmsley road in village of Brandsby.

Great Ayton

FURNITURE MAKING
John Harrison
Farndales Yard, 11 High Street
Great Ayton TS9 6NH
☎ 0642 724236
Designer craftsman specialising in Windsor chairs, dining tables and dressers in English and American hardwoods. Also Bergère arm chairs and Shaker furniture. All furniture is made to order. C/E
Visitors welcome by appointment. (Closed Aug).
Directions: from Stokesley on A173, cross River Leven, turn right into High Street; workshop down first drive on left.

Harrogate

SADDLERY
Seymour Saddlery
(T N Seymour)
Dacre House, Ripley HG3 3AY
☎ 0423 770772
Saddlery work, fancy leather goods, vintage car replacements. Stockists for horse clothing and accessories. C/R&R/W/E. Credit cards.
Mon-Sat 9am-5.30pm.

Hawes

CERAMICS
Wensleydale Pottery
(Simon J B Shaw)
Market Place
Hawes DL8 3QX
☎ 0969 667594

A full range of oven-to-table ware (for dishwashers and microwaves). Commissions taken for commemorative pieces. Many small gifts and souvenirs. C
Open Easter-Oct Mon-Sat 10am-6pm, other times by appointment.

ROPE MAKING
W R Outhwaite & Son
(Dr and Mrs P Annison)
Town Foot, Hawes DL8 3NT
☎ *0969 667487*
Small rope-making firm, started over 150 years ago, now employing 22 people making rope, braid, cord and twine. Suppliers of agricultural rope products, church bell ropes, dog leads, barrier and bannister ropes, horse leading reins etc. Visitors can view traditional rope-making methods of twisting together thin strands of yarn. Admission free. C/W/E
Mon-Fri 9am-5.30pm. Sats Jul-Oct, Easter and Bank Holidays from 10am. (Closed Xmas-New Year).

TOY MAKING
Jigajog Toys
(Martin and Penny Cluderay)
The Penn House, off Market Place Hawes DL8 3QX
☎ *0969 667008*
Hand-painted wooden toys based on traditional ideas. Members of British Toymakers Guild. Also a wide selection of toys by other toymakers. W/E. Credit cards.
Mon-Sat 9am-5.30pm, Sun (summer months) 2-5pm. (Sometimes closed for lunch during winter months)
Directions: adjacent to Market Hall in centre of Hawes.

CERAMICS
Yorkshire Flowerpots
(Gabriel Nichols)
Brunt Acres Industrial Estate Hawes DL8 3UZ
☎ *0969 667464*
Makers of fine quality hand-thrown terracotta garden pots. Range of traditional flowerpots, wall pots, strawberry pots etc. W. Credit cards.
Mon-Fri 8.30am-5pm, other times by appointment. (Closed Xmas-New Year).

WOODWORK
Swallowdale
(Mr and Mrs R H Tully)
Simonstone, Hawes DL8 3LY
☎ *0969 667730*
General woodwork and cabinet making, woodturning, repairs and restoration work. C/R&R/E
Open daily 10am-5pm.

Hellifield

ANTIQUE FURNITURE RESTORATION
Graham Coles
Old Co-Op Hall, Hellifield BD23 9XX
☎ *0729 850573*
A co-operative of several craftsmen specialist cabinet making, restoration work, French polishing, woodturning etc. Sales of restoration materials. C/R&R/W/E
Open daily 9am onwards.

Helmsley

UPHOLSTERY
Peter Silk Upholsterer
18 Bridge Street, Helmsley YO6 5DX
☎ *0439 70051*
Upholstery workshop employing three traditional upholsterers and a seamstress. Restoration work for

Duncombe Park. Member of Guild of Traditional Upholsterers. C/R&R/W. Credit cards.
Mon-Fri 8am-5pm, Sat 8am-4pm, lunchtime closing.

ARTIST
Holly House Studio
(S and D M Barnes)
Main Street, Wombleton YO6 5RX
☎ 0751 31429
Framed original watercolours and limited edition prints (framed or unframed) by D M Barnes.
C/W/E. Credit cards.
Visitors welcome by appointment.

CERAMICS
Wold Pottery
(Jill Christie)
Unit 3, Station Road
Helmsley YO6 5BZ
☎ 0439 71238
Hand-thrown, slip decorated earthenware. Commissions taken; commemorative pots. W
Mon-Fri 9am-5pm, Sat 9am-12noon.

Ingleton

CERAMICS
Bentham Pottery
(Kathy Cartledge)
Oysterber Farm
Low Bentham LA2 7ET
☎ 05242 61567
Small workshop making hand-thrown studio pottery in a variety of glazes. Pottery courses for beginners to advanced students. Also pottery supplies. Holiday accommodation. C/W/E
Mon-Sat 8.30am-5pm. Some Suns 2-pm, advisable to telephone first. Closed Sat-Sun during winter months).

CERAMICS
Ingleton Pottery
(Dick and Jill Unsworth)
Ingleton LA6 3HB
☎ 05242 41363
Studio pottery run by small family concern for 20 years making hand-thrown stoneware pottery. Demonstrations for large or small parties by appointment. C/W/E
Open daily 9am-5.30pm.
Directions: Ingleton is just off A65 between Settle and Kirkby Lonsdale.

Leyburn

DOLLS' HOUSES
Longbarn Enterprises
(Dr Christopher Cole)
Low Mill, Bainbridge DL8 3EF
☎ 0969 50416
18th-century working watermill housing workshop making decorated and furnished dolls' houses, Georgian and modern styles, also available as kits or made to customers' requirements. Working model watermills, castles; dolls' house accessories. C/R&R/W
Wed & Fri 2-5pm July — mid Sept and Bank Holiday weekends. Other times by appointment.

SCULPTURE, CASTING & WOODWORK
Old School Arts Workshop
(Peter and Judith Hibbard)
Middleham DL8 4QG
☎ 0969 23056
Run as a study centre for sculpture, woodwork, modelling and other visual arts. Picture framing, furniture making, repairs and specialist casting in GRP and concrete undertaken. Gallery. Demonstrations Thurs. Residential

courses. C/R&R/W/E. Credit cards.
Open daily 10am-5pm in summer.
Advisable to telephone first in winter.

CERAMICS
Swineside Ceramics
(Martin and Judith Bibby)
Leyburn DL8 5QA
☎ *0969 23839*
Manufacture of novelty teapots
(collectors' items). Eleven people
working in factory with a walkway
enabling visitors to view the work.
Teapots made from slip-cast white
earthenware hand-painted, in a
changing range of styles. Showroom
and tea room. C/W/E. Credit cards.
Mon-Fri 9am-5pm. Sat-Sun summer
time:10am-4pm (winter time
telephone for details).
Directions: 12 miles from A1, on a
small industrial estate in Leyburn.

CERAMICS
Moorside Design
(The Nichols family)
West Burton DL8 4JW
☎ *0969 663273*
Ornamental cats and other animals
made in fine, strong semi-
porcelainous stoneware. Collected
all over the world. Also terrracotta
garden pots. C/W/E. Credit cards.
Mon-Fri 9am-5pm, Sat-Sun 10am-
4pm, (lunchtime closing) by appoint-
ment. (Closed Xmas-New Year).

STAINED GLASS
Tana Stained Glass
(Simon Fitton)
7 Railway Street, Leyburn DL8 5EH
☎ *0969 22715*
Design, construction and restoration of stained glass, working
mainly to commission; windows,
mirrors, light shades, terrariums.

Member of Guild of Master Craftsmen.
C/R&R/W/E. Credit cards.
Mon-Sat 9am-5pm.

CERAMICS
Cottage Pottery
(Mr Jan Ward)
Vine Cottage, Bellerby DL8 5QP
☎ *0969 22184*
Working pottery making hand-
painted fragranced ceramic gifts
including Victorian bottles, boxes,
eggs, teddy bears, miniature
animals and pictures. Also glazed
aromatic oil vaporizers and
aromatherapy oils. W/E
Mon-Fri 9am-6pm, Sat-Sun 11am-4pm.

CABINET MAKING
Thornsgill Joinery
(Colin Gilyeat)
Thornsgill House, Moor Road
Askrigg DL8 3HH
☎ *0969 50617*
All types of high-class joinery and
cabinet making. C
Open at any reasonable time.

FURNITURE MAKING
Wynn Bishop
The Workshop, Bishop's Wynd,
Market Place, Middleham DL8 4NP
☎ *0969 22418/22703*
Cabinet making workshop estab-
lished ten years, specialising in
making replicas of classic pieces
and designing contemporary pieces
Training courses. C/E
Mon-Fri 8am-5pm, lunchtime closing
Sat 8am-12noon.

CERAMICS
Askrigg Pottery
(Andrew Hague)
Old School House, West End
Askrigg DL8 3HN

☎ *0969 50548*
Functional and decorative stoneware
including mugs, lamps, platters,
brush-decorated porcelain. C
Telephone to check opening hours.

Malton

SILVERSMITHING
Wolds Silver
(Vincent and Margaret Ashworth)
Rothay Cottage, Leppington YO17 9RL
☎ *0653 85485*
Silver and gold work, from earrings
to teapots, including commissioned
work. Also residential silverwork
courses (only four pupils at a time).
C/R&R/W/E
*Open daily 9am-5.30pm. (Closed Jan-
Mar).*

SADDLERY
Peter Ward (Saddlery)
5 Wood Street, Norton YO7 9HF
☎ *0653 694561*
2 years' experience in harness
making and repairs to all types of
saddlery, specialising in servicing
racing stables and pony clubs
throughout North Yorkshire and
Cleveland. Also belts made to
individual requirements. C/R&R/W
*Mon-Fri 9am-5pm; advisable to
telephone first.*
Directions: follow main York-
Scarborough road, over railway
crossing. Turn right at roundabout,
take first left; workshop 100yd on left.

CERAMICS
Sophie Hamilton
Beerholme Pottery
High Marishes YO17 0UQ
☎ *0653 86228*
Pottery set up in a converted farm
building with assistance from the

Rural Development Commission.
Decorative functional stoneware
pots, all hand-decorated. C/W/E
*Mon-Sat 10am-6pm. Bank Holidays
& Suns (July-Aug) 10am-6pm.*
Directions: from Malton follow A169
road to Pickering, turn right to High
Marishes; pottery up turning to right.

CLOCK MAKING
Stephen Barrett (Clocks)
Newcroft , West Knapton YO17 8JB
☎ *0944 28813*
New business manufacturing
reproduction grandfather clocks.
C/R&R/E
Open at any reasonable time.

SADDLERY
Hadfield Saddlers
(M A Hadfield)
57 Commercial St, Norton YO17 9HX
☎ *0653 694095*
Family saddlery business with
manufacturing work and repairs
carried out on the premises. Also
stocking horse clothing etc.
C/R&R. Credit cards.
Mon-Fri 9am-5pm, Sat 9am-3.30pm.

FORGEWORK
Sherburn Forge
(C F Cade)
31 St Hilda's St, Sherburn YO17 8PG
☎ *0944 70279*
Two employed producing ironwork
to individual requirements. Also
restoration work. C/R&R/E
Mon-Fri 8am-5pm, Sat 8am-12noon.

Northallerton

CERAMICS
Shire Pottery
(Georgia and Peter Naylor)
Scruton DL7 0RD

☎ 0609 748225
Domestic and decorative stoneware.
Work widely exhibited throughout
the north. C/W/E.
Open Sat-Sun and most weekdays
10am-4pm; advisable to telephone first.

CABINET MAKING
Gordon Hodgson
Cross Lane , Inglesby Cross DL6 3NQ
☎ 0609 82414
Handmade furniture and cabinet
making in solid woods, to customers requirements. Work for home,
office and church. Each piece
marked with a hand-carved bell,
named and dated. C/R&R/E
Tues-Fri 9am-5pm, most Sats 12.30-
5pm. Advisable to telephone first.
Directions: Ingleby Cross is 1 mile
north of Cleveland Tontine between
A19 and A172.

Pateley Bridge

GOLD & SILVERSMITHING
Debby Moxon and Ian Simm —
Jewellery
King Street Workshops, King Street
Pateley Bridge HG3 5LE
☎ 0423 712570
A partnership producing contempo-
rary jewellery in precious and non-
precious materials including titanium,
silver and gold. C/R&R/W/E
Mon-Fri and most weekends 10am-5pm.
Directions: opposite police station,
behind Nidderdale Museum in
Pateley Bridge. Park in town centre.

GLASS BLOWING
Andrew Sanders & David
Wallace — Glassmakers
King Street Workshops, King Street
Pateley Bridge HG3 5LE
☎ 0423 712570

Handmade blown glassware
including scent bottles, paper-
weights, vases and bowls etc.
Specialist and historic glasswork (eg
for Jorvik Viking Centre and
Egyptian glass for BBC TV). C/W/E
Mon-Fri and most weekends 10am-5pm.
Directions: opposite police station,
behind Nidderdale Museum in
Pateley Bridge. Park in town centre.

CERAMICS
Dianne Cross Ceramics
King Street Workshops, King Street
Pateley Bridge HG3 5LE
☎ 0423 712570
Potter making a range of decorated
glazed stoneware pottery for use;
also larger more individual work
and one-off pots. C/W/E
Mon-Fri and most weekends 10am-5pm
Directions: opposite police station,
behind Nidderdale Museum in
Pateley Bridge. Park in town centre.

CERAMICS
The Terracotta Potter
(John Dix)
King Street Workshops, King Street
Pateley Bridge HG3 5LE
☎ 0423 712570
Pottery making a range of frost-
proof terracotta garden ware; also
strongly coloured stoneware. C/W
Mon-Sat 9.30am-5.30pm, (closed
Mon&Tues during school term time).
Sun 10am-5pm.
Directions: opposite police station,
behind Nidderdale Museum in
Pateley Bridge. Park in town centre.

Pickering

CERAMICS
The Pottery
(James Brooke)
Appleton-le-Moors YO6 6TE

☎ 07515 514
Slipware and reduced stoneware
pottery, all hand-thrown. C/W
Open daily 10am-dusk.

Richmond

FORGEWORK
Maurice Stafford
(M and M A Stafford)
Blacksmith's Shop, High Street
Gilling West DL10 5JW
☎ 0748 824373
Village blacksmiths, specialist in
restoration of ironwork to historic
buildings. Makers of general
ornamental ironwork: gates,
lanterns, railings etc. Work mainly
to commission, but some work on
display for purchase. C/R&R/E
Open daily 9am-dusk.

CERAMICS
The Mill Studio
(Mary Weir)
The Mill, Newsham DL11 7RD
☎ 0833 21367
Individual ceramic items made by
slipcasting, ware assembly, decorat-
ing and firing. Instruction. C/W
*Mon 10am-4pm or Tues-Fri by
arrangement.*

FURNITURE
Philip Bastow — Cabinet Maker
Stonegate, Back Lane, Reeth DL11 6TJ
☎ 0748 84555
Small country workshop making
furniture mainly to commission.
Display area with samples of
completed furniture and wood gifts
for sale. C/E. Credit cards.
Mon-Sat 9am-6pm.
Directions: from Richmond cross over
river, take first turn left into Back
Lane; workshop straight ahead.

PICTURE FRAMING
Frame Plus of Richmond
(Graham Saxton)
The Engine Shed, Station Yard
Richmond DL10 4LD
☎ 0748 850505
Picture framing service with four
people in workshop. C/R&R/W/E
Mon-Fri 9am-5pm, Sat 2-6pm.

Ripon

GLASSWORK
Custom Display Cases
(Nigel James Vine)
Wayside, Skelton-on-Ure HG4 5AG
☎ 0423 322633
Manufacture of glass display cases
for models, jewellery, clocks,
taxidermy, trophies, porcelain and
glassware etc. C/R&R/W
Mon-Fri 9am-5pm, Sat 9am-1pm.

CERAMICS
Masham Pottery
(Kathleen and Howard Charles)
Rear of Kings Head Hotel
Market Square, Masham HG4 4EF
☎ 0765 689762
Small studio pottery housed in a
converted barn just off the market
square. Hand-thrown and finished
domestic stonework pottery sold on
the premises. Other crafts for sale. C
Tues-Sun 10am-5pm.

GLASS BLOWING
Uredale Glass
(T J and M Simon)
42 Market Place, Masham HG4 4RF
☎ 0765 689780
Glassmaking business producing
vases, bowls, goblets, paperweights,
lamps etc. Traditional glass blowing
demonstrations.

C/R&R/W. Credit cards..
Open daily Easter-Oct 10am-5pm.
Nov-Easter Tues-Sat 10am-5pm.
(Closed Jan). Glass blowing demon-
strations Tues-Sat 10am-5pm.
Glass blowing demonstration charges:
adults £1, OAPs 50p, children free.
Directions: behind King's Head PH
overlooking Market Place in Masham.

Scarborough

WEAVING
Ankaret Cresswell
(Justin P Terry)
Wykeham YO13 9QB
☎ 0723 864406
Handweavers producing exclusive
fabrics in wool. Made-to-measure
suits using own material. Short
lengths supplied to garment
makers. Credit cards.
Mon-Fri 10am-5pm, lunchtime closing.

FORGEWORK & FARRIERY
E K Readman & Son
(Alan Thomas Readman)
The Forge, High Street
Cloughton YO13 0JE
☎ 0723 870376
General blacksmithing, wrought
ironwork and farriery. Saddlery
sales. C/R&R/E. Credit cards.
Mon-Fri 8am-5pm, lunchtime closing.
Sat 8.30am-12noon.

FURNITURE
David Crews & Co
(David Crews)
Church House, Ebberston YO13 9NR
☎ 0723 859751
Business set up with assistance
from the Rural Development
Commission, employing three
cabinet makers producing all types
of handmade furniture, fitted and

free-standing, mainly in English
hardwoods. Recently commissioned
to design and fit-out the interior of
a luxury schooner. C/R&R/W/E
Open daily 8am-6pm.
Directions: midway between
Scarborough and Pickering on A170.

SPINNING & WEAVING
Woldcraft Woollens
(Sue Bradbury)
Potter Brompton Wold Farm
Ganton YO12 4PH
☎ 0944 70381
Spinning, weaving, felting and
hand-knitted garments made with
wool from own Jacobs sheep. Also
selling locally made pottery and
other craftwork. C/W/E
Open at all reasonable times, but
advisable to telephone first. School
parties welcome by arrangement; charge
of 50p each includes refreshments.
Directions: turn off A64 at traffic
lights in Sherburn, head south signed
Weaverthorpe. After half-mile take
left fork (no signpost) onto single
track road. Half-mile beyond crown
hill turn left down private road.

METALWORK
Ron Field (Craftsman in Metals)
Rowhowe House, Wykeham Lane
Wykeham YO13 9QG
☎ 0723 862640
Restoration and renovation of
church and architectural metal-
work, mechanical antiques. Work
for churches and stately homes,
local authority listed building work
English Heritage listed, featured in
books on restoration work, and on
Guild of Master Craftsmen's list.
C/R&R/E. Credit cards.
Mon-Fri 9am-6pm. Sat-Sun by
appointment.

Selby

CABINET MAKING
Blackfenn Reproductions
(Julian and Lois Crump)
Lodge Hill Farm, Garmancarr Lane
Wistow YO8 0UP
☎ *0757 268644*
Reproduction and restoration of
antique furniture and fire sur-
rounds. Furniture, mainly Victorian
and Regency reproductions, can be
made to customers' specifications.
Work sent overseas. C/R&R/W/E
Mon-Fri 9am-5pm.

ENGRAVING
Abbey Engravers
(Michael Cooper)
3 New Street, Selby YO8 0PT
☎ *0757 210519*
Small business hand and machine
engraving on all materials; watches,
rings, pewter, glass, signs in brass,
aluminium, plastics etc. C/R&R/W/E
Mon-Fri 9am-5.30pm, Sat 9am-5pm.

Skipton

GOLD & SILVERSMITHING
Gemini Studios
(Mrs Sheila Denby & Miss Katie Denby)
1a Main Street
Grassington BD23 5AA
☎ *0756 752605*
Visitors are welcome to watch gold
and silver jewellery being handmade by Katie Denby, Exhibition
silversmith for The Viking Museum
at York. Workshop in a large old
barn, also displaying the work of
other local craftspeople; pottery,
metalwork, knitwear, woodcraft
and original Dales paintings.
/R&R/W
Mon-Sat 10.30am-5.30pm, Sun 1-5pm.

FURNITURE
Old Rectory Kitchens & Bedrooms
(R Gower)
Skipton Commercial Centre
Water Street, Skipton BD23 1PB
☎ *0756 798773*
Cabinet work and one-off furniture
to commission; kitchens and
bedrooms custom built in any
timber, any style, finish etc. Also
small batch work. C/R&R/E
Mon-Fri 8.30am-5.30pm, Sat 8.30am-5pm.

UPHOLSTERY
Revivals
(David Paling)
Units 1&2 Eshton Road
Gargrave BD23 3SE
☎ *0756 749451*
Eight people employed providing
upholstery service, furniture
making, French polishing and
renovation work. C/R&R/W/E
*Mon-Fri 8.30am-5pm, Sat 9.30am-
12.30pm.*

WOOD CARVING
David Tippey
Victoria Lodge
Kirkby Malham BD23 4BS
☎ *0729 830547*
Carved and painted wooden bird
sculpture and decoy ducks. Work
on display in studio and exhibited
in galleries and at birdwatching
events throughout the north west.
C/R&R/W/E
*Mon-Fri 10am-4pm, lunchtime
closing. Sat-Sun 2-4pm. Please
telephone before visiting.*

FURNITURE
Peter Merrell Furniture
Chapel Fold, Grassington BD23 5BG
☎ *0756 752016*
Mainly contemporary furniture

design in hardwoods. One-offs and work to commission, but happy to meet individual requirements; fitted pieces undertaken. C/W
Mon-Fri 9.30am-5pm, Sat 9.30am-1pm.

FURNITURE
Kent Traditional Furniture
(Janet and Paul Kent)
Spring Croft, Moor Lane
Grassington BD23 5BD
☎ 0756 753045
Cabinet makers specialising in solid oak domestic furniture. 14 craftsmen and apprentices working in converted agricultural buildings. Attention given to fine detail, each piece bearing a carved White Rose, the company's trademark.
C/W/E. Credit cards.
Open daily 9am-6pm by appointment only. (Closed Xmas-New Year)

CERAMICS
Little Gallery & Studio
(Ron and Margaret Walker)
Elbec, Litton BD23 5QJ
☎ 0756 770284
Small ceramic workshop producing sculpture and calligraphic plates, with gallery showing original paintings. C
Thur-Sun 10am-5pm. (Closed Nov-Mar and May).

Stokesley

DRY STONE WALLING
Eddie Rowney
94 Sowerby Cres, Stokesley TS9 5EE
☎ 0642 712960
Member of regional branch of Dry Stone Walling Association and tutor in the North York Moors National Park where demonstrations can be seen.

Thirsk

FURNITURE
Kilvington Studio
(K A Wilkinson)
South Kilvington YO7 2LZ
☎ 0845 522328
Furniture handmade to commission; domestic, ecclesiastical and architectural work. C/R&R/E
Open at any reasonable times or by appointment.

FURNITURE
Malcolm Pipes, Fox Furniture
The Old Hall Workshop
Carlton Husthwaite YO7 2BQ
☎ 0845 401359
Visitors are welcome to visit the workshop to view traditional furniture making and joinery. All domestic furniture made in English oak; tables, chairs, dressers etc. Also grandfather clocks and small articles. A fox's mask, trademark of the business, is carved on each item. Furniture is adzed, pegged, mason-jointed and finished with beeswax polish. C/R&R/E
Mon-Sat 8am-5pm. Sun by appointment.
Directions: from A19 Thirsk to Easingwold/York road take turning (north) to Carlton Husthwaite. At turning left to Coxwold, workshop is on corner.

FURNITURE
Design in Wood (Thirsk) Ltd
(Andrew James — Managing Director)
Chapel Street, Thirsk YO7 1LU
☎ 0845 525103
A small group of skilled craftsmen with an established reputation for fine workmanship, designing and making comporary furniture to individual commissions. Domestic,

commercial and ecclesiastical work undertaken including kitchens and other built-in work. Interior design service. Work also for other designers. C/E
Mon-Fri 9am-4.30pm. Sat 9am-12noon by appointment. (Closed Xmas-New Year).

FURNITURE
The Treske Shop
(John Gormley)
Station Works, Thirsk YO7 4NY
☎ 0845 522770
Established 20 years ago, this workshop is growing rapidly in size and reputation, now employing 25 people making a very wide range of furniture. Best known for simple household furniture in solid ash, their work can be found in churches and palaces, with commissions from all over the world. Visitors are welcome to view work in the workshops. C/R&R/E. Credit cards.
Open daily 10am-5pm.
Directions: in old maltings behind a row of houses on north side of A61 where it crosses railway line. Building lies along west side of railway down a road marked with the Treske sign.

FURNITURE
Albert Jeffray
Sessay YO7 3BE
☎ 0845 401323
A wide range of domestic furniture, identified by its carved eagle symbol, handmade in English oak and finished to the highest standards. Special items made to order in a variety of timber. Also small pieces and carvings available. C/E
Mon-Fri 8am-5pm, Sat-Sun 9am-2noon.

Directions: from A1 take A168 by-passing Topcliffe, then turn right to Dalton and continue to T junc, turn right to Sessay. Workshop on right.

FURNITURE
The Old Mill
(W H and M Knight)
Balk YO7 2AJ
☎ 0845 597227
Handmade furniture in a variety of wood; 14 people employed in workshop. Each piece of furniture is made by a single craftsman and bears his own signature. Showroom with full display and other craftwork. C/E. Credit cards. on craftwork sales.
Mon-Sat 9am-5pm
Directions: from Thirsk take A170 to Scarborough for 2 miles, then turn right to Balk. Workshop on right after 1 mile.

FURNITURE
Robert Thompson's Craftsmen Ltd
(The Cartwright family)
Kilburn YO6 4AH
☎ 03476 218
Family run workshop employing 30 craftsmen continuing furniture making and carving in the tradition of Robert Thompson, whose carved mouse trademark became world famous. Their wide range of furniture is on display in his old timbered house; commissioned work is also undertaken. C/E
Mon-Thur 8am-5pm, Fri 8am-3.45pm, lunchtime closing 12noon-12.45pm. Sat 10am-12noon. (Closed Xmas 2 weeks, Easter 1 week).
Directions: 7 miles east of Thirsk, beneath the chalk White Horse.

Whitby

STAINED GLASS
The Stained Glass Workshop
(Joan and Alan Davis)
The Studio, High Street
Lythe YO21 3RT
☎ 0947 83246/820476
Manufacture and restoration of stained glass and leaded lights, specialising in kiln-fired painted work. Small showroom stocking hand-crafted items: roundels, lampshades, terrariums etc. Tuition. C/R&R/W/E. Credit cards.
Mon-Fri 9am-5pm, Sat 10am-5pm.

York

FURNITURE & UPHOLSTERY
W Hutchinson & Son
(W and T Hutchinson)
Squirrel Cottage
Hursthwaite YO6 3SB
☎ 03476 352
Furniture handmade from English oak, bearing a squirrel carved on each item. Also traditional upholstery service. C/R&R
Mon-Fri 8am-5pm, Sat 9am-4pm, lunchtime closing.

CABINET MAKING
A E Houghton & Son Ltd
(B A Houghton)
Common Road
Dunnington YO1 5PD
☎ 0904 489193
Joinery workshop specialising in ecclesiastical furnishings. 30 people employed, including site workforce. Many certificates from York and N Yorkshire Society of Architects. C/R&R/W/E
Mon-Fri 8am-6pm.

CERAMICS
Coxwold Pottery
(Peter & Jill Dick)
Coxwold YO6 4AA
☎ 03476 344
Small established country workshop producing a wide range of useful pots for house and garden; slip decorated earthenware and low stoneware. Commemorative pots etc. Work mostly sold from showroom also stocking a selection of garden pots from other potteries. C/W/E
Tues-Fri 2am-5.30pm. Also Suns in July & Aug. Advisable to telephone first. No charge to individuals. Groups (evenings) up to 20 people, by appointment; tour charge £1.50 pp.
Directions: 19 miles north of York, take A19 to Easingwold, turn right to Hursthwaite and Coxwold.

FORGEWORK
Don Barker Limited
(Don Barker)
Bracken Edge Forge, Plainville
Wiggington YO3 8RG
☎ 0904 769843
Artist Blacksmith with 20 years' experience as a design engineer, employing five other blacksmiths. Worshipful Company of Blacksmiths' Certificate of Merit. C/R&R/W/E
Mon-Fri 9am-5pm, Sat-Sun by appointment.

BASKETWORK & RUSH SEATING
Rosemary Hawksford
St Mary's Rush Seating and Willow Basket Workshop
St Mary's Church
Bishophill Junior, York YO1 1EN
☎ 0904 644788/651407 (home)
Traditional English willow baskets;

a wide selection available and baskets made to order (catalogue available — orders can be sent by post). Also rush seat restoration work. C/R&R
Tues-Thur & Sat 10am-1pm. Other times by request.

CERAMICS
Holtby Pottery
(Mick Arnup)
Holtby YO1 3UA
☎ 0904 489377
Decorated stoneware plates, bowls and vases. Also ceramic letters and numerals. Member of Craftsmen Potters Association (1978); work exhibited in London. Workshop also has work by potters Ben Arnup, Hannah Arnup and animal bronzes by Sally Arnup. C/W/E
Open daily 10am-6pm.
Directions: Holtby is 5 miles from York on A166 Bridlington Road.

CERAMICS
Crocodile Ceramics
(Philip and Anna-Marie Magson)
Windmill House, Sutton Road
Wigginton YO3 8RA
☎ 0904 750054
Small pottery making distinctive hand-thrown terracotta garden pottery with hand-painted decoration. Also stocking a selection of other handmade British pottery, mainly for the garden. W
Open daily 10.30am-6pm. (Closed 2-3 weeks in Oct).

RUG MAKING
Jacqueline James Handwoven Rugs
Rosslyn Street
York YO3 6LG
☎ 0904 621381
rugs and wall hangings, individual

designs and limited collections widely exhibited. Work made to order to co-ordinate with any interior setting. Notable commissions include weaving for York Minster, Wakefield Cathedral, Jorvik Viking Centre etc. C/W/E
Mon-Fri 10am-4pm, lunchtime closing.

LEATHERWORK
Connell Leather Products
(Allan Hewitt & Co Ltd)
Ryders Corner
Crambe YO6 7JR
☎ 0653 81359
Four people employed manufacturing leather goods for industry and the shooting fraternity.
C/R&R/W/E. Credit cards.
Mon-Fri 9am-5pm. Other times by appointment.

PICTURE FRAMING
Wild Cherries Gallery
(Terence and Sue Robinson)
Sutton-upon-Derwent YO4 5BT
☎ 0904 608466
Comprehensive framing service and restoration service in award winning workshops. Also selling a selection of etchings and original prints as well as a large range of popular limited edition sporting and animal prints.
R&R/W/E. Credit cards.
Mon-Wed & Sat 10am-3pm, other times by appointment.
Directions: from York take road to Howden, through Elvington, past Turpins Tavern PH on right. Shortly after, Wild Cherries Gallery on right.

FURNITURE
Adam Jackson
Poplars Farm
Beningbrough YO6 1BY

☎ *0904 470842*
Established furniture making and woodworking business. Individually commissioned furniture in solid hardwoods. Also antique restoration work. Winner of 1989 craftsmanship award for excellence from RIBA.
C/R&R/W/E. Credit cards.
Mon-Fri 8am-6pm, lunchtime closing. Sat 8am-1pm.
Directions: from York take A19 north for 5 miles. In Shipton take 1st left turn signed Beningbrough. After 1 mile take next left signed Beningbrough. After 1.5 miles take next left signed Beningbrough. Poplars Farm is 2nd turn on right in village.

FURNITURE
R Watson & Son
(D P Watson and P R Watson)
Joiner's Cottage
Linton-on-Ouse YO6 2AS
☎ *0347 4233*
Started as a joinery and undertaking concern in 1913, the business now designs and makes domestic furniture including kitchens on a commission basis; joinery restricted to specialised work, chiefly building timber-framed conservatories. C
Mon-Fri 9am-5pm, advisable to telephone first. Sat 1-5pm by appointment.

FURNITURE
Butterfly Furniture
(Andrew Neal Conning)
Tancred , Whixley YO5 8BA
☎ *0423 330580*

Sole cabinet maker and joiner producing fine quality furniture made to customers' own requirements, constructed from a wide range of hardwoods. C/R&R/E
Mon-Fri 8am-6pm, Sat-Sun 10am-12noon.
Directions: half way between Harrogate and York, just off B6265, 1 mile north of Green Hammerton.

FURNITURE
Harry Postill Furniture
The Old Chapel, Fangfoss YO4 5QP
☎ *0759 6209*
Workshop employing three people; pine furniture specialists, in particular quality beds and tables. C/E
Mon-Fri 8am-5.30pm, Sat 8am-4.30pm.
Directions: Fangfoss lies between A1079 and A166, south of York.

WOODWORK
Walpole Woodcrafts
(Mr and Mrs Howard)
Home Farm Cottage
Everingham YO4 4JD
☎ *0430 861276*
Furniture making, carving and turning. Holders of cabinet maker certificates and members of Guild of Master Craftsmen. Also spinning and dried flowers. C/W/E
Mon-Thur 9am-6pm, Fri half-day closing. Most Suns 10am-5pm. (Closed end Sept and Oct)
Directions: From A1079 York/Hull road turn off at Hayton or Thorpe-le-Street to Everingham. Workshop behind Home Farm Cottage.

SOUTH YORKSHIRE CRAFT WORKSHOPS

Barnsley

SCULPTURE
Leonard Bedford (Sculptor)
29 Old Road, Smithies
Barnsley S71 1UE
☎ 0226 285632
Clay modelling and sculpture, cold
casting in resin and candlemaking.
Original sculptures from which
moulds are made and then cast in
resin and painted by airbrush. Also
candles. C/W/E. Credit cards.
Mon-Fri 10am-6pm, lunchtime closing.

FORGEWORK
Wentworth Forge
(Lindsay Gregory)
Old Buildings Yard
Wentworth S62 7SB
☎ 0226 749234
General blacksmithing, light
engineering and fabrications.
Mainly industrial and commercial
work for builders etc. Ornamental
work (flower stands, hayracks,
hanging basket brackets and
weather vanes) mostly to commission. Small showroom. C/R&R/W
Mon-Fri 9am-5pm.

Rotherham

ANTIQUE FURNITURE
RESTORATION
Stubbin Cottage Antiques
(D Smith)
Stubbin Cottage, Low Stubbin
Rawmarsh S62 7RX
☎ 0709 522881
French polishing, staining and
waxing service specialising in
antique furniture restoration,
mainly for private customers, many

overseas. Also sales of antique
furniture, clocks etc restored and
polished. C/R&R/W/E
*Variable working hours; advisable to
telephone first.*

DRY STONE WALLING
Town & Country Walling
(John Lackenby and David Bullivent)
3 Plantation Close, Maltby S66 8JR
☎ 0709 817523
Master Craftsman and four interme-
diate dry stone wallers.

TAXIDERMY
Barbot Taxidermy Studios
(Graham Teasdale)
Barbot Hall Farm
Greasbrough S61 4QL
☎ 0709 563038/364351
Renowned commercial taxidermist
working in old farm buildings.
Expertise acquired through
museum work involving model
making, blacksmithing, tannery and
joinery. Working mainly for
museums and on commission.
Some restoration work undertaken.
Also model making using a large
variety of materials. Founder
member of Museum Taxidermy
Guild; some lecturing and demonstrations given. C/E
*Mon-Fri 9am-5.30pm, Sat 9am-
12noon, or by appointment. Advisable
to telephone first.*
Directions: farm workshop on hillside
below B6089, 1 mile north of Rotherham.

LAPIDARY
Morthen Craft Workshop
(Brian and Sylvia Scott)
Morthen Lane, Wickersley S66 9JQ
☎ 0709 547346

Established business cutting and polishing stones for jewellery; mostly semi-precious stones from overseas, also some local stone. Stone repair work. Some jewellery sold. Group visits by arrangement. Also dried flowers for sale. Workshop is housed in converted farm buildings, also providing showroom and tea room. C/R&R
Open daily 9am-5.30pm; other times by appointment.

SPINNING
Wingham Woolwork
(Mrs Ruth Gough)
70 Main Street, Wentworth S62 7TN
☎ 0226 742926
Hand-spinning and suppliers of a wide variety of fibres for spinning; mohair, cashmere, silk, camel hair, yak etc. Tuition courses for spinners of all levels. Also natural knitting yarns, some hand-dyed. Special spinning demonstrations by arrangement. W/E. Credit cards.
Sat & Mon 1.30-5.30pm, Sun 11am-5.30pm.

Sheffield

DRY STONE WALLING
South Pennine Walling Co
(R Meller)
331 Glossop Road, Sheffield S10 2HP
☎ 0742 738135
Dry stone walling for private gardens, builders and farmers etc.

FURNITURE
M D Finney Fine Furniture Makers
(Mark Finney)
Unity House, Clay Lane (between Eyre Lane and Arundel Gate)
Sheffield S1 2LQ

☎ *0742 755817*
Furniture makers and designers producing hand-crafted and hand-finished solid hardwood occasional furniture, individually made using traditional methods. Tuition courses run for members of the public. Also materials for professional and home woodworkers. Articles written for woodworking publications.
C. Credit cards.
Mon-Fri 9.30am-5pm, Sat 9.30am-3pm.
Directions: from the direction of the Cathedral, turn down Arundel Gate; Clay Lane and workshop on left.

HEDGE LAYING
Jasper Prachek
280 Manchester Road
Deepcar S30 5RG
☎ *0742 885113*
Hedge laying (ATB Instructor); also hedge trimming, woodland management and dry stone walling. Winter demonstrations, details available. Please telephone for viewing in situ work. Also coppice crafts demonstrations given.

HAND-PAINTED GIFTWARE
Knollcraft
(David Tranter)
Unit 4, Rother Valley Craft Centre
Rother Valley Country Park
Mansfield Road, Wales Bar S31 8PE
☎ *0246 416577 (evenings)*
A family business hand-painting on wood, metal, basket ware and leather. Also furniture. C/W/E
Wed-Sun 11am-4pm (7pm in summer)
Directions: from M1 Junc 31(Worksop) take A57(T) to Sheffield. Turn left on A618 heading south, past turn to left (B6059 to Kiveton); Rother Valley Country Park on right.

CERAMICS
Chris Boddy
Rother Valley Craft Centre
Rother Valley Country Park
Mansfield Road, Wales Bar S31 8PE
☎ *0302 366595 (evenings)*
Earthenware clay used to produce originally designed pottery from tiny vases to large jars (75cm/30in high). Table, kitchen and bathroom ware glazed and decorated in contemporary colours.
Summer months 10am-dusk: Winter months 10am-5pm.
Directions: from M1 Junc 31 (Worksop) take A57(T) to Sheffield. Turn left on A618 heading south, pass turn to left (B6059 to Kiveton); Rother Valley Country Park on right.

CERAMIC SCULPTURE
Antonia Salmon
20 Adelaide Road, Nether Edge
Sheffield S7 1SQ
☎ *0742 585971*
Contemporary ceramic sculptures, vessels and dishes; white stoneware thrown and handbuilt, carved and incised with burnished surfaces. The pieces are high bisque fired, then slowly smoked in sawdust before final wax polishing. Work sold in shops and galleries throughout UK. Fellow of Craft Potters Association. C/W/E
Visitors welcome by appointment only.
Directions: from A625 into Sheffield turn right into Brincliffe Edge Road, take 3rd turn left into Union Road, take first right and right again into Adelaide Road; workshop on corner.

CANDLE MAKING
C Brooks — Candlemaker
South View, Ridgeway Cottage
Industry Centre, Ridgeway S12 3XR
☎ *0742 478591*
Hand-carved candles made to customers' requirements and colours. Two people employed. C/R&R/W/E. Credit cards.
Open daily 10.30am-5pm.

WEST YORKSHIRE CRAFT WORKSHOPS

Bradford

DRY STONE WALLING
Ian E Barraclough
Little Hill Farm, Pit Lane
Thornton BD13 3ST
☎ 0274 833010
Dry stone walling, working with others on large projects. Master Craftsmanship Award from Dry Stone Walling Association.

SILK WEAVING
Seshan Textiles
(Aneeta Milligan,BAHons)
South Square Gallery, South Square
Thornton BD13 3LD
☎ 0274 835076
100% pure silk textiles hand-woven on a traditional wooden dobby loom; scarves and shawls in brilliant jewel colours inspired by Mexican costumes and crafts. Business set up with assistance from the Crafts Council. C/W/E. Credit cards.
Mon-Fri 10am-6pm.
Directions: from Bradford follow Thornton Road beside Alhambra Theatre/Cinema for several miles to Thornton village. South Square on left beside New Inn PH.

Halifax

CERAMICS & FORGEWORK
Jim Cooper Pottery
(James and Gill Cooper)
103A Oldham Road
Ripponden HX6 4EB
☎ 0422 822728/823565
Mainly slip-cast ceramic giftware. Old forge with lineshaft and stationary engine; traditional blacksmithing and ornamental ironwork. C/R&R/W/E
Open during normal working hours. Advisable to telephone first.

CERAMICS
Throstle Nest Pottery & Gallery
(Pat Kaye)
Old Lindley
Holywell Green HX4 9DF
☎ 0422 374388
Pottery producing thrown and hand-built decorative and functional stoneware. Also gallery housed in old Yorkshire stone barn selling a wide variety of regional crafts and paintings with regular exhibitions. C
Open daily 10am-5pm.

Hebden Bridge

DOLL RESTORATION
Sue Rouse
The Dolls House
Gladstone Buildings, Hope Street
Hebden Bridge HX7 5AG
☎ 0422 845606
Restoration of dolls and dolls' clothes; most types of old fashioned antique and collectable dolls. Dolls clothes sympathetically made. Advice on restoration and identification given. Some dolls and accessories for sale. C/R&R
Thur-Sat 12noon-5pm, advisable to telephone first. Other times by appointment.
Directions: from A646 through Hebden Bridge turn down Hope Street opposite Picture House Cinema Workshop is situated down a short passage between library and Collectors' Corner.

FINE ART & SCULPTURE
John Hawkwood
The Turret, Valley Road
Hebden Bridge HX7 7BZ
☎ *0422 846199*
Portraits and illustration work;
drawings of landscape and figures.
Well established in the field of
architectural drawings and artist's
impressions working from architectural plans. Also figurative sculpture suitable for garden settings. Sat
and evening courses in drawing
and scuplture. Gallery with
changing exhibitions. C/R&R/W/E
Wed-Sat 10am-5pm, Sun 2-4pm.
Other times by appointment.

CERAMICS
Mill Pottery
(Jan Burgess and John Kerrane)
Bridge Mill Workshops
St George's Square
Hebden Bridge HX7 8ET
☎ *0422 844559*
Hand-thrown domestic stoneware
pottery and decorated plates. A
range of practical and functional
domestic ware at reasonable prices.
Also decorative and commemora-
tive plates and bowls. C/W
Mon-Fri 9am-5pm, Sat 10am-5pm,
Sun 1-5pm. (Approximate hours).

Keighley

FURNITURE PAINTING &
DECORATION
Ann & Richard Oxley
Thomas Whitaker Ltd, Greenfield
Joinery Works, Eastburn
Keighley BD20 7SR
0535 653281
Visitors welcome to workshop
producing hand-painted and
decorated furniture, both antique

and modern. Members of Guild of
Master Craftsmen. C/R&R/E
Mon-Sat 10am-4pm.

FURNITURE
Thomas Whitaker (Eastburn) Ltd
Greenfield Joinery Works
Main Road, Eastburn BD20 7SR
☎ *0535 653281*
Established company of craftsmen
joiners making traditional furniture
in English oak and other hardwoods; specialist joinery and
specialist casework. A full range of
living, dining and bedroom
furniture tailored to suit individual
requirements. C/R&R/W/E
Directions: from Keighley on Main
Road Eastburn, pass General Hospital
on right, White Bear PH on left; take
next turn left to workshops.

METALWORK
Freegate Metal Products
(Tom Ashby)
Freegate Mill
Cowling BD22 0DJ
☎ *0535 632723*
Six people using traditional foundry
methods reproducing Victorian
garden furniture and decorative
castings. C/R&R/W/E
Mon-Fri 7.30am-4.30pm.

ENGRAVING
Cravengraving
(John T and E Mitchell)
73 Main Street
Crosshills BD20 8PH
☎ *0535 634001*
Engraving (computerised and
manual) on wood, plastics, metals
etc. Industrial and commercial
work, signwork, nameplates,
trophies etc. Computer aided
design engraving. Member of

League of Professional Craftsmen.
C/R&R
Mon-Fri 9am-5.30pm, Sat 9am-12noon.
Directions: on A6068 Keighley-Colne
road (Main Street) at Crosshills.

DRY STONE WALLING
Stuart Walker
5 Meadow Lea
Sutton-in-Craven BD20 8BY
☎ *0535 636972/0860 614673(mobile)*
Two employed in dry stone walling
business; registered Master Craftsman, Dry Stone Walling Association. Specialists in miniature dry
stone walls 2.5in high.

Leeds

ANTIQUE FURNITURE
RESTORATION
G W and J Woodcock
Ledston Lodge
Ledsham, South Milford LS25 5LU
☎ *0977 683234*
Rural business employing three
people mainly restoring furniture
for private customers and antique
shops. French polishing, re-
upholstery, carving etc. Some
furniture made to commission.
Unusual garden furniture made.
Showrooms. C/R&R
Mon-Fri 9am-5.30pm, Sat-Sun 10am-5pm.

Otley

TILE MAKING
The Medieval Tile Company
(Ian Stewart Hargreaves)
Woodnook Farm, Blubberhouses LS21 2PQ
☎ *0943 880370*
Three people employed in tile-
making business producing wood-
fired handmade tiles for floors,

walls, patios, conservatories etc.
C/R&R/W/E
Mon-Fri 8am-5pm, Sat-Sun by appointment.

Todmorden

DRY STONE WALLING
Peter Walker
1 West View, Todmorden OL14 8BN
☎ *0706 818353*
Dry stone walling contractor. Also
courses in dry stone walling.
Examiner for Craftsman Certification Scheme.

ARTIST
Shade Studio
(Bohuslav Barlow)
262 Rochdale Road
Todmorden OL14 7PD
☎ *0706 814247*
Trained artist, member of Manches-
ter Academy with studio and
gallery. C/R&R/E
Mon-Fri 9am-5pm, by appointment. Sat 10am-5pm. (Closed fortnight in July).

Wakefield

GLASS ENGRAVING
Wadeland Hand Crafted Glass Engraving
(Elizabeth Sockett)
120 Northfield Lane, Horbury WF4 5H
☎ *0924 279426*
Working mainly to commission,
glass engraving on pieces ranging
from crystal pendants and earrings
to large bowls and trophies.
Member of the Guild of Glass
Engravers. C/R&R/E. Credit cards
Open daily 9am-5pm; advisable to telephone first.
Directions: on A642 Wakefield-
Huddersfield road, south of Horbury

FOR SKILL
&
INTEGRITY

THE GUILD
OF
MASTER CRAFTSMEN

INDEX TO CRAFT WORKSHOPS

Roger Heaton, Lincs
Roger Mills Picture Framer, E Sussex
Strides Gallery, Humber
Taggart Gallery, Cambs
The Bart Luckhurst Gallery, Notts
The Stables, Leics
The Studio, Oxon
Townsend Picture Framers, Derbys
Village Workshop, Northants
Webster & Hill Fine Art Studio, Som
Wild Cherries Gallery, N Yorks

PIPE MAKING
David Cooper, Hants
Tilshead Pipe Co Ltd, Wilts

PLASTER CASTING
Casting Images, Dorset
Neil's Crafts, Warwicks

**PRINTING/ETCHING/SILK SCREEN
PRINTING**
Alan Stones, Cumbria
Aquarius Designs, Dorset
Bacon & Bacon, Northum
Capelli Printmaking Studio, Warwicks
Fidgen Design, Norfolk
Frances St Clair Miller, Heref/Worcs
Left Bank Arts, Lincs
Periwinkle Press, Kent
Sarah Ringrose, Devon
The Black Swan Press, Oxon
The Crafty Printmaker, Notts
The Orchard Press, Shrops
Timberland Art & Design, Lincs

**PUPPET/THEATRE COSTUME
MAKING**
Doreen James Parsons (Puppets),
Norfolk
Pat Thompson Designs, Northum

PYROGRAPHY
Close Connections, Lancs
Thorpe Hall Carpenter's Shop, Cambs
Woodway, Dorset

RESTORATION WORK GENERAL
Michael Wallis, Bucks

ROPE MAKING
Outhwaite, W R & Son, N Yorks
The Historic Dockyard, Chatham, Kent

**RURAL SKILLS/FARM/FOREST
MANAGEMENT**
Clifton House Craft Group, Lancs
Dethick Crafts, Derbys

Gary Rowlands, Avon
Living Wood Chairs, Devon
Wilderness Wood, E Sussex

RUSH & CANE WORK
Abbey Cane & Rush Seaters, Leics
Antique Cane Restoration, Notts
Basketry World, Warwicks
Broadwindsor Chairs & Crafts, Dorset
Bryant Crafts, Berks
Cane & Woodcraft Centre, Hants
Cane Corner, Devon
Cane Workshop, The, Som
Chiaroscuro, Wilts
Country Chairmen, Oxon
Country Crafts, Leics
Hammacott, A W, Dorset
Ian D F Sim, Glos
Joan Gilbert, Derbys
John Harriman, Surrey
Lomas Pigeon & Co Ltd, Essex
Middleway, Warwicks
Mona Leckie, Norfolk
Paul Spriggs, Glos
Payne & Poole, Glos
Rob King — English Willow
Basketworks, Norfolk
Rosemary Hawksford, N Yorks
S & M Cane & Rush Seating, Staffs
Springs & Things, Oxon
Stephen Hill, Glos
Sturgess, B T & M, Derbys
The Chair Lady, Cumbria
Twists & Turns — Spinners, Norfolk

SADDLERY see also Leathercraft
Acorn Saddlery, Devon
Bart J Snowball, Northum
Birdsall, R, — Saddler, Derbys
Blackmore Vale Saddlery, Dorset
Blisland Harness Makers, Cornwall
Boots & Saddles, Kent
Calcutt & Sons Ltd, Hants
Clothes-horse, Northum
Dragonfly Saddlery, E Sussex
Ermington Mill Saddlery, Devon
Evans, J C, Shrops
Glenn Hasker, Hants
Godden Saddlery, Dorset
Hadfield Saddlers, N Yorks
Hide-Horn, Cumbria
Hunt, C P, Saddlers, Norfolk
Hurst Saddlers, Leics
John Hayes Products, Glos
Kestan Horses, Hants
May, A D R (Saddler), Warwicks
Mark Bushell — Saddlers, Lincs
McCoy Saddlery & Leathercraft, Som

ENGLAND'S REGIONAL TOURIST BOARDS

These official Tourist Boards will be happy to supply you with further general information on their areas.

Cumbria Tourist Board
(Cumbria)
Ashleigh, Holly Road,
Windermere, Cumbria LA23 2AQ
☎ 05394 44444

Northumbria Tourist Board
(Cleveland, Durham, Northumberland and Tyne & Wear)
Aykley Heads, Durham,
County Durham DH1 5UX
☎ 091 384 6905

North West Tourist Board
(Cheshire, Greater Manchester, Lancashire, Merseyside and the High Peak District of Derbyshire)
Swan House, Swan Meadow Road,
Wigan Pier, Wigan WN3 5BB
☎ 0942 821222

Yorkshire & Humberside Tourist Board
(North Yorkshire, South Yorkshire, West Yorkshire and Humberside)
312 Tadcaster Road, York,
North Yorkshire YO2 2HF
☎ 0904 707961

Heart of England Tourist Board
(Gloucestershire, Herefordshire, Worcestershire, Shropshire, Staffordshire, Warwickshire and West Midlands)
Woodside, Larkhill, Worcester,
Worcestershire WR5 2EQ
☎ 0905 763436

East Midlands Tourist Board
(Derbyshire, Leicestershire, Lincolnshire, Northamptonshire and Nottinghamshire)
Exchequergate, Lincoln,
Lincolnshire LN2 1PZ
☎ 0522 531521

Thames & Chilterns Tourist Board
(Oxfordshire, Berkshire, Bedfordshire, Buckinghamshire and Hertfordshire)
The Mount House, Church Green,
Witney, Oxfordshire OX8 6DZ
☎ 0993 778800

East Anglia Tourist Board
(Cambridgeshire, Essex, Norfolk and Suffolk)
Toppesfield Hall, Hadleigh,
Suffolk IP7 5DN
☎ 0473 822922

London Tourist Board
(Greater London area)
26 Grosvenor Gardens,
London SW1W 0DU
☎ 071 730 3488

West Country Tourist Board
(Avon, Cornwall, Devon, Somerset, West Dorset, Wiltshire and The Isles of Scilly)
60 St David's Hill,
Exeter, Devon EX4 4SY
☎ 0392 76351

Southern Tourist Board
(Hampshire, Northern & Eastern Dorset, and Isle of Wight)
40 Chamberlayne Road,
Eastleigh, Hampshire SO5 5JH
☎ 0703 620006

South East England Tourist Board
(East Sussex, Kent, Surrey and West Sussex)
The Old Brew House, Warwick Park,
Tunbridge Wells, Kent TN2 5TU
☎ 0892 540766

Tourist Information Centres
There are over 800 Tourist Information Centres throughout the United Kingdom. Look in your local telephone directory under 'Tourist Information'. TICs can give you details about local attractions, events and accommodation

VISITING
CRAFT
WORKSHOPS
IN THE ENGLISH COUNTRYSIDE

**An essential guide to the craftsmen and women
who work in rural England**

All over England craftsmen and women produce a wide range
of objects — both beautiful and useful. This book provides a
wealth of information about where to find everything from
pottery to quilting. It guides you to where you can watch these
craftspeople at work in craft centres and small workshops.

Furniture makers, gold and silversmiths, leatherworkers,
spinners and weavers, toymakers, clockmakers, glass
engravers, basket makers, blacksmiths and many more are all
to be found here. This book will solve many problems and
provide new and interesting ideas for gifts.

ISBN 0 86190 195 9

£5.99

9 780861 901951

MOORLAND
PUBLISHING
Co Ltd